Transportation Security

Clifford R. Bragdon

AMSTERDAM • BOSTON • HEIDELBERG • LONDON
NEW YORK • OXFORD • PARIS • SAN DIEGO
SAN FRANCISCO • SINGAPORE • SYDNEY • TOKYO

Butterworth-Heinemann is an imprint of Elsevier

Transportation Security

Butterworth-Heinemann is an imprint of Elsevier
30 Corporate Drive, Suite 400, Burlington, MA 01803, USA
Linacre House, Jordan Hill, Oxford OX2 8DP, UK

Library of Congress Cataloging-in-Publication Data
Transportation security / [edited] by Clifford Bragdon.
    p. cm. – (The Butterworth-Heinemann homeland security series)
Includes bibliographical references and index.
ISBN 978-0-7506-8549-8 (hbk. : alk. paper) 1. Transportation–Security measures. I. Bragdon, Clifford R.
HE194.T75 2008
363.12–dc22

                                                                                            2008017447

British Library Cataloguing-in-Publication Data
A catalogue record for this book is available from the British Library.

ISBN: 978-0-7506-8549-8

For information on all Butterworth–Heinemann publications
visit our Web site at www.elsevierdirect.com

Transferred to Digital Printing in 2012

Acknowledgments

I would like to thank several individuals who have contributed to the development and preparation of *Transportation Security*. Mary Dyer, my Administrative Assistant at Florida Tech, has assisted me throughout this manuscript process. Her devotion to organizational detail, administrative coordination, the necessity of meeting important deadlines throughout this process, and most importantly, her positive attitude, provided important motivation to complete any task at hand.

Another major contributor was my wife Sarah Bragdon, who was actively involved in examining the chapters that I prepared, providing review and recommendations. This important perspective gave me thoughtful and helpful feedback from a person whose area of competency includes English composition as a former teacher.

An important mentor who planted the seed for *Transportation Security* was Major General Mike Sumrall, Assistant to the Chairman of the Joint Chiefs of Staff for National Guard Matters. Mike inspired me to address the problem of national preparedness, recognizing there is no easy answer but a compelling need to protect this country and the ideals and traditions for which it stands.

My entire family gave me encouragement to pursue the book to the entire end, knowing the task was formidable but rewarding. All three daughters, Katherine Bragdon, Rachel Rose, and Elizabeth Hole, along with my sister, Peggy Shepley, thoughtfully listened and made suggestions.

Lastly, I want to thank all the authors who contributed chapters for their hard work and their important insights into the subject covered by this book. It was truly a team effort.

Clifford R. Bragdon, Ph.D.

Contents

Contributors

John C.W. Bennett, J.D.

Dr. Bennett is currently chief executive officer of Marine Protective Services, a provider of ISPS Code–MTSA consulting services and training (certified by the U.S. and UK governments). In 2004 he established Asset Tracking Logistics and Security, LLC, to enter the field of supply-chain security and visibility with technologies allowing real-time tracking and condition monitoring worldwide. While on active duty as a career officer in the U.S. Navy he completed three deployments in support of the U.S. Antarctic Research Program and obtained an LL.M. in Law and Marine Affairs from the University of Washington (1981). He served 6 years at the Pentagon, where Dr. Bennett became head of the Law of the Sea and Law of Air and Space branches of the Office of the (Navy) Judge Advocate General, including 3 years as an oceans policy planner on the Joint Staff, and participated as a representative of the U.S. at several international treaty negotiations, nongovernmental organization conferences, and bilateral talks. While assigned to a NATO command, a U.S. combatant command, a Navy fleet headquarters, and a submarine force staff, he became proficient in U.S. and NATO rules of engagement (ROE) and participated in numerous war games, counterterrorism and similar exercises, and nuclear incident response drills. He negotiated revisions to the NATO ROE on behalf of the Supreme Allied Commander, Atlantic.

He graduated from Swarthmore College with a B.A. (with high honors) in economics in 1970 and obtained a J.D. degree in 1973 from Georgetown University Law Center, where he was an editor of the *Law Journal,* following a year spent clerking for a federal appeals court judge. Dr. Bennett received certifications as CSO, SSO, and PFSO from Maine Maritime Academy in 2004. In 2006, he was appointed a Distinguished Visiting Research Professor at the University College, Florida Institute of Technology. He is a member of the American Bar Association, American Society of International Law, the U.S. Naval Institute, and the Military Officers Association.

Thomas A. Bruno

Mr. Thomas A. Bruno is president and founder of Bruno Associates Incorporated (BAI) since 1994 and has over 35 years of commercial, military, and medical logistics support experience. His firm specializes in microcircuit technology in support of logistics applications consisting of bar code technology, optical memory cards (OMCs), contact memory buttons (CMBs), radio frequency identification (RFID), and biometric and wireless communications. BAI provides consulting service, technical evaluation of RFID systems, independent validation and verification, and system integration for the Department of Defense (Battelle and Oak Ridge) and for commercial industry.

One of the original coauthors and editor of the U.S. Army's Logistics after-action report after the Gulf War, Mr. Bruno also was one of the concept developers for implementing radio frequency identification technology (RFID) within the Department of Defense (DOD). He wrote the Army's initial automatic identification and data capture (AIDC) and automated identification technology (AIT) strategies for the Army's deputy chief of staff for logistics. Mr. Bruno developed the concept of operations for the Defense Logistics Agency (DLA) and Defense Advanced Research Projects Agency (DARPA) to evaluate radio frequency identification "intelligent labels." Bruno Associates wrote the executive summary, developed the funding, and programmed the RFID intelligent label project for the U.S. Department of Defense, entitled "Advanced HAZMAT Rapid Identification Sorting and Tracking (AHRIST)." He was the principal writer of the concept of operations (CONOPS) and test criteria for electronic article surveillance tags and the RFID passive intelligent labels. Mr. Bruno also was a member of the White House antiterrorist technology advisory group in 1998 with Oak Ridge National Laboratory.

Corey A. Cook, LTC (Ret.)

Corey A. Cook currently serves as the program manager of Theater Opening and Sustainment for Lockheed Martin. Formerly a lieutenant commander in the U.S. Navy, he enlisted as a senior electronics technician until his selection to Officer Candidate School, where he received his commission as a naval supply officer. His introduction to radio frequency identification (RFID) and automatic identification technology first came during a 2-year special assignment with the U.S. Army, when he served as naval liaison and action officer for South American and Pacific operations, assisting in the integration of RFID asset tracking for material transport.

LCDR Cook later served as the director of operations for the Navy Expeditionary Medical Support Command, responsible for humanitarian assistance and natural disaster and wartime medical response. He oversaw the deployment, activation, administration, procurement, assembly, and logistical support of eight 500-bed fleet hospitals worldwide, multiple rapid-deployment expeditionary medical facilities, and numerous forward-deployable preventative medical units responsible for disease and contagion detection. A staunch advocate of RFID and AIT technology, he actively oversaw their integration into medical asset identification and packaging, biomedical equipment maintenance, civil engineering fleet management, medical warehouse real-time location systems, and iridium tracking for pharmaceutical transportation. During his tenure, he directed command disaster relief response and humanitarian assistance service for Hurricane Ivan, the Indonesian tsunami, Hurricane Katrina, and the Pakistan earthquake. Mr. Cook received his bachelor of science degree in business from Christopher Newport University, followed by a master's degree in logistics from Florida Institute of Technology.

Sarah R. James

Ms. Sarah R. James, Executive Director of SOLE - The International Society of Logistics has actively promoted the development, advancement, and recognition of logisticians since 1983. As a member of SOLE's board of directors, as chair of the Society's awards board, as the Society's Vice President of Finance, and as its President, she has nationally, as well as internationally, provided leadership for logistics excellence and development of incentive programs supporting recognition of logisticians. Her efforts were recognized by the Quartermaster of the U.S. Army when Ms. James was awarded the

Distinguished Order of St. Martin for her long-term service for and on behalf of the Army's logisticians. This unique recognition to an individual outside the U.S. Army is matched through the awarding of SOLE's Distinguished Service Medal, which recognized—among other things—the broad international footprint she has developed, for both SOLE and herself. While in government Ms. James was the recipient of many honors and awards, including the Commander's Award for Civil Service, the Meritorious Civilian Service Medal, and the Superior Civilian Service Award. She is currently working with defense and other government officials, industry leaders, and a broad spectrum of academe—both internationally and in the United States—to develop graduate curricula and models for development of whole-life–life-cycle logisticians and humanitarian and disaster relief logisticians. Through her efforts, SOLE's professional certifications (Certified Professional Logistician, Certified Master Logistician) are recognized and used for both career promotion and assignment consideration, by both governments and the private sector. Florida Institute of Technology has appointed Ms. James a University College Fellow. This is first time Florida Tech has had a formal association with an individual recognized nationally and internationally for logistics excellence.

Tom Jensen

Mr. Tom Jensen is the chief executive officer and chairman of the board of the National Safe Skies Alliance and National Safe Waterways & Seaports Alliance, headquartered in Alcoa, TN. In 1997 Mr. Jensen was instrumental in founding National Safe Skies Alliance as a nonprofit organization in cooperation with the U.S. Federal Aviation Administration (FAA). Under his leadership the organization administers a multimillion dollar annual grant from the Transportation Security Administration (TSA) to conduct operational testing and evaluation projects in airports nationwide. In 2003 he founded National Safe Waterways & Seaports Alliance to address the security concerns of our nation's waterways, ports, dams, and locks.

To date, Safe Skies has conducted over 100 operational testing and evaluation projects at more than 35 civil aviation airports nationwide. These projects are in the areas of passenger security checkpoint, checked baggage and cargo access control, biometrics, and perimeter. In addition, Safe Skies preoperationally performs tests and evaluations at its Alcoa, TN, facility. Safe Skies also conducts special projects and applied research, such as site surveys and vulnerability assessments. It also has hosted an annual conference with several federal agencies in Washington, D.C., addressing airports and security systems, technology, and related issues. Presentations have been made by airport and aviation safety and security experts from throughout the world.

Mr. Jensen was a member of the Tennessee House of Representatives from 1966–1978 and served as minority leader for 8 years. He was responsible for passage of 85% of the legislation introduced on behalf of Governor Winfield Dunn. In 1977 Mr. Jensen served as president of the National Conference of State Legislatures, representing the nation's 8,000 state legislators. For 15 years he served on the board of commissioners of the Metropolitan Knoxville Airport Authority.

Ralph V. Locurcio, Brig. Gen. (Ret.)

Brigadier General (retired) Ralph V. Locurcio is a professor of Civil Engineering and director of the Undergraduate and Graduate Construction Management Program in the College of Engineering at Florida Institute of Technology. He is a professional engineer

(PE) with over 39 years of experience in the U.S. Army Corps of Engineers and private industry related to planning, engineering, design, and construction of military facilities and civil infrastructure. Most notable in his military career was his leadership of the Kuwait Emergency Reconstruction Operation (KERO). He formed and led a team of 125 engineers in the $650 million reconstruction of civil infrastructure in war-torn Kuwait, following the 1991 Gulf War. In 1996 he left the military to become senior vice president for STV Inc., as director of federal programs involving eight offices and 1,300 based personnel. He served as director of public works for a community of 21,000 and has worked on literally hundreds of military facility projects and civil projects. These include power systems, substations, highways, bridges, flood control, navigation, hydropower, recreation, water supply, and environmental and emergency response to natural disaster initiatives. A graduate of the U.S. Military Academy at West Point and the College of Engineering at Purdue University (M.S., Engineering), he also graduated from the U.S. Army War College, the Wharton Strategic Management Program at the University of Pennsylvania, and the JFK National Security Program at Harvard University. Professor Locurcio has published and lectured widely. Due to his demonstrated international leadership in large-scale reconstruction management he has become an advisor and consultant to the Corps of Engineers, among others, in disaster mitigation and recovery, including rebuilding Iraq and New Orleans for the United States government.

Charles P. Nemfakos

Mr. Nemfakos, newly elected Fellow of the National Academy of Public Administration, was the former Deputy Under Secretary, Department of the Navy. He was responsible for the formulation, presentation, and execution of the U.S. Navy's budget; directing the department's base closure process; providing executive-level continuity for the department in areas of institutional management and strategic planning; and supporting privatization initiatives, incentive structures, and right-sizing efforts. Finally, Mr. Nemfakos was the department's Chief Financial Officer. During the last decade of his career he played a central role in the transformation of the department after the cold war.

Following this federal career phase Mr. Nemfakos was an executive with Lockheed Martin Corporation, Naval Electronics and Surveillance Systems, directing efforts aimed at rationalizing product lines and providing program focus to enhance competitive strategies. He also led Nemfakos Partners, LLC for a period of 4 years.

A member of various professional associations, Mr. Nemfakos has lectured at the Naval Postgraduate School on public policy in resource allocation, at Georgetown University on national security issues, at Indiana University on public administration policy, and at the Defense Acquisition University on public–private entity relationships. He has served as a Senior Fellow at the Center for Naval Analyses and an adjunct at the National Defense University; and he currently is the Chair of the Humanitarian and Disaster Relief Logistics Board of Advisors of SOLE - The International Society of Logistics.

He has been recognized by presidents of the United States with both the Presidential Rank Award of Meritorious Civilian Executive (two awards), and the Presidential Rank Award of Distinguished Civilian Executive (two awards). He was selected by American University to receive the Roger W. Jones Award for Executive Leadership for 2000. In 2004, the Secretary of Defense honored Mr. Nemfakos as one of only nine Career Civilian Exemplars in the 228-year history of the U.S. Armed Forces.

William S. Pepper IV

Bill Pepper is the project manager and system architect for the Harris SafeGuard™ situational awareness command and control system. Mr. Pepper initially joined Harris in 1990 after successfully performing engineering management and systems architectural duties for the Operations and Control and Communications subsystems on the Grumman Joint-STARS aircraft. Mr. Pepper has been responsible for SafeGuard™ product marketing, demand generation, and business development activities. He has developed several conceptual designs for CBRNE-Cyber protection in support of the Department of Homeland Security's Science & Technology Directorate.

Prior to this he developed the biometric-based Passenger Authentication Security System (PASS) and supported the Airport Access Control Pilot Program. As the product manager for the STAT Neutralizer™ Intrusion Prevention computer security product, he led all design and development efforts and was directly responsible for successful certification of the STAT Scanner by the U.S. Army. He is the author of numerous technical papers and has made numerous presentations contained in the proceedings of national conferences related to security. These technical articles cover a host of themes related to situational awareness associated with various transportation modes (e.g., airport and maritime ports), communications interoperability, and biometrics, as well as other software engineering subjects related to security. A veteran of the United States Air Force, Mr. Pepper is an active member of the Safe Skies Alliance, the Association of Naval Aviation, the IEEE, and the National Defense Industrial Association. He performed undergraduate studies at the Wichita State University, University of Arizona, and State University of New York, and graduate studies at the Wichita State University and holds a master's degree in computer resources and information management and computer science from Webster University.

Peter V. Radatti, Ph.D.

Dr. Peter Radatti is the president and CEO of CyberSoft Operating Corporation, which he established in Conshohocken, PA, in 1998. He wrote the first Unix antivirus software adopted in the industry (i.e., military and civilian) and has 21 patents pending or issued in the computer industry. With an academic and professional background in chemistry, electronics, and astrophysics, Dr. Radatti worked at General Electric Space in its military programs department for 13 years designing and developing early fiber-optic computer networks. His work in the computer security field is now a standard utilized in both Unix and Linux applications primarily used by the U.S. Department of Defense and all branches of the service. Due to this unique intellectual property and diverse working experience involving secure information technology–based environments, he is a national and international consultant to the United States government. As a major area of interest, Dr. Radatti is detecting and preventing targeted hostile software in homeland security. He is now developing a new wave anti-virus product for McIntosh, which is unusual in its simplicity for the computer user. His interest and competencies are diverse, as evidenced by the invention of the world's first all dietary fiber bake mix; it utilizes no grains or carbohydrates, which can be a significant assistance to people with medical issues that are treatable with dietary fiber. Currently Dr. Radatti has also been appointed as Distinguished Visiting Research Professor at Florida Institute of Technology, Melbourne, FL.

Ms. Jo Ram

Ms. Jo Ram is chief operating officer and vice president of Indusa Global, headquartered in Atlanta, GA. She joined the company in 2000 and has focused all of Indusa's technical and financial resources on developing tourism and national security solutions for clients worldwide. Ms. Ram was an integral part in implementing the first fully biometric-based ePassport with facial scans and fingerprints in the Americas for the government of the Bahamas. This contract with the Bahamas Ministry of Foreign Affairs began December 22, 2006. Beside ePassports, it has included machine-readable visas, eIDs, and a deportation and border management system for 38 locations. Indusa Global is also working for the Bahamian Ministries of Tourism and Labour & Immigration, as well as other countries (e.g., Jamaica, Singapore, India, and the United States). She is responsible for increasing Indusa's revenues from $250,000 to over $10 million in 7 years.

As an independent management consultant, Ms. Ram also had large corporate and governmental clients. Her engagements include developing quality control management software for Ritz Carlton and Hilton Hotels; strategy and financial consulting for development funds for the United Nations Development Program (UNDP); and a business plan for the Caribbean Institute of Technology (CIT) in partnership with Indusa in Montego Bay, Jamaica. Before Indusa she had over 10 years of consulting, financial, and accounting experience. At MCI Worldcom she was involved in setting up financial processes for all of MCI's international operating units. Ms. Ram has also worked at Accenture (formerly known as Andersen Consulting) on the Symphony Alliance engagement, a partnership. Ms. Ram graduated in 1995 from the University of Georgia in Business Administration, majoring in Accounting.

Dr. Robert Sewak

Dr. Robert Sewak is the managing director of Viasat Systems, LLC, located in Delray Beach, FL. Prior to this he served as executive vice president and director of Education and Special Projects for AEGIS Technology Companies, Deerfield Beach, FL. His primary focus has been organizing and developing for AEGIS maritime and intermodal cargo security, tracking, surveillance, and monitoring. This has included instituting the PILOT, performance, integration, logistics, and operations test, which represented one of the first comprehensive field trials of end-to-end 24/7 tracking and monitoring of intermodal cargo performed on a real-time basis. This technology is now being utilized for multiple transportation modes as part of safety and security protocols to prevent transportation-related terrorism.

Dr. Sewak has also directed the activities of the Neuro Acoustic Research Centre, where he is responsible for the discovery, creation, and development of a unique audio modality to aid in the betterment of the human condition. He has written two books, contributed chapters to four other books, published 22 professional articles, and participated in many national conferences related to transportation safety and security. Dr. Sewak holds the title of Distinguished Research Professor, in the University College at Florida Institute of Technology. His Ph.D. was completed at the University of Memphis.

L. David Shen, Ph.D.

Dr. L. David Shen is a professor of civil engineering and director of the Lehman Center for Transportation Research Engineering (LCTRE) at Florida International University.

He is also a senior technical advisor to the United Nations Development Programme (UNDP). Dr. Shen is a member of three national committees of the American Society of Civil Engineers (ASCE) and the university representative for the Transportation Research Board (TRB). He is a registered professional engineer in Florida, Maryland, and California. He has extensive industrial experience, which includes current consulting to international and Florida transportation firms. His former position was with the District of Columbia Department of Transportation. He is also a senior technical advisor to the Beijing University of Technology (BJUT) in China.

Dr. Shen is the principal investigator and project director of two U.S. Federal Transit Administration–sponsored research projects. His recent research projects have been the examination of factors influencing successful implementation of intermodal guideway public transit systems and the impact of technological and demographic trends on future transportation system efficiency. Dr. Shen has an extensive list of publications and numerous papers in various journals. He is the author or coauthor of over 130 papers, books, and articles. He has received several distinguished awards for excellence in research scholarship. Dr. Shen teaches in the areas of highway planning and design, transportation and land development, highway capacity and control systems, mass transit planning, and airport planning and design. Currently he is also the FIU coordinator for the University Consortium for Intermodal Transportation Safety and Security. He received his Ph.D. in civil engineering from Clemson University.

Michael Workman, Ph.D.

Dr. Michael Workman received his Ph.D. in organizational development from Georgia State University. Following a distinguished 22-year career in the computer industry working in network applications and international security for Bank of America and Telecom/Equant, Dr. Workman began an academic career. Initially he was an assistant professor in information science at Florida State University, and more recently he came to Florida Institute of Technology as an associate professor in the College of Business. His academic and research focus is examining how human perceptions and technologies interact and affect performance, particularly in team-based work. Currently he is the director of the Center for Immigration and National Security, as well as of the Security Policy Institute. He has served as a national security advisor on biometric passports to foreign governments. Dr. Workman has 30 publications, including three books, and has made presentations at 16 international conferences with proceedings and published reports. A scholarly reviewed journal, *Journal of Global Security and Preparedness,* is being developed jointly with Elsevier. He will serve as senior editor of this Florida Tech–based journal. His most recent research focus is the issue of mobility-based security, technological applications, and the relationship to human behavior (i.e., both individual and group dynamics). Dr. Workman is co-principal investigator on the Army Research Laboratory's $2 million grant, Biologically Inspired Security for Mobile Ad Hoc Networks. He will also be initiating a professional development series on visual semantic communication, beginning this year.

Dedication

The book is dedicated to Ronald R. Polillo, a visionary and globally-renowned transportation security expert in anti-terrorism and force protection. He established national leadership in development of the Aviation Security Technology Integration Plan and its subsequent implementation supporting the United States Aviation System. Ron's contribution is duly recognized by both governmental and business leaders worldwide, and his presence will be sorely missed.

Cliff Bragdon

Foreword

Transportation Security represents a refreshing and interdisciplinary approach to understanding and addressing global preparedness from a transportation perspective. The authors have very capably set the table to understand the essential interrelationship among natural disasters, human disasters, and sustainable infrastructure that must be *collectively addressed* if we are going develop effective solutions to this worldwide problem.

Dr. Bragdon has introduced extremely important concepts that fill voids in this complex subject while offering breakthrough insights. First of all he has properly defined intermodal or integrated transportation as it should be: the safe, secure, sustainable, and efficient movement of people, goods, and information by air, land, sea, and space. Second, transportation has not been narrowly defined as dealing with just physical modes of movement, but also the electronic communication of information, which addresses interoperability. On this basis, he has introduced the concept of "transcommunication" as a bridge between the physical and electronic world of mobility, which the UN identified at one time, but has lain fallow since Habitat. Third, mobility, which has a cultural lifeline to civilization, appears to have both a genetic, as well as psychosocial basis. This means that the concept of movement is a fundamental component of the human species and its operative functioning system. If movement becomes impaired from a human perspective, there is a reduction in comfort, enjoyment, and biophysical functioning, including the potential onset of stress, and suboptimization. This is a profound observation that Dr. Bragdon hypothesizes, and it should be rigorously explored.

Dr. Bragdon has very capably dealt with many of the historical and institutional impediments that have resulted in a stovepipe approach to the subject of security that requires systems integration. Initially institutional biases were advanced by individual modal advocates, at the exclusion of a holistic and integrated transportation model. This laid the foundation for a piecemeal and disjointed problem-solving management approach.

Personal private transportation (the private car) historically was advocated by Robert Moses in New York as a national model, with no role for public transit. The consortium of General Motors, Standard Oil, and Firestone, through their "National City Lines" company, discouraged rail-based transit, substituting GM buses through the predatory purchasing of municipal trolley and street car systems (over 400). This was followed by the establishment of Highway Departments for planning, designing, and managing road systems at state and local government levels and the development of a federal highway system as the primary means of national mobility. It was only much later that there were attempts at having a more diversified approach, or as Dr. Bragdon states, a "total modal approach." Now with Departments of Transportation in play, followed by legislation using highway trust funds beyond merely supporting road systems (pavement), a more balanced transportation system has begun to appear. The book also discusses the fallacy that the federal government alone is the most effective manager for addressing natural and manmade disasters. It examines the important tri-part teaming of government with business and nonprofits.

Clearly one of the most provocative discussions involves Dr. Bragdon's forecasting that energy-based reliance on fossil fuel for personal transportation will become a nonissue by 2020. "Carhenge" will now become liberated from a petroleum-based fuel supply, potentially reducing the threat of terrorism. The next formidable hurdle will be the growth and size of transportation systems, which could put a squeeze on the finite space available for human habitation. This means we should institute a spatial management, "diet cities" approach, to optimize urban space three-dimensionally, rather than using two-dimensional "land use" planning as our urbanized world expands to 10 billion persons before 2050. Super-sizing our infrastructure, a drive-thru society dominated by SUVs, crossovers, and trucks, matched with larger homes ("McMansions") and personal appetites is the wrong formula, even without the necessity of fossil fuel. Spatial gridlock and restricted access will impair needed transportation response to any disaster and will make urban society assets more vulnerable.

The magnitude of impact of all these disasters is outlined, and the economic consequences, especially in terms of the world's GDP, are profound and, according to Dr. Bragdon, equal to 5%. The financial cost of 9/11, estimated at $2 trillion, and Katrina at $800 million are at a magnitude no economy can support on a long-term basis. The frequency of both manmade and natural disasters is growing, and the author recognizes the financial escalation as our civilization becomes increasingly urbanized, with greater infrastructure and valued assets located in vulnerable coastline locations.

Organizationally, *Transportation Security* follows a logical sequence. The first section of the book addresses the importance of transportation, the need for a security systems approach (that is presently missing in transportation), the importance of logistics, and then security behaviors (which to date have received little attention). These modal discussions are especially helpful to gain an understanding of several role players (i.e., aviation, maritime, and roadways). The second section then describes the primary modal elements of security, followed by the various technologies currently available. There is an excellent discussion of applied modal container tracking technology as well as command and control. This section investigates interoperability and the situational awareness the military has experienced, and their applicability to civilian counterpart operational centers, including cyber security.

The reader is not left in limbo about solutions and case studies that outline best professional practices, including future approaches. The third section of *Transportation Security* discusses ways in which transportation is now addressed. Automatic Identification Technology (AIT) is being used extensively by the military and is described in detail, with applicability to civilian logistics as well. It is important to note that much of the AIT and related tracking systems were first applied in the military, and subsequently have been used in other governmental and commercial markets. The important case study of rebuilding Kuwait is presented to demonstrate the complex but organized approach that was used by the U.S. Army Corps of Engineers following the Gulf War. This critical path methodology is now being employed as "lessons learned" to assist the post Katrina

Hurricane rebuild. One of the most topical chapters addresses immigration and national security, including the newest multifaceted technologies that are being incorporated into passports. Protecting borders with biometrics, RFID implants, surveillance measures, and national identification systems are being examined by every country. This chapter presents secure technologies that just have been implemented in the Caribbean, and specifically, the Bahamas.

In the concluding chapter, Dr. Bragdon builds on the Army experience in discussing the Fast Integrated Response Systems Technology (FIRST) and how that could be used in emergency response and recovery incidents that cover a large geo-political, multistate area. Humanitarian disaster relief logistics is in its infancy and must grow in sophistication and applicability to meet the logistical challenges. Whatever approach is used, the author encourages that an integrated approach take place, with stakeholders from civilian, military, business, and nonprofits utilizing an effective interoperable system that is cyber secure. Dr. Bragdon makes an important case for establishing a global systems approach, and he outlines the Global Center for Preparedness (GCP) and the multiple missions that need to be in place to be more effective worldwide. The GCP already is attracting interest among businesses, governments, and academia as we must focus on global preventative solutions.

The contributing authors have written complementary chapters that make this book an excellent one-stop primer for practicing professionals who are actively involved in preparedness activities at public and private sector levels. Their backgrounds are diverse, and these chapters address essential components to understanding and applying solutions to this problem impacting most nations of the world. The book also has an important place in college and university curriculums that deal with homeland security, national preparedness, and disaster response and recovery.

Dr. Jay Stein, FAICP
Provost and Vice President of Academic Affairs
SUNY Plattsburgh, New York
Former Dean, College of Construction, Planning, and Architecture
University of Florida

About the Author

Clifford R. Bragdon, Ph.D., AICP, FASA

Dr. Bragdon is the associate provost, dean of the University College, and Distinguished Research Professor at Florida Institute of Technology. He is also the director of the Global Center for Preparedness, which is an international think tank focused on natural and human disaster prevention and sustainable planning and management, with public and private sector and nonprofit partners. Prior to this he established and was executive director of the University Consortium for Intermodal Transportation Safety and Security (UCITSS). Endorsed by Congress and signed into law by President Bush, this was a $10 million federally funded center involving 12 public universities, coordinated by the U.S. Department of Transportation (DOT), and based at Florida Atlantic University. It was the largest grant ever awarded to a university by DOT. The consortium dealt with safety and security issues for intermodal transport systems (highways, airports, seaports, rail-transit, utilities, and communications–IT).

While in New York, Dr. Bragdon was dean and vice president of the National Aviation and Transportation (NAT) Center at Dowling College. He was the first dean in the United States for a school of aviation and transportation devoted to the integration of all modes of transport by air, land, and sea. At the NAT Center he invented the first intermodal transportation simulation system (ITSS) for performing virtual simulation using all transportation modes. The ITSS was invented by Dr. Bragdon and patented by the U.S. government. It was rated the ninth most important new U.S. invention for the next 100 years by *Newsday* in 2000. Previous to this Dr. Bragdon served as professor, associate dean, and associate vice president at Georgia Institute of Technology, Atlanta, GA. While at Georgia Tech he was also the executive director of AMCEE, a consortium of engineering-based universities involved in distance learning, including MIT, Stanford University, Purdue University, University of Florida, and Arizona State University, among others. Dr. Bragdon also taught for 13 years at Emory University's College of Medicine, and the School of Public Health.

Dr. Bragdon's specialty is the field of transportation, land use–space use, simulation, environmental planning, sustainability, and safety and security. He has published five books as well as over 100 articles. He has lectured widely on these general subjects throughout the world and been an invited lecturer and distinguished speaker at over 70 universities (including Harvard, MIT, Columbia University, University of Pennsylvania, University of North Carolina, Moscow State University, Peking University, and Tianjin University). The United Nations and their United Nations Development Program (UNDP) invited him three times to participate as distinguished speaker in Turkey, Singapore, and

Egypt regarding sustainable environments and strategic envisioning; similar invitations have come from the mayor of Moscow and the Chinese National Academy of Science.

A distinguished professor and researcher, Dr. Bragdon has been a principal investigator in over $60 million of funded contract research, as well as a consultant to over 150 governmental agencies, governors, the U.S. Congress, the Office of the President, the United Nations, NATO, and major global corporations. Dr. Bragdon has been invited to the Office of the President twice due to his international reputation in intermodal transport safety and security. He also was a consultant to the Office of the Mayor, New York City, and developed a real-time 3-D computer simulation and analysis of the World Trade Center incident. A national transportation and land use planning and security expert on television and radio networks, Dr. Bragdon has appeared on NBC, CBS, ABC, FOX, CNN, Cablevision, National Public Radio, and WNYC. Print media activity has included the *New York Times, Newsday,* the *New York Daily News, USA Today,* the *New York Post, Traffic World,* and *Commerce Business Daily.*

Dr. Bragdon is also a Fellow in the Acoustical Society of America (FASA) and a charter member of the American Institute of Certified Planners (AICP), the American Planning Association (APA), and the Association of Energy Engineers (AEE). He has been honored with the Engineer Achievement Award of the Year for New York by 11 professional societies and by the Federal Aviation Administration (FAA) for Excellence in Education. He has been given the Citation for Technological Excellence by both Suffolk County and Nassau County, New York, as well by the governor of Georgia. He is listed in *Who's Who in the World, Who's Who in America, Who's Who in Science and Engineering, Who's Who in Finance and Industry, Who's Who in Environment and Energy,* and *Who's Who in American Education.* Dr. Bragdon serves on SOLE - The International Society of Logistics advisory board for Humanitarian and Disaster Relief Logistics and the National Academy of Science transportation research board. He has recently been appointed Managing Editor for the Journal for Global Preparedness, published by Elsevier, Amsterdam.

Academically, Dr. Bragdon holds an A.B. degree in political science and sociology from Westminster College. His master's degree (M.S. in urban planning) was obtained at Michigan State University, while his Ph.D. in city planning was completed at the University of Pennsylvania, Philadelphia, PA. Prior to his academic career, Dr. Bragdon was a captain, United States Army, Medical Service Corps.

THE BUTTERWORTH-HEINEMANN HOMELAND SECURITY SERIES

Visit **http://elsevierdirect.com/security** for more information on these titles and other resources.

Introduction

Transportation Security and Its Impact

Clifford R. Bragdon, Ph.D., AICP, FASA

Objectives of This Chapter:

- Provide an overview of transportation security and its societal impact
- Explain the purpose of this text, *Transportation Security*
- Describe the text's organizational framework and four distinct sections
- Discuss modal aspects of security
- Discuss technology applications
- Discuss transportation security solutions

> *"Where there is no vision, the people perish."*
> —Proverbs 29:18, King James Bible

Overview

Mobility represents the cultural lifeline of civilization throughout human history, comprising all methods of transport for both economic and social survival. It is the basis by which civilization has supported the character and lifestyle of its population, surviving and evolving through time. The level of effective mobility is directly related to the transportation support system. Ideally a transport system is based on the safe, secure, sustainable, and efficient movement of people, goods, and information utilizing air, land, sea, and space. It is characterized by two mobility components: physical (e.g., nonmotorized transport, aviation, roadways, maritime, rail, transit, etc.) and electronic (e.g., utilities, satellites, distance communication, information technologies, etc.). This concept can be referred to as *transcommunication* (Figure 1.1). Transcommunication was collaboratively developed as part of the United Nations Conference on Human Settlements by a Habitat II task force examining the basis for an effective human habitat, in which this author participated (Habitat II, June 1996, Istanbul, Turkey).

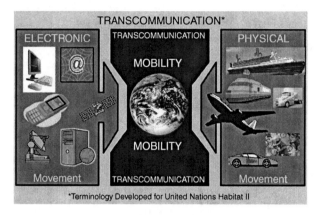

FIGURE 1.1 Transcommunication

It appears that the concept of mobility may have both social and physical character-istics. Not only is movement necessary for cultural reasons (e.g., economic and social well-being), but there also are indicators that it may have fundamental genetic roots. In other words, as humans we physiologically and psychologically require mobility as part of a comprehensive human life support system. A loss or decline in this mobility sys-tem individually and collectively brings about diminished enjoyment and freedom, while also impairing both economic and social well-being. This means that reliance on a secure system for moving people, goods, and information seamlessly is absolutely essential. However, society is at risk, because we do not have a securely integrated mobility system in place. This is evidenced by our response to both natural and man-made disasters in a growing world already containing over 6 billion people.

Today we are at risk because we have used a stovepipe management approach, addressing each transportation mode separately and independently from one another, with limited communications technology and organizational integration. Consequently, gridlock and logistical inefficiency prevail. Underoptimized movement is reducing our world's Gross Domestic Product (GDP) annually by approximately 5% (Figure 1.2). This

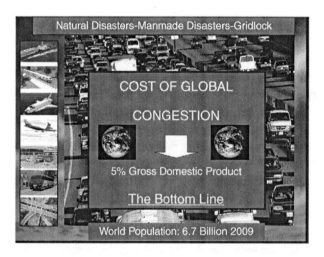

FIGURE 1.2 Cost of Global Congestion: GDP

general GDP percentage has been receiving validation reported in both the United States and Latin America. Transportation as a support system for the economy is further being exacerbated by a growing number of natural and man-made disasters that are affecting the global effectiveness of all transportation modes. Not only is there an increasing number of disasters, but also the magnitude of damage and costs inflicted on our infrastructure is rapidly rising. The infrastructure attack on the World Trade Center by terrorists commandeering commercial aircraft on September 11, 2001, represented a global wake-up call. This event demonstrated to the world that transport systems themselves could inflict profound damage upon our infrastructure and population.

Worldwide the total annual cost is estimated to be between $2.5-$4.0 trillion, if natural disasters, man-made disasters, and associated transportation gridlock are combined (Figure 1.3). The basis for this financial estimate involves combining both direct and indirect costs of impact. Sizeable numbers of the population and associated infrastructure have been severely impacted by natural disasters inflicted by hurricanes, tsunamis, cyclones, tidal waves, and earthquakes throughout the world. Hurricane Katrina has now become the costliest natural disaster in United State's history, calculated to be over $86 billion.

The loss of life from these naturally induced events is substantial. Tsunamis occur mostly in the Indian and Pacific Oceans (85% of the time) and disturb heavily populated coastal areas, with the total number of victims from the most recent such event exceeding 200,000 fatalities. Damage was inflicted across two continents and 12 countries. Tsunamis are not new phenomena, dating back to 1480 B.C. with the destruction of the city of Knossos, which was the capital of the Minoan civilization. The Myanmar cyclone in Spring 2008 has accounted for 70,000 lives, while in China the toll is reaching 50,000. The incidence of natural disasters appears to be on the increase, and the magnitude of their economic impact is growing substantially. The world's buildup of more densely populated coastal communities with increasingly valuable assets (i.e., dwellings, vehicles, roads, bridges, utilities, etc.) is escalating the financial magnitude of damage.

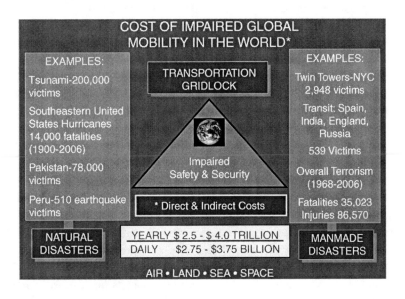

FIGURE 1.3 Estimated Cost of Impaired Global Mobility

Today there are global examples of terrorist acts inflicted on transportation and associated infrastructure in four continents and over 40 countries worldwide. Major world cities, including New York, London, Moscow, Tokyo, Paris, Bombay, and Madrid, have been victims of physically inflicted terrorism. Certainly 9/11 and the loss of 2,948 lives caused by the use of terrorist-directed commercial aircraft gave more attention to the problem. All transport modes are at risk. It appears that public transit has become a new operational theater for terrorists, especially involving rail and bus intermodal stations. From 2003 to 2007 there have been 539 transit fatalities in the five largest terrorist attacks, with 3,363 passengers and crew injured.

Since September 11, 2001, the nation's seaports have also been increasingly viewed as potential targets for terrorist attacks. Security experts are very concerned that the ports can be an entry point for the smuggling of weapons and other dangerous materials into the United States, and cargo and cruise ships could present potentially desirable terrorist targets as well. The ports are gateways for the movement of people and goods and are industrial hubs located very close to population centers, presenting additional opportunities for terrorists. A coordinated port security program is critical to protecting the American people. The 9/11 Commission stated, "... while commercial aviation remains vulnerable it appears that ports are an even greater risk." Maritime piracy near ports is also on the rise, up 20%, compared with comparable 2007 figures. The highest number of attacks occur in the West African nation of Nigeria (22% of the total attacks, according to the International Maritime Bureau: Piracy Reporting Center), or 11 of the 49.

Clearly the entire transportation system is vulnerable to terrorism. The federal government, with congressional support, has initiated efforts, generally by mode of transport, to address this problem, primarily through the Department of Homeland Security. It has been estimated that 1,500 sites (e.g., airports, seaports, rail, transit, etc.) need to be "hardened." Airports have received the most financial attention and come closest to comprehensively addressing terrorism. However, vulnerability remains with all transportation modes. In terms of modal vulnerability, air transport has addressed the possibility of terrorism to the greatest extent but has not eliminated the problem (e.g., incomplete screening of cargo). Air transport also received by far the largest share of federal dollar support, compared with all other transportation modes (i.e., nearly 80% of the total civilian security budget). Transit, maritime, and intermodal modes of movement appear to have the highest risk potential from a possible terrorism incident perspective (Figure 1.4), since their terrorist prevention efforts still remain in the earliest stages of development and effectiveness. This could be partially explained by the budgetary resources allocated, compared with those going to aviation. In 2005 airports received $18 billion ($9 per passenger) for security, in contrast to $250 million ($0.01 per passenger) for transit and rail security, from the federal government. Beginning in 2006–2007 there began to be a slight shift in the financial allocation by transport mode, reducing the percentage given to aviation. Railroads (e.g., Amtrak) and mass transit are now beginning to show some interest in security.

All modes of transportation play an important part in the logistical system supporting movement. For example, the U.S. maritime ports offer multiuse facilities for both commercial and military transportation activities. Approximately 95% of our nation's trade, valued at nearly $1 trillion, enters or leaves our 361 seaports annually. Port safety, security, and sustainable operation are essential to the nation's economic well-being. Maritime operations and the movement of people, goods, and information

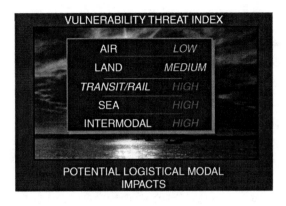

FIGURE 1.4 Vulnerability Threat Index

do not occur in an isolated manner. A cross-modal system of road and rail is an important feature in this transportation infrastructure, along with electronic communication interface. Similarly airports, which now move higher-end commodities (5% by weight but over 35% in value), along with over 50% of all commercial passengers, cannot exist in physical isolation. Landside operations are equally important to airside operations at an airport.

This nation's lack of energy independence is impacting how we pursue needed energy resources, which remain dominated by the use of fossil fuel and petroleum-based energy resources. Until that pattern of demand is significantly modified, the United States will remain influenced by foreign interests, including terrorists who seek retribution (60% of the fuel consumed in the United States is imported, with the percentage rising annually). Record fuel costs, increasing global demand, and ethanol based fuels that are contributing to an increase in food costs all are adversely impacting non–petroleum-based countries. Energy conservation and sustainability are at the forefront of problems that need to be addressed along with global warming.

It appears that through technological innovation, the transportation system can be sustained with a combination of electro-hybrid alternatives by no later than 2020. Such a change would reduce foreign energy dependence and enhance the economy, while at the same time improving this nation's air quality and reducing the heat island effect.

A hydrogen-based system, supplemented by regenerative electric and lithium–polymer recharge systems, will become the dominant player in terms of energy delivery, thereby reducing this societal concern of fuel-based mobility (Figure 1.5). A "carhenge" petroleum-based society will be replaced by energy independence. Gas stations will be replaced by "energy centers." From cellulosic ethanol to plug-in hybrids, and ultimately hydrogen-powered fuel-cell vehicles, a permanent sustainable alternative to fossil fuel depletion is on the horizon in little more than a decade. A potential positive by-product may well be a reduction of global warming and the carbon footprint, if a concerted international initiative can be universally advanced and implemented among 211 nations worldwide.

Transportation demand and the use of space for movement (i.e., air, land, and sea) will be the next societal issue due to the supersizing of our urban habitat (Figure 1.6). This will also bring onto the transportation scene supersized modes of movement including the world's largest aircraft (the Airbus A380 with a passenger capacity of 853); and the world's largest commercial ship (Royal Carribean International Project Genesis with a passenger capacity of 5,400). All these new modes, and their respective supporting

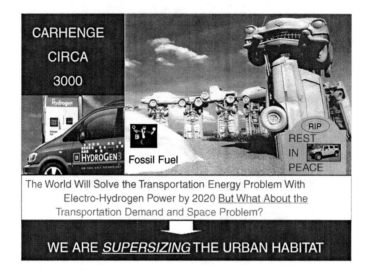

FIGURE 1.5 Carhenge Circa 3000

infrastructure, need to address safety and security, logistics, and emergency response and recovery. Supersizing could bring about spatial gridlock with the bloating of our city infrastructure.

The global population will also continue growing in size, with our habitat becoming more urbanized. By 2009 the world's population will be 6.7 billion, with 50% of the earth's inhabitants living in urban areas. Urbanization will be accelerating by 2050, with 70% of the population being city dwellers and the population reaching 9.2 billion inhabitants (a 37% increase in our total population). Urban areas, according to the United Nations Population Division, will absorb all the growth, with 27 megacities having more than 10 million inhabitants. These populated corridors are becoming more financially vulnerable to natural disasters, whose costs appear to be doubling every decade.

Regardless of societal transitions, the acquisition of world resources, and advancing technology to deal with globalization, the world will remain at risk as long as terrorists

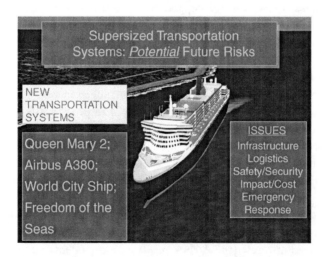

FIGURE 1.6 Supersized Transportation Systems and Potential Future Risks

believe there are cultural inequities and opportunities for advancement. *Transportation Security* identifies the foundations of the problem, how the various modes of transport are underprepared to address the issue, the role of technology to assist with this problem, and various examples for developing solutions to mitigate the elements of natural disasters and man-made disasters at the present time. Approaches for creating smart, sustainable solutions to enhance the safety and security of our urban society in the future will also be addressed. At least one country, Saudi Arabia, is examining how they can rebuild their country using safety and security principles. This is being proposed with the development of King Abdullah Economic City (KAEC) in the Kingdom of Saudi Arabia (KSA). This need is consistent with the passage from the Old Testament, Proverbs 29:18, "Where there is no vision, the people perish."

Purpose and Organizational Framework

Transportation Security is intended to be a one-stop resource for educators, students, and the practicing professional community of planners, engineers, architects, managers, logisticians, and public officials who are involved or interested in any aspect of public and private transportation security risk assessment, situational awareness, management, and control. It should also be appealing to both the private sector and nonprofits that are important players in identifying, addressing, and developing collaborative solutions.

Section I: Introduction

Organizationally this book is divided into four distinct sections. The Introduction (Section I) consists of four chapters that provide the overall framework to *Transportation Security*. Chapter 1 includes the general status of transportation security and its impact on our logistically based global society. This chapter addresses terminology that needs to be applied to understand the total modal integrated approach. It also analyzes the collective economic and social impacts associated with both natural and man-made disasters in a changing global environment. Certain energy improvements are projected and discussed that could make our society more sustainable; however, this may be somewhat offset by how urban space is perceived (three-dimensionally: aerial, surface, and subsurface) and managed in an ever-increasingly populated world. Bloated cities could have an unfavorable impact on addressing potential disaster response and recovery.

The importance of logistics in the transportation security process is the theme of Chapter 2. It presents a strategic perspective of logistics transformation that will allow industry to succeed in the atmosphere of hardened supply-chain management. Experiences are presented working with a European Union port in Greece and the challenges faced with integrating new technology into an existing workforce. The chapter's last section deals with the military in the area of integration and command and control of resources, including initiatives using a risk–reward analysis. Chapter 3 emphasizes the need for a transportation systems approach, which is now conspicuously absent, thereby making nations more vulnerable than they should be. Impediments to achieving both horizontal and vertical integration are described. This is followed by identifying all the modal elements in a transportation systems approach. Chapter 4 explores for the first time a social–psychological perspective by examining security behaviors and intentions that include the use of RFID, biometrics, and a national ID system that is rapidly evolving. The implications of omnipresent surveillance and monitoring of individuals and the impacts these procedures have on psychosocial functioning are discussed from a behavioral security perspective. George Orwell's "big brother" concept from his book *1984* could become a reality in a continuous surveillance-based urban society.

Section II: Modal Aspects of Transportation Security

Section II focuses on selected modal aspects of transportation security, involving three chapters. These chapters include "Road Transportation and Infrastructure Security" (Chapter 5), which addresses roadways, bridges, and parking; "Aviation Security" (Chapter 6), which describes airside and landside aspects of airports including passenger, freight, and support personnel; and "Maritime Security" (Chapter 7), which discusses international and national maritime law, as well as vessel and port measures for passengers and freight.

The first chapter in this section (Chapter 5) describes the National Highway System and the critical role it plays, supporting interstate and defense needs, including evacuation. There are also discussions about security issues and threats to the entire road network and its related infrastructure of tunnels, bridges, and traffic operation centers. Although the probability of a terrorism occurrence is less likely than traffic collisions and vandalism, it must not be dismissed. The history of each of these modes is the subject of discussion in Section II, along with problems associated with terrorist incidents, risk assessment and security planning, programs, technologies that appear to have potential benefits for mitigation, and the general trends for funding.

In Chapter 6 the evolution of aviation security from skyjackings (1961) to the 9/11 incident is discussed, recognizing that the total cost of the attack in September 2001 was approximately $2 trillion. The federal government has taken the lead, since interstate commerce airspace is the responsibility of the Federal Aviation Administration (FAA). Security-related subjects are the focus of the Department of Homeland Security (DHS) and the Transportation Security Administration (TSA). Airport security procedures and technologies are the theme of the remaining portion of this chapter. Maritime security, both at sea and in port (Chapter 7), concludes Section II. Piracy and its evolution to present-day terrorism are presented in a historical context. Since ships can operate on open seas, they are subject to international law, including methods of protection, which are now applicable to every seaport that wants to deal with trade. New security regulatory regimes are now in place both internationally, through the International Maritime Organization (IMO), and nationally, through the U.S. Maritime Administration, (MARAD), and the U.S. Coast Guard.

Section III: Technology Applications to Transportation Security

This section emphasizes the use of technology applications to transportation security, involving three chapters. Cybersecurity and the use of the computer-based targeted hostile software associated with viruses, worms, bots, Trojan horses, and bombs, among others, is discussed in the context of "Computer and Transportation Systems Security" (Chapter 8). Cybersecurity is a major threat to the electronic side of transportation systems, and this chapter addresses this problem in a practical manner.

The expanding base of new technologies associated with real-time intermodal transportation surveillance, covering the entire logistical process, is the theme of "Intermodal

Transportation Security Technology" (Chapter 9). Cargo containers, which carry 95% of the world's international trade, are a principal way of transporting goods and material globally. Recognizing the potential threat to maritime commerce, the Department of Homeland Security issued its "Comprehensive Approach to Port Security," which applies to 361 seaports. The total logistical supply-chain process must be part of the security surveillance process, covering the container passage by ship and the associated intermodal transport process. Container tracking involves a number of handoffs including manufacturers, shipping lines, ports, marine vessels, dray operators, and other players in the supply chain. The need for intelligent containers should remain the goal of an effective maritime system, which is essential for maintaining the security of world trade and commerce. Prior to port vessel and cargo screening to combat piracy and terrorism is being developed using off-shore platforms and smart buoys.

The fact that any means of transportation can become a potential weapon of mass destruction requires a situation awareness system technology advanced by the military (Chapter 10). Experiences of the military in situational analysis, intelligence gathering, information dissemination, network connectivity, and data fusion provide important insights for civilian applications to transportation security and the important surveillance of operational space. Military technologies can assist in advancing command and control communications applications to effectively manage incidents. Now available are third-generation product suites that are network-based Situational Awareness Systems that can be adapted to peaceful uses. The interoperability of military and civilian means of secure communications is essential.

Section IV: Transportation Security Solutions

Case studies that can be applied to the task of developing transportation security solutions are the theme in Section IV. The utilization and distribution of medical facilities in a time-sensitive manner by the U.S. military is presented as a case study in Chapter 11: "AIDC-The Foundation of Military Transportation Logistics." The military deployment of assets worldwide requires an effective Automatic Identification Technology (AIT). As of November 2007, according to the House of Representatives Fact Sheet (House Speaker Nancy Pelosi), there have been 65,354 steel shipping containers lost in transit to Iraq and Afghanistan while in the U.S. Central Command logistics system. An AIT suite of technologies enables the automatic capture of source data, thereby enhancing the ability to identify, track, document, and control deploying and redeploying forces, equipment, personnel, and sustainment cargo. The Naval Expeditionary Medical Support Command (NEMSCOM) instituted an effective medical automated tracking system for rendering medical assistance through its military medical warehouse facility on a global basis. Regrettably FEMA did not avail themselves of this medical assistance resource during Hurricane Katrina, when offered by NEMSCOM.

Chapter 12, "Infrastructure Recovery Initiatives: A Retrospective Assessment," is an important approach to transportation security. This chapter describes the notable reconstruction of Kuwait, which is now being used as a lessons-learned training model by the U.S. Corps of Engineers for the rebuilding of New Orleans due to Hurricane Katrina. Chapter 12 proposes a more responsive recovery system, based on the Kuwait experience, and the development of multistate regional recovery centers that are not limited by geopolitical boundaries (e.g., city, township, county, or state). Often these political boundaries inhibit emergency response and recovery. Rather than reinventing an infrastructure recovery system, lessons should be applied from previous military and civilian initiatives.

Chapter 13 addresses immigration and national security as this subject relates to proper documentation and legal entry. This chapter covers the subject area of biometrics; RFID and implants; surveillance devices; use of national identification systems and machine-readable travel documents (MRTD) such as ePassport, eVisa, eID; and deportation solutions. It discusses the importance of immigration and national security issues that countries must focus on to protect their boundaries. The chapter finishes by providing a background on best practices implemented in the national security arena through examples. Currently the most advanced passport documentation systems in use are outside of the United States. The recently adopted Bahamas passport system is described.

The final chapter in the book (Chapter 14: "Fast Integrated Response Systems Technology [FIRST] and Establishing a Global Center for Preparedness") is based on an integrative concept. The GCP is gaining increasing interest from stakeholders (e.g., Joint Chiefs of Staff, Lockheed Martin, Raytheon, Harris, EMC, Pearson Learning Solutions, etc.). It collectively links the public–private–nonprofit organizations together in addressing any regional–national–global natural or man-made disasters in a sustainable and efficient manner. This combines the comprehensive interest of natural and man-made preparedness, response, and recovery with the concern for establishing a smart, sustainable infrastructure. All too often the "bureaucratic rebuild" does not address a more preventable and sustainably designed community that should be a model for the inhabitants of the future. Preventable green design and more secure community well-being mean we do not want to return to do another ill-conceived "makeover." Since the cost of reconstruction is reaching astronomical heights (e.g., Twin Towers over $2 trillion) we need to rebuild intelligently. The general content of this book, *Transportation Security*, has been shared with the Office of the President, Washington, D.C.

References

Associated Press. "Ship big enough to carry a park." Florida Today, March 16, 2008.

Berkowitz, Carl, and Clifford Bragdon. Advanced simulation technology applied to port safety and security. *Conference Proceedings, the American Society of Civil Engineers*, June 2006.

Berkowitz, Carl, and Clifford Bragdon. Advanced analysis: How can newly developed technology be applied to improving passenger terminal safety and security? *Passenger Terminal World*, November 2006.

Blake, Scott. "Local Company's Offshore Platforms to Combat Piracy." Florida Today, May 15, 2008.

Bragdon, Clifford R. Emergency systems used in transportation systems. *Proceedings, Segunda Feria Internacional de Transporte Masivo* (Second International Mass Transit Fair), Bogota, Columbia, November 8–9, 2006.

Bragdon, Clifford R. Integrated mobility-based transportation system for logistical optimization. *Logistics Spectrum*, 40(3), July–September 2006.

Bragdon, Clifford R., and Stephen Lee Morgan. New visualization technologies for port security. *Cargo Security International*, 2(6), December 2004.

Bragdon, Clifford R., David J. King, and Mathew Hyner. Virtual multi-sensory planning and technology applied to intermodal transportation safety and security. *The New Challenge of*

International Transportation Security. Institute of Traffic and Transportation, National Chiao Tung University, 2003.

Bureau of Transportation Statistics. *Pocket Guide to Transportation 2007*, U.S. Department of Transportation, Washington, D.C., Government Printing Office, January 2007.

Cohen, Charles, and Eric Werker. The political economy of natural disasters. Harvard Business School, January 16, 2008.

Farazmand, Ali. Learning from the Katrina crisis: A global and international perspective with implications for future crisis management. *Public Administration Review, Special Issue*, 148–158, 2007.

First Annual Conference on National Preparedness. *Proceedings, Global Center for Preparedness*, Florida Tech, December 14, 2007.

Flynn, Stephen, and Daniel B. Prieto. Neglected defense: Mobilizing the private sector to support homeland security. *Conference on Foreign Relations*, Washington, D.C., 2008.

Government Accounting Office (GAO). *Transportation Security Research and Development.* Washington, D.C., GAO-04-890, 2004.

Habitat II Conference on Human Settlements, Second United Nations Conference, Istanbul, Turkey, June, 1996.

Howitt, Arnold, and Jonathon Makler. *On the Ground: Protecting America's Roads and Transit Against Terrorism.* Washington, D.C., Brookings Institute, April 2005.

Hubler, Eric. The fittest and fattest cities in America. *Men's Fitness*, March 2008, 85–91.

International Maritime Bureau (IMB) Piracy Reporting Center, April 28, 2008.

Lan, Lawrence W. (Ed.). *The New Challenge of International Transportation Security*, Institute of Traffic and Transportation, National Chiao Tung University, 2003.

Mass Transit System Threat Analysis (Unclassified). Transportation Security Administration, Office of Intelligence, Washington, D.C., February 29, 2008.

Orwell, George. *1984.* New York: Harcourt, Brace & Company, 1949.

Transportation for tomorrow. Report of National Surface Transportation Policy and Revenue Study Commission, Washington, D.C., January 2008.

Younes, Bassem, and Carl Berkowitz. Guidelines for intermodal connectivity and the movement of goods for Dubai. *Logistics Spectrum, 40*(3), July–September 2006.

Zlotnik, Hania. *2007 Revision of World Urbanization Prospects*, United Nations, UN Population Division, February 2008.

2

Transportation Security Through Logistics Transformation

Charles P. Nemfakos
Sarah R. James

Objectives of This Chapter:

- Provide an overview of the complex issues involving the relationship between transportation security and a nation's physical and economic security
- Introduce a framework to explore the relationship between national security and the global economy
- Explore the relationship between enhanced productivity and economic security
- Discuss the impact of logistics transformation on security
- Introduce the concept of "dual outcomes" as it relates to logistics transformation scenarios
- Provide an overview of defense logistics transformation
- Examine the correlation between logistics transformation, transportation security, and national security

> "In today's global market, the threat to the modern nation state's economic viability is far greater than the risk of physical attack. That threat can be redressed through logistics transformation using multiple solution sets."
> —Charles P. Nemfakos, April 2007

Chapter Overview

From a strategic perspective, logistics transformation is not just a critical factor to achieving transportation security, it is the lynchpin that will allow industry not only to survive but also to thrive in today's atmosphere of hardened supply-chain management. There are aspects of this need in both the commercial domain and the national security domain: The RFID (radio frequency identification) technology that was initially introduced as bar coding in retail activities has subsequently evolved into the more recent application to container identification and tracking. We now are in a more demanding era. In order to ensure economic security for the future, nation states and businesses need to take a systems approach to the introduction and application of logistics transformation processes and technologies. This is particularly critical in the area of transportation.

The authors, who are currently working with an EU (European Union) port in achieving security through both capital investment in logistics transformation and human capital development, will draw upon their experiences to provide broad perspectives on these applications. The challenges faced by the port community in integrating new technology into an existing workforce are many. For today's logistics operations, multiple functions are automated and integrated at a level much higher than that with which the existing workforce is familiar. In addition, there is a technology gap in the workforce as older workers with the highest seniority have, for a large part, the least familiarity with the current state of information technology and instructional techniques. Maintaining technological proficiency was not previously a high priority. The desired security system automated many functions formerly done manually, integrated across agencies and information sources, and interfaced with agencies that in the past had not been required to share information or receive regular reports. International sharing of information has been automated to a level not previously seen in the global military and police forces.

Finally, the authors will address the critical applications of logistics transformation in dealing with lightening the burdens of defense spending—particularly as it relates to both stability operations and humanitarian and disaster relief. The largest impact of the United States (U.S.) Department of Defense's (DoD) current transformation continues to be in the area of integration and command and control of resources. Formerly highly stovepiped, this command and control function is being disseminated across levels of command and, in some cases, to new forces and commands with limited experience and operational understanding. Through understandings developed as a result of their work in conducting forums identifying possible "roads ahead" for the U.S. DoD's logistics transformation, the authors will provide insight on the paralysis that frequently grips companies and organizations when faced with the enormity of the totality of transformation. They will suggest that in order to combat such paralysis and enable the transformation to begin, it is highly effective to look at the various components and transformation initiatives as one would undertake a risk–reward analysis. When grouped by level of risk and degree of reward, overlaid by the element of time, initiatives can be phased such that traumatic changes in organizational and workforce culture are either eliminated or minimized. This approach would enable a systems engineering attack on innovations as varied as using the Internet to process military orders for material; producing financial reconciliations and fiduciary controls; developing a robust modeling and simulation capability; designing front-end engineering so as to improve reliability and consumption efficiency; implementing international supply-chain processes; and so on.

With this as a framework, the chapter will make clear that *logistics transformation* is an all-encompassing term that describes both the changes in the logistics processes and the impact of logistics considerations in their development, as well as the application of logistics technology to the broader operational areas, in this case national and global transportation security.

Introduction

The movement of goods in an efficient manner is fundamental to the effective operation of the global economy. This characteristic of the transportation system is mandatory for

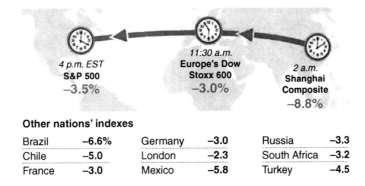

FIGURE 2.1 28 February 2007: China Sneezes and the World Catches Economic Flu

modern supply-chain operations. Anyone who doubts that we truly do have a global marketplace—a marketplace that operates with a compression in the speed with which events are manifest—need only look at events in the past year where the Chinese stock market had a correction. As the news of the correction progressed time zone by time zone across the globe, as shown in Figure 2.1, every other exchange reflected the impact of that correction. The fact of the matter is that the global economy is upon us, and we have to realistically tailor our actions to accommodate the reality of its international and national, as well as local, impact.

Two decades ago business communities worldwide saw the introduction of a series of productivity measures that led to economic expansion while avoiding inflation. In truth, there is evidence that that growth in the U.S. economy came about in an unexpected manner. Many productivity gains were the result of innovative approaches to logistics operations, specifically in supply-chain management (as in the often used examples of Wal-Mart and Dell). To the extent that security concerns disrupt productivity, the effect on the global economy can be significant. An aspect of security concerns not often discussed is the impact that catastrophic or disastrous events can have on the efficiency and effectiveness of the marketplace. Logistics planning can be an effective hedge against the turmoil that could result in the global marketplace as a result of such an event. (The most current example of this set of relationships is the effect of the 2007 earthquake on Japan's automobile industry.)

From a strategic perspective, logistics transformation is not only a critical factor to achieving transportation security, it is also the lynchpin that will allow industry to both survive and thrive in today's atmosphere of hardened supply-chain management. By *logistics transformation* we not only mean the deliberate application of continuous process improvement, advanced technology, and organizational redesign and/or restructure to existing processes and products to increase productivity and capability; we also mean the thoughtful application of change in such a manner as to achieve multiple goals. It is the evolution of this change process to incorporate the concept of "dual use" that is critical for the achievement of the next generation of productivity gains from capital investment. What we saw happening previously was an ability to absorb, in a somewhat random manner, transformational technologies and capabilities and then adapt them so as to extract economic value. For example, the RFID technology that was initially introduced as bar coding in retail activities has subsequently evolved into the more recent application of container identification and tracking. We are now in a more demanding era: In order to ensure economic security for the future, nation states and businesses need to take a systems approach to the introduction and application of logistics transformation processes and technologies. Using our RFID technology example, to extract its value for defense the technology really has to be placed in the larger framework of end-to-end logistics embedded into the larger defense enterprise. Accordingly, if one develops a construct that identifies the broadest capabilities of the logistics spectrum, at the end of the day one will have not only physical security but—more importantly—economic security as a result of the investment in the infrastructure for broader change.

In this chapter we rely, to a large degree, on the challenges faced by port communities in integrating new technologies into an existing workforce. This allows us to use perspectives in logistics transformation and human capital development in achieving the security of a key transportation node. To extract the greatest economic gains from logistics transformation requires both a systems approach to infrastructure investment decisions and a holistic approach to investment in human capital development. A critical aspect to achieving transformation of any kind is the tailoring of human capital development to the proposed investments in transformative systems.

As we have said, many of the productivity gains achieved in the world's industry have stemmed from or are based on transformation in the broad spectrum of activity that we call *logistics management*. Critical to successful logistics management is having a clear view of the outcomes that one wants to achieve. Logistics as a science permeates all endeavors of the modern industrial state, as well as the modern national security infrastructure. One absolutely needs to be able to identify the benefit of the success achieved; that is, what value is added to an industrial enterprise or to the operational effectiveness of forces by the application of effective logistics management. Having said that, we also address the planned use of logistics transformation to lighten the "defense burden" at all levels.

As in any scientific endeavor there are some basic principles that should apply to logistics management. These principles can be summed up in the concept that the logistics product or service needs to be available at the right place, time, and cost—and reliably available for whatever continuing operation it is intended. As may be seen, this touches upon the entire total life cycle of any product line. For the U.S. DoD it has meant identifying those innovations that would, on one hand, support the concept just outlined, while at the same time enhancing what we would call the *enterprise value*. This strategic approach to logistics transformation increases operational effectiveness and achieves economies and efficiencies necessary for the public good.

The Global Economy and National Security

"In today's global market, the threat to the modern nation state's economic viability is far greater than the risk of physical attack. That threat can be redressed through logistics transformation using multiple solution sets."[1] In the case for Greek port security the point was made further that logistics transformation is the keystone to commerce,

---

[1] Speech made by the author (Nemfakos) on "Logistics Transformation and The Global Economy," Unit A: Globalization and Logistics at the Warehouse Transport/Logistics 2007 International Forum, 28 April 2007, Athens, Greece.

port security, and economic stability.[2] In the Summer 2007 issue of *Logistics Spectrum*, the need for extracting both economic and physical security from logistics transformation was identified.[3] Also identified were the four significant areas where logistics transformation shows its greatest promise in extracting value: "1/ a systems approach to decisions; 2/ a holistic approach to human capital development; 3/ sound approaches to work training; and 4/ agility and flexibility at arriving at logistics solutions."[4] Central to the discussion is the notion that "Globalization is the term used to explain disparate events such as the emergence of Starbucks Coffee Company stores in remote corners of the globe, the blending of cultures as evidenced in Bollywood and other foreign film industries, and the cross-national bonds that have come to permeate social movements."[5]

Clearly, forms of globalization have been with us since the times of the Greeks and Phoenicians: Both established nodes of commerce, with factories across their colonies producing finished products. But they didn't work in real time and lacked both transparency and speed of knowledge, which is antithetical to the hallmark of today's globalization, reflecting a speeding up and intensification of connections.

Visibility is also important for the military logistician. The attendant need for better knowledge of requirements and available resources arrayed in time and space is consistent with the private sector view of the commodity-chain function.[6] The "who," "why," "what," and "where" of military visibility needs correlate to perspectives on the global city, and the need for a systems approach.

With the demise of the bipolar security environment (i.e., the United States and its allies balanced against the Soviet Union and its clients), the nature and focus of the meaning of *security* have evolved. Increasingly, the notion that economic security is just as integral to the security of any nation as physical security has found resonance in national security planning. For any nation, the collapse of its markets, banking system, and critical infrastructures (e.g., electricity, communications, etc.) is as profound a concern as vulnerability to missile or aircraft strikes, or land invasion. What is significant is that this view of economic security inevitably leads planners to view events in the global economy with as much interest as the latest developments in weaponry reflected in Jane's. In point of fact, often economic security concerns trump physical security concerns. An example of this dichotomy is reflected in recent decision making by the Association of Southeast Asian Nations (ASEAN)[7] member states, where regional security interests were trumped by more economically rewarding bilateral economic ties.[8]

---

[2] Presentation made on 29 April 2007 by the author (Nemfakos) to port interests of Patras, Greece.

[3] "Notes About the Theme," author (Nemfakos), *Logistics Spectrum*, Vol. 41 Issue 3 (July–September 2007), p. 2, SOLE - The International Society of Logistics: Hyattsville, MD.

[4] *Ibid.*

[5] Sowers, Elizabeth A., "Mobility, Place and Globalization: Toward a Sociology of Logistics," *Logistics Spectrum*, Vol. 41 Issue 3 (July–September 2007), p. 12, SOLE - The International Society of Logistics: Hyattsville, MD.

[6] Christianson, LTG Claude V., "In Search of Logistics Visibility: Enabling Effective Decision Making," *Logistics Spectrum*, Vol. 41 Issue 3 (July–September 2007), p. 18, SOLE - The International Society of Logistics: Hyattsville, MD.

[7] Association of Southeast Asian Nations, www.aseansec.org.

[8] DHL commissioned research by the *Economist* Intelligence Unit entitled "Trading Up: A New Export Landscape for ASEAN and Asia," November 2007.

A further example of this linkage of economic activity with nation-state security concerns was raised in December 2007 by the Greek minister of the Interior. Reiterating comments made earlier in the year by the Greek foreign minister (Dora Bakoyannis, in ministerial meetings in Italy and the United States[9]), he especially highlighted the dual threats to Greek and western European security posed by illegal immigration from potential terrorists, and the smuggling of goods and human cargo for illicit purposes.

There is tension between the needs of a developed nation-state for cheap labor, goods, and materials and the nation's assurance of the safety of its citizens and culture. There is corresponding tension between the needs of developed nation-states and the desires of individuals from developing nations to participate in the economies of developed nation-states, both as producers of goods and as migratory providers of labor. While these tensions are most visible today in the European Union and, to a certain degree, the United States, this new form of human migration may be the Achilles' heel of achieving the fruits of the global economy while ensuring national security.

This evidence of the linkage between the global economy, with its attendant movement of goods and people across national borders, and concern for national security, with its need for knowledge of ongoing activity and its control, is manifest in a broad array of capital investments and international agreements. All the investments and agreements have similar fundamental goals: transparency, visibility, information security, and improved functional response times. These goals are not unique either to the global economy or national security. Rather, they illustrate the blending of the requirements of each.

FIGURE 2.2 Greece: A Strategic Crossroads in the Global Economy

Greece serves as an interesting microcosm for our purposes because it is, and has been since the outset, both an EU member state and a founding member of NATO, the

---

[9] As cited on http://news.pseka.net/index.php?module=article&id=6631&print=1, Ms. Bakoyannis made the following comments in her meeting on 22 March 2007 with Italy's foreign minister, Massimo d'Alema: With respect to terrorism, "it is clear that we cannot be complacent, nor can we limit ourselves to isolated actions. On the contrary, it is necessary that we be vigilant and cooperate. … Our geographical position in the Mediterranean and our role as 'access gateways' to Europe attract waves of migrants to our two countries, rendering cooperation a necessity." She further said that the EU countries have come to realize that in order to confront illegal immigration and terrorism, cooperation was necessary among all the states, including those from which the migrants originate.

security umbrella of the economic union. Because Greece's economy lags behind some of the other EU states—particularly in the areas of infrastructure investment—it will serve us well in our examination of productivity investments, logistics transformation, and transportation security. This last aspect is particularly useful, given Greece's geographic position and the amount of its economy that is tied, either directly or indirectly, to secure and reliable transportation (see Figure 2.2). The effects of the lack of a reliable transportation system are readily seen in the enormous dislocations that occurred in China in January and February of 2008.[10]

Economic Security and Enhanced Productivity

Through the middle of the first decade of the 21st Century there has been sufficient capacity to effect change through the introduction of technology, process improvement, and training such that nation-states could allow initiatives in these areas to take hold with little or no advanced planning. Serendipitously, and as a result of after-the-fact observation, it was determined that these unplanned productivity gains had produced a dampening effect on global inflation for close to two decades. Because of the methods by which they were introduced, these gains occurred in individual silos of endeavor. The fact that this source of economic security (i.e., productivity gains) has been largely consumed is evidenced by the effects on the global economy of the removal of the artificial support for growth that occurred in the last half of 2007 and the first half of 2008. When paranormal financial mechanisms as engines of growth (such as the financial vehicles that contributed to the collapse of the U.S. mortgage market) were removed from the global economy, it became obvious that differing forms of productivity would be needed to contain worldwide price inflation. That is particularly true in any environment where fuel and other commodities are in a high demand–low supply situation. This differing form of productivity is one that looks across multiple silos of endeavor to achieve improvements while limiting capitalization demands.

This enhanced productivity is fundamental to achieving economic security in the new global market paradigm that is emerging. Logistics transformation is a critical enabler in allowing the capability to look across silos to achieve this latest range of essential productivity gains. In a way, the Japanese experience in reviving the country's economy after the "bubble" burst is a testament to this thought. The Japanese prime ministers (most notably Keizo Obuchi and, to a lesser extent, his predecessor Yoshiro Mori) formulated to the Japanese Diet that logistics transformation was a critical element to reviving the country's economic growth and achieving economic security.[11] It is noteworthy that logistics transformation shares with banking reform the requirement for the same factors to enable their achievement: transparency, visibility, information security, and agility. It is

---

[10] The challenges facing the People's Republic of China as it modernizes its economy and transforms its infrastructure are extraordinary topics worthy of study by themselves. While the authors recognize that much can be learned from the Chinese experience of this decade, frankly, the subject matter is so overwhelming that doing it justice in this chapter would overwhelm the larger lines of thought being pursued.

[11] The streamlining of business operations—particularly those within the logistics enterprise—has been a key element of the many initiatives Japan has embarked upon since it first undertook a program of economic reforms in 1996. Beginning with its "Program for Economic Structural Reform" in 1996, and continuing through the May 1997 adoption of its "Action Plan for Economic Structural Reform," the "Industrial Revitalization Program" of January 1999, and the convening by Prime Minister Mori of the "Industrial Rebirth Council" in July 2000, the government has been dedicated to strengthening the cost competitiveness of doing business in Japan by improving the efficiency of logistics and promoting reform to reduce Japan's high cost structure.

also noteworthy that these precepts are similar to those driving capital investments and international agreements.

We can see, therefore, that it is no longer sufficient to say that "I have developed a warehouse system [or a trucking control mechanism, or the best procurement system] in the world." We need to take a systems engineering approach to assure that investments in productivity enhancement are melded in a manner so as to achieve the highest gain from the range of capital investment made. The expression of capital investment is an expression that, for our purposes here, encompasses infrastructure investment, material investment, technology investment, process improvement investment, human capital investment, and training investment.

Given the extraordinarily broad range of the coverage of these investments, and the demands of any scientific application of systems engineering to the problems of logistics transformation and the ensuing enhanced productivity, one must focus on outcomes. For our purposes, the outcome desired is to assure the largest measure of productivity in the whole enterprise. By *whole enterprise* we can mean the nation-state, an industry, a segment of the population, or even a profession. In ensuring that the right materials are at the right place, at the right time, and for the right cost, it may well be that while the best warehousing system may serve supply-chain management best, it may behave in a manner that actually serves as an impediment to achieving the greatest amount of productivity in the "whole enterprise." It is for this reason that the authors focus on the science of logistics as opposed to the practice of supply-chain management (which is a subordinate element of logistics).

The continuing growth of globalization, even in the face of widespread disenchantment with its effects, places an extraordinary burden on logistics transformation to enable economic security. The seminal study on the effects of globalization on security, commissioned by the U.S. Department of the Navy[12] and performed by the U.S. National Defense University in 2001[13]), highlighted both the security repercussions resulting from changes in the global economy and the transformational impacts—both economic and national—of those changes. From a different perspective, there are two significant examples of the linkages between the global economy and economic security: the events surrounding the 2004 strike in the U.S. West Coast ports, and the stock market correction in February 2007 triggered by the drop in the Shanghai Composite Index. As Figure 2.1 illustrated, extraordinary speed characterizes the global marketplace.

The "perfect storm" in the Los Angeles, California, area ports illustrated the capability to dam the "river of commerce" when technology failures, insufficient device and process reliability, and information and communication network failures intersect at a single point of failure.

---

[12] The project was initiated by then undersecretary of the Department of the Navy, Jerry MacArthur Hultin (now president of Polytechnic University), with deputy undersecretary of the Department of the Navy, Charles P. Nemfakos (chapter co-author). Their personal involvement reflected the emerging understanding of the forces the global economy was exerting on the national security issues of the U.S., with which they had to deal on a daily basis.

[13] Ellen L. Frost and Richard L. Kugler, editors, *The Global Century: Globalization and National Security* (two volumes), Institute for National Strategic Studies, National Defense University: Washington, D.C., June 2001. A summary report of the project on globalization and national security can be found on the National Defense University Web site at http://www.ndu.edu/inss/books/Books_2001/Challenges%20of%20the%20Global%20Century%20June%202001/CHALENG.PDF.

July 2004 –
LA "Perfect Storm":
and the economy freezes

   • *Failure of technology, insufficient*
   device reliability and shipping
   network failures
   • *No cameras for the World Series,*
   Marshall Fields' sale fails for lack of
   merchandise
   • *Dammed the "River of Commerce",*
   affecting $1B/day in goods and
   2.8 million jobs

FIGURE 2.3 Los Angeles, California, 2004 "Perfect Storm"

Enhanced Productivity and Logistics Transformation

The collapse of the logistics infrastructure at the Port of Long Beach, California, with its attendant inability of the global economy to recover, is illustrative of the inter-relationships that exist in today's markets (see Figure 2.3). A more recent example (albeit with less evident impacts, because it certainly didn't impact the U.S. World Series) is the effect on automobile production by earthquake damage at a Japanese small auto parts manufacturer.

Introducing relevant technology, ensuring against device or process failure, and building redundantly reliable communications networks present a high cost burden to any economy. In the Japanese example, maintaining multiple sources of supply erodes manufacturing economy.[14] These cost burdens, not dissimilar to notions of the defense burden, are large and present market and political difficulties in sustaining support for incurring them over time. In the market and in defense, the theory of "dual use" has been advanced as a way to reduce cost burdens while increasing capability. However, "dual use" merely represents an effort to distribute overhead costs of item production to multiple users. The more appropriate formulation would appear to be one of achieving "dual outcomes." *Dual outcomes* is the process of affecting transformative activity in a resource-constrained environment by achieving several desired outcomes through a single capital investment.

The premium on the most efficient and effective application of talent, time, tools and capital—which we will subsequently refer to as "T^3C"™—results from a combination of applying a systems engineering approach to logistics transformation, and expending political capital to reduce the defense burden (i.e., physical security) while securing another desired outcome (economic security) from capital investments made. Achieving

---

[14] Of note is the fact that a proposed, but not implemented, consolidation of Japanese automaker multiple parts inventories into a single common inventory would actually serve as an aggravating event in these circumstances. Instead of achieving manufacturing economy, one would achieve—with a single point of failure—manufacturing gridlock, with its attendant cost increases.

dual outcomes is particularly important in the global marketplace as it has evolved, where economic security may be a more important aspect than physical security.

Multiple Scenarios of Logistics Transformation

It could be said that a nation-state's logistics energy and resources are applied without boundaries across its multiple and disparate political and economic segments. In approaching logistics transformation, two types of strategies could thus be employed: either an approach that aligns logistics segments to market segments, or one that customizes the delivery of logistics services to the individual customer of the services. In addition, a case can be made for the importance of dual coordination: coordination between the organization and its customer, and coordination among a firm's different services.[15] Arriving at the correct decisions for T^3C^{TM} application—which represents a form of risk–reward analysis—the notion of taking a systems engineering approach to making logistics management decisions becomes apparent.

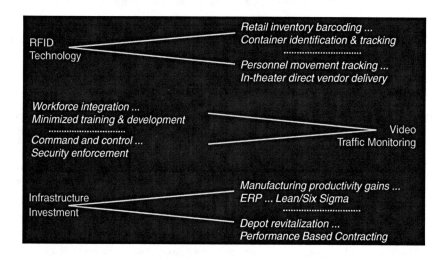

FIGURE 2.4 Examples of "Dual Outcome" Logistics Transformation Initiatives

By *dual outcome*, we mean avoiding making capital investment decisions with a focus of the cost–benefit analysis being made within a single silo or swim lane. So, for example, (as illustrated in Figure 2.4) one could see RFID technology as a solution to inventory tracking. Alternatively, it could be seen as a solution for personnel location and performance management. A similar example is the use of video traffic monitoring to minimize training and development requirements through better workforce integration, while—at the same time—achieving physical security enforcement through enhanced command and control.

The employment of a dual outcome approach automatically forces the development of logistics transformation scenarios that apply to not only the logistics enterprise but

[15] Sendel, Marie-Pascale, Ph.D., "Innovation in the Logistics Field: Collaborative Practices between Manufacturers and Retailers," *Logistics Spectrum*, Vol. 37 Issue 4 (October–December 2003), p. 22, SOLE - The International Society of Logistics: Hyattsville, MD.

also the corporate enterprise. While many of the scenarios are technology driven, many others are the result of process improvements that, in turn, drive the development of transformational technologies. Some scenarios that have had or are currently having a significant impact on logistics transformation are highlighted as follows:

> *Enhanced management of range and depth.* In a logistics transformation roundtable discussion hosted by SOLE - The International Society of Logistics, Steven Garrett, founder of the Phoenix Genesis Consortium (an internationally focused group of technology and security providers), made the important point that in the global economy, there is a higher value than that derived from "just-in-time" delivery. To fully leverage life cycle support, one must actually have knowledge "... down to the split second of where things are and where things move. Your sphere of control is total, and your knowledge of the handling and the handlers goes down to give actual responsibility to each and every stage of shipping."[16] This implication for the need for speed and transparency is, in fact, at the heart of e-commerce, with the key being increasing speeds of information transfer and access from all levels—operational to strategic—throughout the enterprise. To sustain the speed, obviously, it requires a shared database that supports the entire enterprise from development through manufacturing to delivery to the market.

> *Web-based commerce.* In the past two decades, the Internet has gone from a curiosity and personal tool of convenience to a critical infrastructure element used both internally (e.g., with suppliers) and externally (e.g., with customers) by businesses. The exchange of information between manufacturers and their suppliers, retailers, and shippers is critical to inventory management and order fulfillment. More and more retail sales are being handled directly between the consumer and the manufacturer, distributor, or retailer. Further, retailers are using electronic point-of-sale systems to capture and analyze emerging sales trends.[17] Customers routinely make use of "made to order" online ordering systems for everything from computers (e.g., Dell) to denim jeans (e.g., Levi). In the U.S. Department of Defense, the requisitioning process has moved from punch card and paper to the Defense Logistics Agency's "DoD Emall" online ordering.

> *Life cycle support.* The same principles that apply to range and depth apply to life cycle support. The introduction of best practices through logistics engineering allows for the integrated management of systems from initial design throughout the entire life cycle of the system (whether data, process, or material). This results in such diverse elements of efficiency and economy as the reduction in the cost of design development, production engineering, fabrication, and unutilized excess material; the reduction in the time necessary to effect those elements; and the improvement of control, flexibility, and risk management.

---

[16] "Security Through Logistics Transformation," SOLE - The International Society of Logistics Roundtable, Washington, D.C., February 2007.

[17] Sendel, Marie-Pascale, Ph.D., "Innovation in the Logistics Field: Collaborative Practices between Manufacturers and Retailers," *Logistics Spectrum*, Vol. 37 Issue 4 (October–December 2003), p. 25, SOLE - The International Society of Logistics: Hyattsville, MD.

Vehicle health management. While often described in terms of specific "whats" (e.g., onboard diagnostics and prognostics, offboard integrated diagnostics, autonomic system monitoring), vehicle health management is a technology and process that allows for the reduction of redundancies (i.e., systems, spares, support equipment, and personnel). Further, with implementation it reduces the total cost of ownership and improves system reliability, with the net effect that "... continuous monitoring, analysis and [system] improvements will cost less than the added cost of not implementing ... over the life of a product. Besides the tangible cost-benefit, customer loyalty, reliability and safety are important factors that make an [integrated] vehicle health management program a must-have requirement."[18] It is the very type of technology that enables the visibility of where things are and where things move. In fact, this may be the clearest example of applying technology investment to achieve dual outcomes.

The preceding four illustrative examples are the beginning of a process that will ultimately require a systemic and systematic application of logistics engineering to pursue dual outcomes as an intended practice, as opposed to a serendipitous consequence of arriving at a particular solution set because it is an inevitable result of the transformation process. What should be clear is that the next step in achieving both economic security and physical security through logistics transformation is the ability to harvest gains from investment by achieving dual outcomes.

Defense and Logistics Transformation

What is obvious is that there are common characteristics that support all logistics systems' transformations. Those are agility, transparency, and robust communications. These threads of the global economy and enhanced productivity, addressed in earlier sections of this chapter, represent a consistent theme that is reflected in any endeavor in the modern logistics enterprise.

The duality of purpose in making capital and workforce development investments is central, on many different levels, to achieving efficiencies and economies. In the security domain the mandates for dual outcomes are generated by disasters, human devastation, terrorism, and political instability. These are more prevalent and, in the global economy, felt more pervasively than ever before. Accordingly, businesses need to improve their physical security capabilities while—at the same time—creating corporate value. These, then, need to be integrated across national economies, starting from remote entities (such as vendors in call centers) and concluding in core business operations. The relationship between corporate value propositions and physical security—whether reflecting disaster or political instability—are exemplified by the recent events in China as a result of the instability effected during the 2008 Lunar Year celebrations and a little less recently in the Japanese earthquake of 2007.

Logistics transformation is essential if national security, both economic and physical, is to be responsive to the vagaries of the global economy. Today's social and economic structures inform and influence the business of logistics. In defense, the United States arrived at a defense logistics "tipping point" post-1989 (see Figure 2.5).

---

[18] Hampson, Todd C., "Integrated Vehicle Health Management Requires Data and Integration," *Logistics Spectrum*, Vol. 39 Issue 2 (April–June 2005), p. 18, SOLE - The International Society of Logistics: Hyattsville, MD.

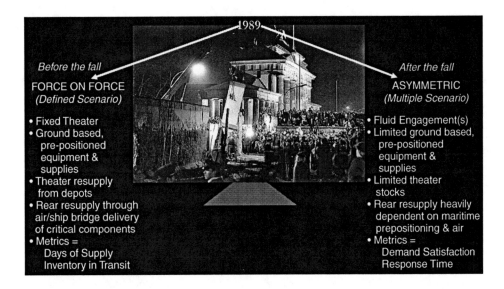

FIGURE 2.5 U.S. Department of Defense Logistics Tipping Point

Prior to the fall of the Berlin Wall in 1989, the U.S. Department of Defense dealt with defined scenarios that could be effectively supported through pre-positioning, with resupply effected through air–ship bridge delivery from central depots at the rear with visibility being expressed in terms of "days of supply" and "in-transit inventory." The migration to multiple scenarios supported by limited pre-positioning and requiring resupply heavily dependent on maritime pre-positioning and air delivery changed the metrics to "demand satisfaction" and "response time." This required a transformation in military logistics that reflected both the technology of weapons and forces as well as the social structure (e.g., dealing with terrorists, coalitions, and nongovernmental organizations) at the modern point in time. Recognizing the extraordinary dynamics attendant to dramatic transformation and the need, therefore, for an acculturation of defense logistics transformation, the U.S. Department of Defense initiated a series of forums to examine certain aspects of required changes. We will briefly discuss the results of those forums and draw relationships to activity in the civil sector and interrelationships created across the logistics management spectrum.

These forums had participation from thought leaders and practitioners from various DoD components, the private sector, and academia. Total participants numbered in the hundreds, and the range of propositions addressed fully reflected the characteristics of this large and diverse group of participants. The various ideas were winnowed through the use of a decision matrix that characterized ideas proposed in terms of benefit to risk, both economic and cultural. A representative sampling of those ideas that would entail capital investment included the following:

- *"Webify" the requisition process*: Management of material ordering, delivery, and return through portal management.
- *Develop and source dynamic decision aids*: Introduction and use of decision analysis tools (e.g., expert systems and artificial intelligence).

- *Establish integrated systems for enterprise-wide fiduciary controls and financial reconciliation*: Development and implementation of functionally integrated enterprise-wide resource systems.

- *Develop a robust simulation capability*: Use of modeling and simulation for both training and process re-engineering.

- *Develop flexible and adaptive training curricula*: Design and promulgation of standards and certifications.

- *Acquire required sensing and monitoring*: Development and use of autonomic systems.

- *Develop open-ended and integrated structures*: Increased design and use of dual outcome solution sets for use in a net-centric enterprise.

- *Design and build configurable and flexible facilities*: Investment in enabling adaptive reuse of physical infrastructures.

- *Upgrade IT infrastructure and architecture*: Modernization of backroom functions.

- *Develop an adaptive, versatile, diverse, and integrated leadership and workforce*: Transition to new generations with new skill sets for success through change.

- *Develop and support meaningful mechanisms for recognition of competency and excellence*: Development of "sincere" incentive structures tied to performance and outcomes.

- *Design and implement a joint logistics value net*: Investment in systems to increase whole enterprise visibility and transparency.

While there were other significant ideas, they were more tactical in nature or did not require the same level of capital investment.

What is interesting to note is that many, if not all, these ideas for DoD logistics transformation have approximations in the civil sector. For example, if one were to examine the Dell approach to computer manufacturing, delivery, and support of "build on demand" items, one would find many of these same principles being applied—not to achieve physical security, in the national defense sense, but to achieve economic benefit (i.e., economic security). If one were to examine Wal-Mart's approach to supply-chain management, even though it does not entail the full range of logistics management, one would find many of the same ideas discussed in the reduction in the need to reduce the size of the DoD's logistics footprint. The difficulty in, and critical importance of, effecting this cultural change (i.e., reduction of the logistics footprint) is evidenced in the ongoing Iraqi and Afghani campaigns.

Interestingly, both the Dell and Wal-Mart experiences have a strong element of the transportation functionality transformation embedded in them. In both those experiences, as in the U.S. DoD, capital investments had to be made with the view of dual outcomes, because both physical security and efficiency (i.e., economic security) were at stake. The value of that stake can be perceived by examining the amount of U.S. DoD spending on transportation, as illustrated in Figure 2.6.

Transportation Security Through Logistics Transformation

The transportation network is comprised of many nodes. The large economic values devoted to transportation support the notion that examining any one of those major nodes would allow us to address the issue of achieving transportation security through

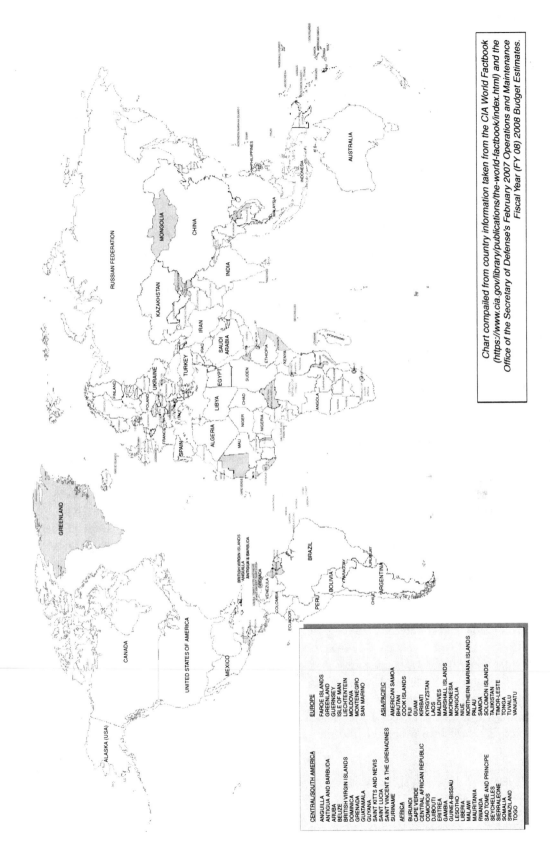

FIGURE 2.6 U.S. DoD FY 2006 Transportation Expenditures: Greater Than 62 Country GDPs

logistics transformation. And, as we pointed out previously, this security has dual aspects: security from an economic perspective and security from a physical perspective. It, therefore, becomes a useful stage from which to see the pursuit, during investment decision making, of dual outcomes.

The authors have done much work with various Greek maritime interests on the challenge of investing in and achieving port security. The reasons for this concentration by the authors were that there were enough Greek ports to satisfy the needs of systematic examination, and some of the Greek ports were of sufficient size so as to be able to draw comparisons with and analogies to other major ports in the global economy. In addition, they reflected a national coherence because of the size of Greece and its geographic relationship to many of the elements of activity that we addressed earlier as corrosive to security (e.g., terrorism, political instability, and disasters). Finally, the nature of governance in Greece makes examination easier than, for example, looking at ports in the United States, even though the issues may well be the same.

An interesting aspect to our examination of capital investment needs is the decision on the part of the current Greek government to seek privatization of the port node in its transportation system. Specifically, this privatization represents an effort on the part of the government not dissimilar to the efforts in many of the U.S. states, which have turned over to private investors the ownership and/or management of elements of their transportation network (e.g., toll roads and bridges that support interstate commerce[19]). In the United States, the privatization was done to effect efficiency, generate new capital investment in modernizing the infrastructure, and remove from the shoulders of the state the burden of infrastructure management. One could surmise that the Greek government would welcome privatization for the same reasons, and in addition, for the purpose of relieving the Greek government of the responsibility for investments in and development of the workforce.

This need for investment in human capital is fundamental to achieving any large transformation. It is particularly the case when the workforce views its jobs as a right and, thus, views any proposed transformation as a threat to its individual personal interests. Accordingly, to be successful, the workforce must have a clear understanding of the goals of the transformation, an acceptance of the incentives that are being offered to encourage change, and a belief that management's expression of intent and purpose is sincere. This all represents an initial cost that the enterprise must be willing to bear. The cost is a result of both economic incentives and the expense of educating and retraining the workforce. This education and retraining activity needs to be done in a manner that fully conveys the sincerity of enterprise management to the transformational change. To say it somewhat differently, training courses and seminars need to be given over periods of time and with the frequency that is consistent with the workforce's understanding of the degree of difficulty represented by the transformational change.[20]

The view, in both the United States and Greece, is that the private investment community can deliver these transforming ideals in a more rapid and agile manner than can any government entity. In a way, this is an instance of government enlisting the private sector to enhance economic and physical security through investments that would

---

[19] As a specific example, the Chicago Skyway Bridge is managed by a private Australian and Spanish consortium that collects $3 each time a car crosses the 7.8-mile span.

[20] "Security Through Logistics Transformation," SOLE - The International Society of Logistics Roundtable, Washington, D.C., February 2007.

result in the transformation of major nodes of the nation's logistics system. The reason for this is that the global projection of the sums required for all infrastructure investment suggests that it is larger than the entire gross domestic product of most nations.[21]

If we examine the functional aspects of the areas the authors have recommended to the Greek authorities for consideration in managing their port enterprise, one can readily recognize many of the transformational constructs that we already discussed, in both the defense sector and the general national civil sector. At a very high level these recommendations deal with making investments in real-time, indoor and outdoor asset management technologies. These, coincidentally, also provide visibility into manufacturing processes. Recommendations have been made to introduce these transformational products and processes on a sector-by-sector basis, thereby enhancing opportunities for scalability. The authors introduced the notion of economic enhancement with the economy of material handling being the engine to deliver new and higher levels of return on investment, and with the dual outcome of physical security enhancement being a by-product of the investment. Finally, recommendations were made to fully integrate sensor-based technology, which on the one hand enables physical security and on the other hand increases economic benefits by creating customer value through visibility. To one degree or another each of those fundamental ideas was addressed in the series of DoD Logistics Transformation forums. While because of the unique requirements of U.S. national security they may not have emerged as the most prominent transformation ideas, all the recommendations previously discussed were also areas of interest in the forums.

While there was general acceptance of the proposition that logistics transformation could affect transportation security, the Greek experience has been muted by the dynamics of civil–political events. However, even though the Greek port security issue is—in the minds of the authors—the exemplar for affecting security through logistics transformation, the idea of the use of this transformation in the port node is widespread. For example, in the summer of 2007 the American Association of Port Authorities (AAPA) addressed the application of high-dollar value grants for the purpose of enhancing facility security at America's seaports. In their view, the financing was needed to pay "... for access control systems, personnel training, waterside security and interoperable communications, as well as the costly facility requirements related to implementing the new Transportation Worker Identification Credential."[22] These types of investments are both consistent with the logistics transformation needs addressed throughout this chapter and similar in the attribute of dual outcome, as each contributes to the physical as well as the economic security of the port authorities and the nation.

Ports, of course, are only one—albeit, in the United States, a large one—node in the transportation network. However, if one expands beyond the port node into the railway node that supports movement to and from the port, one can see similar needs for improvements in functionalities. Interestingly, some of the technological innovations that are becoming available would appear to advance the notion of dual use beyond the

---

[21] "Investors are waking up to the sound of jackhammers worldwide. In the next 20 years, the tab to build and maintain roads and bridges, and to create and maintain systems that deliver electricity, water, sanitation and telecom services will swell to $30 trillion. Rich countries must upgrade decades-old infrastructure, and developing nations must build it to make their economies competitive." Frick, Bob, "Infrastructure, for Stability," Personal Finance (Kiplinger), *The Washington Post*, February 24, 2008, p. F3.

[22] "AAPA Hails Additional Port Security Funding Announcement," *Maritime Executive Newsletter*, 16 August 2007.

economic and physical security aspects that we address in this chapter. For example, the emerging use of Global Positioning System (GPS) technology by rail shippers enables successful network integration between shipping companies. Obviously, that interconnection has both economic and physical security benefits for the shipping companies. If that network is interconnected with port authority asset management visibility through the use of container RFID, it enables delivery of a real-time location and asset condition monitoring capability to the commodity user. Particularly when one considers the network's application to the movement of hazardous materials, one can readily see the national security aspects of this capability. This manifestation assumes increasing importance when dealing with the security mandates of terrorism and political instability addressed previously.

Transportation Security and National Security

At a conference hosted in July 2007 by the International Cargo Security Council (ICSC) in San Diego, California, that focused on port security with an emphasis on combating cargo theft, the attendees discussed measures and preparedness to counter this global security threat, which has both economic and physical security aspects. This concern is not limited to the United States: In March 2008 the government of Turkey hosted an international conference dealing with border security and global terrorism, where the concerns were the same as those discussed in San Diego in 2007. While the topics covered national counter-terrorism activity, the program focused on how to use technology to improve screening and enhance port security. It is clear that the vulnerability of any country's logistics network is a global concern. The notion of the "global city" and its implications for economic security are integral to any discussion of economic security in the global economy. The application of technology both to enhance asset visibility and to increase the transparency of transactions and the visibility of threats is at the heart of reducing logistics vulnerability, thereby enhancing both transportation security and national security.

Issues as diverse as maintenance and piracy serve to highlight this notion of reducing vulnerability through logistics transformation. Autonomic systems (which we briefly discussed in the paragraph on vehicle health management) would have been of material use to the British Royal Navy in January 2008. At that time, it was reported that one of their few remaining aircraft carriers had to return to port because a refrigeration unit used to store meat broke down shortly after the ship left Portsmouth to join multinational operations in the Indian Ocean.[23] That deficiency in the logistics management chain could have had significant national security implications, rather than just the embarrassment of the French reporting on the inadequacies of the British. A similar type of event, on any of several parts of the logistics management chain, could just as easily have severe economic or environmental repercussions. While we have not focused in this chapter on environmental implications, the breakdown in supply-chain shipping in the last year in Asia, Europe, and the Americas highlights the damage incurred to national economies as a result of disasters of an environmental nature.

At the other end of the continuum lies the highly critical issue that piracy poses to a nation's economic and physical security. The capital investment or capital asset application needed to combat this threat to logistics management is phenomenal, both because of the asymmetric nature of the threat and because of the lack of systems

---

[23] "Faulty Fridge Sends Warship Back to Base, " Agence France-Presse press release, 23 January 2008.

(e.g., communications, sensors, platforms, information) dedicated to dealing with it. For example, 20 different countries—including the United States—are chasing a variety of pirates off the coast of Somalia, who are attacking the supply chains of a broad variety of East African countries.[24] The degree of difficulty in providing asset visibility, interconnectivity (e.g., people, ships, and countries), and command and control in a threat and counter-threat environment such as this can only be imagined. And yet, to ensure physical and economic security, solutions must be found. This problem has become so prominent that the resolution of it is now listed as a goal of the U.S. Navy's maritime security operations.[25] In the summer of 2007, the *Smithsonian* magazine devoted six pages to a colorful description of this piratical activity, in the process making the point that "No vessel seems safe, be it a supertanker or a private yacht."[26] The article makes the point that, with 95% of the world's trade traveling by ship, there is no shortage of targets.[27] Consequently, the potential for damage to the global economy is extraordinary.

Summary and Conclusions

At the national level, one can develop a set of desired activities that can be seen as an expression of the idea of dual outcomes. These goals can be supported by one or more outcomes of transformational tools that could be applied to the process. In our example, the tools would produce visibility as one of several possible outcomes that would support national goals. These national goals have aspects of both economic and physical security.

A similar construct to that reflected in Figure 2.7 could be made at the local level, or for an enterprise. While the names of the goals would change (e.g., "Industrial Capability," "Global Stewardship," "National Aspiration," and "National Security" in Figure 2.7), the construct would remain constant.

As previously discussed, there are sets of solutions that if applied correctly would allow the achievement of the construct depicted in Figure 2.7. In our examination of Greek port security, the solution sets identified were in three specific areas: physical, economic, and—not surprisingly—political. Physical security solutions included underwater, surface, perimeter, and personnel security initiatives. In addition to the obvious economic benefit from the integration and protection of the national supply chain, other aspects of the economic security solution sets included the development and use of trusted shippers, improved cargo visibility (a variant of supply chain's asset visibility), and inventory transparency. Finally, the most difficult to accomplish was the political solution set composed of centralized coordination, global integration, human capital development, and stability of the "Politeia."

---

[24] "20 Navies Join Forces to Ensnare Pirates off Lawless Somali Coast," AP Wire release reported in the *Arizona Daily Star*, 2 December 2007.

[25] "Navy Maritime Domain Awareness Concept 2007," disseminated to the field on 29 May 2007 under cover letter from Admiral M. G. Mullen, Chief of Naval Operations. The document " ... provides overarching guidance for the development and application of Maritime Domain Awareness (MDA) across all levels of command for the United States Navy. It provides the conceptual framework to prioritize MDA efforts across Navy, ensure alignment with external MDA initiatives, and inform the Fleet MDA Concept of Operations (CONOPS)." The issue of piracy is addressed in the section entitled "Enabling Naval Operations," as called out in the strategic objective "Strengthen existing and emerging alliances and partnerships to address common challenges." A .pdf copy of the concept can be found on the Navy's Web site at http://www.navy.mil/navydata/cno/Navy_Maritime_Domain_Awareness_Concept_FINAL_2007.pdf.

[26] Raffaele, Paul, "The Pirate Hunters," *Smithsonian*, August 2007, Vol. 38 Issue 5, p. 41.

[27] *Ibid.*

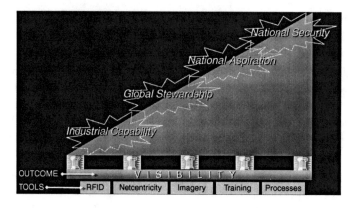

FIGURE 2.7 Security Through Logistics Transformation

Just as we observed in the discussion of Figure 2.7, while these solution sets address national perspectives, they can just as easily—and effectively—address enterprise needs. In this case, we don't even have to change the goals, just some of the characterizations of where and how the solution sets would be applied. For example, underwater security would not be required for an enterprise that was not contiguous to a body of water. But it *would* be required for an enterprise that was supported by a massive system of water and sewage distribution (e.g., a power plant). Our conclusion, then, is that achieving economic security (by means of effective efficiency) and national security (by means of enhanced physical security) is fundamentally achieved through the use of logistics transformation. Our further conclusion is that transportation security, both economic and physical, can be achieved only through transformation of the logistics enterprise, which in turn transforms the "whole enterprise." Finally, given the competitive demands of the global economy, all investment in transformation (both infrastructure and human capital) must strive to achieve—through the application of the principles of logistics engineering—the "dual outcome" approach to ensure the appropriate application of T³C™.

3

The Need for a Transportation Systems Approach

Clifford R. Bragdon, Ph.D., AICP, FASA

Objectives of This Chapter:

- Convey the importance of a systems approach to transportation security
- Point out impediments to an integrated system of movement
- Provide examples of failure when systems integration has not been effectively employed
- Discuss security and threat level where there are intermodal transportation infrastructure facilities
- Describe characteristics of an integrated system of transport or intermodalism
- Discuss trend analysis and the potential for achieving a systems approach

> *"We need to build resiliency into the systems that make modern American life possible—transportation, communications, trade, basic infrastructure and government agencies."*
> —Stephen Flynn, author of *The Edge of Disaster: Rebuilding a Resilient Nation*

Introduction

Transportation is the backbone of our urban society. It is the mechanism by which all movement occurs. Whether human society is agriculturally or industrially based, representing either a pastoral food-gathering culture or one that relies on mechanization and production, transport has been an integral element in its survival and evolution. The rise and development of all civilizations have relied on mobility for social and economic commerce. Depending upon its historical point in time, land travel has utilized either nonmotorized (i.e., walking, bicycles, carriages, animals, etc.) or motorized (i.e., automobiles, trucks, motorcycles, buses, aircraft, etc.) movement. Consequently, transportation activity is directly related to both human and economic survival, the fundamental tenets of civilization. No civilization has ever survived without an effective system of movement incorporating land, sea, and/or air.

An integrated intermodal system of transportation involves the safe, secure, sustainable, and efficient movement of people, goods, resources, and information traveling by air, land, and sea. This complex system of integrated or holistic transport consists of both physical modes of movement (classically involved with roadways, rail, transit, airports, maritime transportation, utilities, and pipelines) and electronic modes of movement (associated with communication, electronic data interchange, related information technology, satellite, and digital and fiber optic connectivity and interoperability). However, this desirable systems approach, seamlessly integrating physical and electronic modes of movement, is virtually nonexistent at present. The result is a gridlock condition that underoptimizes our economic potential and social responsiveness and that impairs the total effectiveness of emergency preparedness, response, and recovery. Interoperability is lacking and a fusion of physical and electronic mobility must be seamlessly developed before we have optimized systems integration.

It is becoming evident that mobility is an integral part of a physical and logistical system supporting our economic way of life, as well as a possible genetic physiological necessity that our human population needs and desires. Impaired mobility has direct effects on the emotional, social, and physical well-being of the individual's health (utilizing the United Nations definition of health being not merely the absence of physical disease but also the emotional and social well-being of the individual).

Constrained transportation-related conditions involving both drivers and passengers result in health-related impacts ranging from nuisance to hazard potential. Neural–humoral stress response in traffic-related conditions can include elevated blood pressure and cardiovascular and circulatory effects. Performance efficiency can also be impaired, resulting in mental health–related stress outcomes whose by-products are road rage, aggressive behavior, and general annoyance. The transportation management approach to initiate "traffic calming" in surface transport design is recognition of this growing problem and the need for physical design solutions in urban traffic-impacted environments. There is also a need to improve traffic management from a human factors–social psychological perspective. It appears that preserving the comfort and enjoyment of the physical environment surrounding the transportation milieu restores normalcy and improves human health and well-being. The United Nations Educational Scientific and Cultural Organization (UNESCO) through its 2000 Millennium Declaration has established a set of Millenium Development Goals (MDG) for health. Goal 7, "Ensure environmental sustainability and access" is a recognized objective throughout the world. It addresses the problem of providing adequate infrastructure serving all the population.

In addition to the health consequences of transportation mobility on driver and passenger, congestion affects the efficiency of movement and therefore gridlock becomes a major factor in preparing for or responding to emergencies. Accident rates involving emergency response vehicles rise yearly. The Texas Transportation Institute (TTI) in its National Mobility Study conducted annually since 2001 indicates that the delays in metropolitan travel as well as excess fuel and congestion costs continually are on the rise. Certain urban locations are at a greater risk than others, with the largest populated metro areas demonstrating the highest risk. The most traffic-impacted metropolitan areas are Los Angeles (1st), New York (2nd), Chicago (3rd), Dallas (4th), and Miami (5th), averaging 72 to 50 hours in traffic delays annually per person. New York is ranked 23rd due to its public transit system, which 39.3% of all workers use. However, if we analyze only those who commute by car, then New York is rated near the top in terms of congestion, as reported by The Tri-State Transportation Campaign in its bulletin, "Mobilizing the Region." Congestion often results from an imbalance of modal choices, characterized by a high preference for one way to travel over another (e.g., private vehicle versus public transport, or land transport versus marine transport). The consequences are impaired movement in terms of traffic volume or flow rate, and a reduced level of service as well as accident potential and health impacts, especially among drivers.

Responding to emergencies or developing effective evacuation routes is an ever-increasing problem with population and vehicle demand rising disproportionately to instituting surface-based transportation improvements. Transportation infrastructure and needed improvements or expansion are always behind the demand curve. This also becomes a threat to improving energy efficiency and energy independence and subsequently increases the demand for foreign-based and -produced petroleum resources. The necessity for importing fuel to support the United States demand (now at about 65%) compromises our independence and encourages the threat of terrorism.

Impediments to an Integrated System of Movement

There is a series of impediments that impair an integrated approach to transport movement. Understanding these constraints should lead to eliminating the problems and working toward an effective and seamless system of integrated transportation before the end of the 21st century. Furthermore, a nation based on global preparedness requirements must be effectively positioned to be responsive to any natural or man-made disasters by utilizing an integrated system of mobility. The challenge is there but so are the impediments that must be addressed to maximize this goal of integrated systems movement. The following eight impediments need resolution before we can achieve a safe, secure, and effective integrated system of transportation movement:

Modal Bias

Each mode of transportation has evolved through an inventive technological process, followed by a series of governmental assurances (e.g., agreements, patents, and initiated public policies reflecting territorial exclusiveness). In the New York region it first began in the 18th century with maritime development initially involving sailing vessels naturally powered by wind, followed later by steam and subsequently diesel engine power. The railroad became a second major player in the 1880s when public rail service to Long Island was established within an exclusive right-of-way (the first in the United States). Subway development occurred in nearly the same time frame in New York, Paris, and London. Roadways and highways supporting motor vehicles were next (including establishment of a federalized U.S. interstate highway system based initially on national defense and known as the Dwight D. Eisenhower National System of Interstate and Defense Highways, approved in 1956). This was closely followed by the introduction of commercial aircraft service using a network of public-use airports. The national aviation system and navigational airspace became federalized in 1958, with the establishment of the Federal Aviation Administration (FAA), since aviation principally involves interstate commerce.

In each instance these transportation modes were supported and protected by their own business and governmental interest groups. They each had a core business and supporting organizations (i.e., trade and professional) as well as lobbyists, including separate advocates and entities "working" the federal government and Congress. Typically these transportation interest groups operated in isolation, protecting their self-interest (roadways, railroads, transit, marine, and airport interests) individually and independently of one another. Joint public use facilities were nonexistent. The planning, designing, constructing, and operating transportation infrastructure was an insular system with no provision for its parts to be linked together in any manner. This is a perfect definition of stovepiping, with individual interests prevailing over collective interests.

Robert Moses, for example, designed an extensive parkway road network for the development of Long Island, New York, which intentionally excluded streetcar and railroad rights-of-way, as well as bridge clearances for buses. He did not want his model parkway system, designed for the "automobile age," contaminated by public transit. Responsible for $125 billion worth of public works projects, Moses felt that mass transit was other people's business, as described by Robert Cairo (1974) in *The Power Broker*. Chauffeured all his adult life, the master builder led efforts to remove streetcar track beds (1940s) and to ensure that the Long Island Railroad (LIRR) did not interfere with his grand plan for the automobile: recreational driving to enjoy parkways and ultimately parks and beaches. No roadway project he oversaw during his 40-year "reign" was ever permitted to include any access for public transit within the planned parkway right-of-way. What was good for New York was considered worth emulating for other metropolitan areas worldwide. Consequently, in the U.S. model, automobiles prevailed, isolated from any public or mass transit.

Institutional Protectionism and Self-Interest

The evolution from a trolley- or streetcar- (light rail) to a bus-oriented society was not based only on evolving technology; this transition also was assisted by certain institutional self-interests led by the motor vehicle industry. Even though there is now a countertrend, which is encouraging subsidized public transit, in 1936 three important motor vehicle–related industries (i.e., General Motors, Standard Oil of California, and Firestone) allied themselves in a partnership. They established a new holding company, National City Lines (NCL), created to encourage trolley-based transit companies to consider either modifying or replacing their fixed-rail fleets with buses (initially designed to be powered by overhead electrical wires), built exclusively by General Motors and subsequently powered by GM diesel engines. NCL purchased over 100 electric-surface transit systems in 45 cities including New York, Philadelphia, Baltimore, St. Louis, and Detroit (Bradford Snell and Jane Holtz Kay).

There were various advantages and incentives advanced by National City Lines, which were somewhat compelling to these transit operators who often were struggling financially. However, the ultimate goal of this NCL enterprise was the elimination of the trolley system, to be replaced by diesel-powered General Motors buses using Firestone tires and fueled by Standard Oil petroleum products. The decline of the trolley infrastructure, including the mandatory paving of these rail beds, primarily with asphalt, became a major trend, encouraged if not required by National City Lines. Gradually the streetcar rail system, used in both interurban and city service throughout the United States, was abandoned for bus service. In 1936 there were 40,000 streetcars, and by 1965 only 5,000 remained.

In many instances these abandoned rail beds were buried under asphalt road pavement, such as in Atlanta, Georgia. Atlanta is now rethinking the possibility of restoring a portion of this streetcar alignment in its downtown as part of its center city master plan. The idea is to enhance tourism and improve commuting, which New Orleans, Portland, San Francisco, and Tampa, among others, have done. Washington D.C. ceased streetcar operations in 1962, and New York Mayor LaGuardia proudly announced their demise with the introduction of a more "technologically advanced transportation system," buses. It is interesting to note that a large number of the streetcars sold by New York City at that time were purchased in Austria, and today (more than 50 years later) they remain in municipal service.

Ultimately National City Lines and the three companies, General Motors, Standard Oil, and Firestone, were taken to court. Investigated by the FBI, they were eventually found guilty in federal court of criminal conspiracy to replace electric streetcars, and in 1949 they were fined (less than $100,000). However, they received no additional penalties. By that time the streetcar grid and infrastructure, including the inter-urban one that served many metropolitan regions, was effectively dismantled in most areas in the United States Rubber-wheeled transportation, including extensive use of buses and personal use of the automobile, began to reshape the density and patterns of residential settlements, as well as the overall design and form of American cities.

Social Conditioning: A Drive-Thru Society

It is now readily apparent that the motor vehicle has become the preferred mode of transport, which affects our ability to have a balanced system of transportation. Nearly 92% of the U.S. population use private vehicles for personal trips, while just 2.1% use public transit, according to the Center for Transportation Excellence. The social conditioning and acceptance of the motor vehicle has created a drive-thru mentality that is beginning to challenge the form of our human settlement patterns. It is becoming increasingly common to be born, married, or even buried in your vehicle (see Figure 3.1).

Collectively the drive-in wedding chapels in Las Vegas, Nevada, annually perform over 100,000 marriages in cars. A wedding of this type typically involves a preacher or justice of the peace. The length of service may be as short as 5–15 minutes with the bride and groom seated in their car and the engine idling for a quick departure. Valentine's Day is the most popular day for couples to be married at one of the curbside wedding facilities.

Motor vehicle access is not limited to just wedding venues, since in Montgomery, Alabama, among other cities, there is now a drive-thru funeral home. Here you can pay your last respects to a loved one, sitting in your personal car the entire time. A video screen selects the appropriate room to view the deceased for the automobile occupants and to pass on condolences (via car speakerphone) to the mourners inside the funeral home. Vehicle centrist activities now even address divorce issues and burial options. There is at least one business that advertises a drive-thru divorce service, in Atlanta, Georgia. Several states now permit a person to be buried in their favorite automobile or SUV and placed in their preselected cemetery plot, individually designed for their permanent home. The gamut of cars designated for funeral burial arrangements goes from classic cars to creature-comfort old cars driven a lifetime to upscale recent-vintage foreign imports (e.g., Ferrari, Rolls Royce, Aston Martin, BMW, Mercedes, and Bentley).

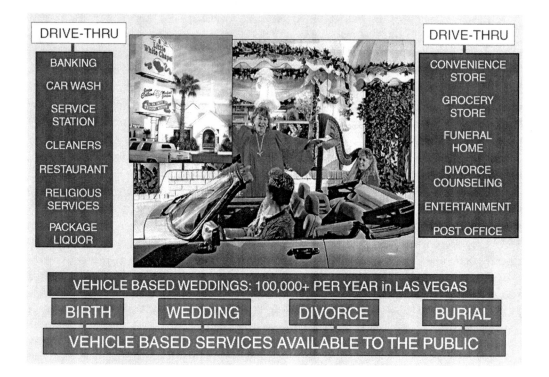

FIGURE 3.1 United States: A Drive-Thru Society

This American preoccupation with automobiles is contributing to a growth of vehicle ownership (the United States has the highest in the world with more motor vehicles than licensed drivers). There are also associated design trends that are shaping urban space, influenced by the growing demand for owning or leasing more personal vehicles. One impact is the growth of three-car garages (20% of all new single-family homes have them). Parking spaces are now 1.5 feet wider than they were 20 years ago due to the influence of larger personal vehicles and their expanded dimensions. This is changing many space-restricted residential neighborhoods. In major cities coast to coast, more property owners are paving their front yards for storing their cars. This has been referred to by *New York Times* writer Patricia Leigh Brown (2002) as "chroming" the front yard.

Vehicles are equipped with an increasing number of creature comforts, orchestrated through the use onboard computers. The distinction between cars and homes is becoming blurred. Cars now have adopted home-type features including more onboard real-time communication systems; enhanced audio, video, and geo-positioning technologies; air-conditioning and pollen controls; safety human factor devices; and now refrigerators, washing machines, and even microwaves. The automobile is becoming a home away from home (see Figures 3.2-3.5). With over 30 computers now integrated into automobiles, the opportunity for additional homelike creature comforts in vehicles is nearly technologically unlimited. Human factors (i.e., driver processing) and vehicle weight may become the limiting factors to their use. There is a question as to how popular these may become, as well as what the market demand is; however, manufactures are trying to entice the vehicle owner with "needed" options.

FIGURE 3.2 The Automobile with Microwave and Washing Machine Appliances

FIGURE 3.3 Multimedia Entertainment Systems

FIGURE 3.4 Navigational Communication Technology

Maytag has introduced traditional household appliances created especially for vans, SUVs, cross-over vehicles, or station wagons (see Figure 3.2). The company has designed and manufactured prototype microwave and washing machines capable of being installed in the back of these vehicles so passengers can take care of certain chores while on the road. The loss of vehicle storage capacity and weight may become a limiting factor to the popularity of these features. Consumer demand is presently an unknown.

FIGURE 3.5 Mobile Offices

Automobile entertainment centers have the characteristics similar to what is available in home-based multimedia theaters. Satellite radio networks, iPod connections, and television–DVD systems are becoming more common, especially for families with children (see Figure 3.3). The television monitors are either imbedded in the headrests of vehicles or flat screens drop down from the vehicle's headliner, frequently automatically. Surround sound speaker systems and headphones complement these visual devices for enjoying TV or favorite DVDs.

Global Positioning System (GPS) technology, first used by the military, has aggressively entered the consumer market. Navigational positioning and tracking travel to predetermined destinations are rapidly replacing paper maps in both commercial and private vehicles. They are appearing both as original equipment and in aftermarket sales. In Europe, telematics is advancing the scope of GPS to include restaurants, menus, and reservation capabilities that can be selected by city, food preference, rating of the restaurant, and meal price (e.g., Michelin guides are electronically online for passengers). Lodging, historical sites, cultural amenities, and all the interactive logistics can be part of the vehicle concierge package. Traffic monitoring matching road conditions with driving options, as well as real-time interface with roadway support services, is being offered as well.

The concept of the journey to work may well be fading, as an increasing number of personal vehicles are becoming partial or complete offices on the road (Figure 3.5). Business and travel may well be integrated into a mobile platform designed for your automobile. Beginning with the wireless phone system, more workers are expanding their business hours by using the car phone while driving. Onboard computers, combined with fax, scanning, and reproduction machines, keep the worker involved with their business practice, while others use their personal digital assistant (PDA) for text messaging, video teleconferencing, etc. Again human factor issues, including multitasking while driving, are becoming a safety issue. Vehicle accidents are on the rise since more people are interfacing with their electronic onboard technologies (e.g., handheld cell phone, etc.) while driving. Cell phones, along with other communication devices, are distractive to the driver, and that is why in many states, counties, and other political jurisdictions personal cell phones are restricted by law while a car is in motion.

As vehicles grow dimensionally our residences are also increasing in size. Garage sizes are expanding to compensate for the larger vehicle footprint and storage-capacity demands. Physically the average single-family house has grown from 1,645 sq. ft. in 1975, to 2,434 sq. ft. in 2005; this represents an increase of 789 sq. ft., or a spatial increase of 32% in home size in the past 30 years. Even though houses are significantly larger, the number of people per household has declined from 3.11 persons in 1970 to 2.60 persons per household in 2005. This means the space per household member has grown from 529 sq. ft. per person in 1970 to 936 sq. ft. per person in 2005, representing an increase of 43% in residential space per person. In summary, homes are larger but the total number of occupants are fewer, resulting in more square footage per inhabitant. Slowly but surely our personal vehicles are incorporating amenities found in our homes, which is beginning to blur the distinctions between the attributes in our homes and those in our personal vehicles.

Dominance of Motor Vehicles in Governmental Planning and Policy

Historically, all states established highway departments with their primary mission being supporting the planning and development of a highway network. Fortunately, in more recent years, not only have these agencies been renamed as departments of transportation, but their responsibility at the state level now also deals with multiple transport modes, rather than just road networks for vehicles.

In the development of the Long Island Transportation Plan (LITP) 2000, the New York Department of Transportation (DOT) did make a concerted attempt to include all transportation modes in the planning process. However, the LIRR rail system was too inflexible to cooperatively serve the region's needs, as reported in *The New York Times,* during the development of the eastern Long Island transportation plan. For example, the present public transit system was not permitted to have stops on the LIRR station property, and the former president of the LIRR felt they were competing with the commuting railroad service. Consequently, LITP 2000 recommended a rubber-wheeled rapid commute vehicle system (buses) operating within existing and expanded highway right-of-way.

Recent planning decisions appear to reflect a more conciliatory position of the LIRR, which now is willing to be part of an integrated network of transportation providers. For example, they are modifying this earlier decision by endorsing inter-hamlet rail connection using the Long Island Railroad right-of-way. This was one of the recommendations contained in the Southampton Intermodal Transportation Study (SITS), issued in 2003, but in 2008 is only now being considered. Joint use of the LIRR for handling freight along with passenger commuters still remains possible, but the conditions of use placed upon freight operators offering the service is heavily restrictive. Consequently, this limits the possible use of the railroad for such joint-use purposes. Over 96% of all shipping to and from eastern Long Island utilizes trucks and the limited road network, not the railroad corridor.

With the passage of the federal Intermodal Surface Transportation Efficiency Act (ISTEA) in 1991 by Congress, a new opportunity was created to begin addressing and planning for transportation in a comprehensive integrated manner. This rethinking is a very important subject for improving response times relative to both natural and man-made disasters as well as emergency preparedness. Initially this resulted in a financial shift of approximately $20 billion from federal highway programs to support nonhighway transportation projects (*Transportation Reform and Smart Growth: A Nation at the Tipping Point*, Funders' Network for Smart Growth and Livable Communities, July 16, 2001). Subsequent federal legislation (reauthorized in 1998 as the Transportation Equity Act, TEA-21) and later the SAFETEA-LULU (Safe, Accountable, Flexible, and Efficient Transportation Equity Act—A Legacy for Users) now provides a major infusion exceeding $247 billion for reauthorization over 6 years. Some progress now is beginning to be made to at least address integrated transportation issues, along with supporting funds to be used exclusively for such projects.

There are existing indicators of this imbalance toward road-based solutions and associated parking trends. The number of miles driven daily on our U.S. roadway network has risen by 85%, between 1995 and 2005 (4.8 billion miles). Roadways are growing, with the largest now being 26 lanes wide and containing a 400-ft. right-of-way. The average number of commuters per car is decreasing, with 77% being single occupancy, which increases traffic generation and associated congestion. There are now approximately eight parking spaces for every private vehicle, which translates to 37 square miles of land devoted to parking per additional 1 million persons.

Despite this funding to create a more balanced surface transportation system, there remain many obstacles, including the need to put the nation's infrastructure in a state of good repair in the most efficient and cost-effective manner. This is the number-one recommendation by the National Surface Transportation Policy and Revenue Study (NSTPRS) Commission, established by Congress, and contained in their final report, *Transportation for Tomorrow* (January 2008). There remains a tendency for each transport mode to avoid collaboration, and the primary funding sources presently rest with the Highway Trust Fund, which is unable to keep pace with all the transportation requirements.

We can ill afford uncoordinated efforts with a series of disconnects resulting in less than a comprehensive and integrated transportation planning system. The Commission in their report reinforces this need, "The concept of mobility is so fundamental to the American dream, integral to our national character, and necessary to our economic well-being, that it is imperative that our surface transportation, in all its varied modes, be the best in the world. The Commission believes that the Nation's leaders must provide for them—free of parochial interests, cognizant of energy sustainability and environmental impacts, and providing the needs of all who use it and depend upon it." This remains a noteworthy goal that needs universal support, but up till now it has not been achieved.

Spatial Management and Diet Cities

The next serious issue demanding global attention is the management of space, since space is a finite resource, like fossil fuel. Property consists of three dimensions: aerial, surface, and subsurface (see Figure 14.4), but we are primarily addressing land only as a two-dimensional surface resource. We are practicing "Titanic planning," (see Figure 14.5) since we refer to urban space as "land," having just two rather than three dimensions. Consequently, we are considering only the surface, rather than also addressing aerial and subsurface development opportunities. The preoccupation with land as only a surface feature reduces our consideration of water-based development, along with more traditional land-based developments.

Rooftop utilization, air rights, subterranean master planning, and even aquatic cities such as those being designed and built in Dubai, UAE, are not part of critical spatial planning within the United States. We continue to follow a two-dimensional, land-use–based model, which dimishes our spatial options. In terms of access during times of emergency response and evacuation, air, land, and sea routes are vital corridors, which were underutilized during the Katrina evacuation. For example, New Orleans access for evacuation to the Houston area relied solely on Interstate 45, which proved to be a gridlocked, congested corridor reducing valuable evacuation time, with no other intermodal support (e.g., rail- and water-based modes were not used, and only limited air evacuation support was given). During this time the average backup was approximately 100 miles, road rage flourished, motorists ran out of gas, and there was a bus fatality. A partial contraflow system was finally instituted on I-45 by the governor but with limited success.

Furthermore, population demand for housing and transportation is creating spatial excess, resulting in a nearly uncontrollable appetite. We are "supersizing" our urban environment, bringing about unplanned sprawl and "fat cities." Spatial excess and "upscaling" have replaced spatial conservation. A hint of this problem is now being identified in the annual rating of fit and fat cities, based on certain personal health parameters, which *Men's Fitness* began reporting annually in 1998. Their methodology includes a sampling each year of the 50 largest metropolitan areas. Beside personal consumption and lack of exercise, transportation habits are now part of the rating system.

Arlington, Texas, for 2008 is rated second on the fat city list, in part due to increased sedentary (single occupancy) commuting by the dominant use of personal vehicles. The average commute time has doubled over the past 5 years in Arlington, according to the 2008 *Men's Fitness* magazine survey (March 2008). *Forbes* has also identified obese cities and commuting as unhealthy when drive time is combined with air pollution and traffic accident indices. In *Forbes*'s 2007 rating the Arlington–Dallas–Ft. Worth metro area ranks eighth as the most dangerous commute in the United States. The sample size included the 50 largest metropolitan areas in the United States.

Weight gain and body size, due in part to decreased personal exercise, expanded appetite and a more sedentary urban lifestyle, have also contributed to the physical increase in the size of vehicle systems. For example, passenger seats on aircraft are being enlarged in the next generation of aircraft, to respond to increased body weight and girth. According to the FAA, this is also affecting the gross weight of aircraft, which adversely impacts the fuel consumption and burn rate of planes, as well as the profitability to the airlines.

Clearly, a diet-cities strategy is needed, as advanced by this author (dietcities.com). Our present-day urban society has established an unhealthy appetite that represents bloated congestion (it is estimated that transportation gridlock is equivalent to 5% of the gross domestic product, (or approximately $0.5 trillion in the United States), while human health weight gain is now costing our economy nearly $100 billion annually in additional health care services. These two trends appear to be interrelated; therefore, a diet-cities initiative in the redesign of our personal lives and the form of the city has multiple benefits, with reduced congestion and improved quality of life. Improvements in these areas will enhance both national preparedness and response to natural and man-made disasters.

No Consensus in the Definition of Terms: Intermodal

There is not a consistent definition when we discuss the need for a seamless system of transportation, whereby all modes (i.e., air, land, and sea) act in unity. The term *intermodal* was considered to be the preferred word for describing a holistic transportation system. This author defines intermodal transportation as the safe, secure, and efficient movement of people, goods, services, and information by air, land, sea, and space. This term is not universally found in languages around the world. For example, in both Russia and China the word *intermodal* is not found in their respective languages. This was observed by this author on trips to both countries on several occasions. Not only is the word missing, but so is this process of using intermodalism as an urban planning and engineering concept to improve congestion planning and management. Meetings with transportation specialists, architects, and planners in China always revolved around the discussion of a single mode (highway, air, rail, etc.). In their minds, addressing a roadway, transit system, or airport had to be the focus of any discussion. They were comfortable discussing transportations problems from a single mode of transport. In their academic institutions there were no courses, certificates, or degree programs in transportation related to intermodalism. When there were signed memoranda of understandings (MOU) or letters of agreement (LOA), *intermodal* was left in English, without translation in either the Chinese or English versions of the documents.

Even in the United States there is not consensus as to the definition of *intermodal*. Depending upon the group involved, it can have either a narrow or a broad definition.

In discussions with rail and trucking industries, *intermodal* is defined narrowly. They refer to intermodal as the transfer of goods between maritime, rail, and truck modes of transport, typically involving freight, including distribution or warehouse centers. The Intermodal Association of North America (IANA) uses this term but excludes air transportation from this use. Other professions apply a broader definition, incorporating all transport modes; for example, the Transportation Research Board (TRB) of the National Academy of Science.

Overseas and particularly in Europe the term *intermodal* is not used. Operationally, the Europeans have substituted the term *integrated transport* for *intermodal transport*. In their countries they are aggressively addressing this concept with the creation of joint-use transportation infrastructure facilities. They include both freight and passenger services with all their airports. Paris Orly and Charles de Gaulle International Airport provide RATP buses, urban rail to Paris RER, and at Charles de Gaulle high-speed rail, The Grand Vitesse (TGV); along with motor vehicles, motorcycle and bicycle accommodations are interfaced with aircraft. They have also raised the bar by offering code share ticketing, allowing a customer the convenience of an integrated ticket that combines both air and ground transportation fares, to facilitate integrated customer services throughout France. Integrated baggage handling using RFID tags is also becoming another service between modes. Private companies are now offering luggage tagged processing services. A traveling businessman can have his bags checked onto the plane from his hotel without going through the airport baggage process.

Governmental Responsibility Usurping Citizen Responsibility

The world has become a nation of laws and governance that has instilled a system of public administration and management to keep order. With this administrative management system, urban institutions of government have evolved to offer increasing levels of services to their citizenry, provided through a taxation process and/or fee for services (e.g., police and fire, street maintenance, utilities, waste management, etc.). Frequently this has deterred or displaced citizen involvement. Money for services is not a replacement for citizen responsibility and public participation. Responsibility of the citizen is slowly being supplanted by government being the substitute provider. Consequentially, there is a philosophical and social change in attitude and sense of responsibility of our urban-based society to become involved. The sense of community and associated responsibility of all citizens to be active participants is therefore diminishing. Governmental substitution for citizen duty and involvement can have serious implications. This impedes the nations of the world to be responsive to natural and man-made disasters as part of global preparedness.

This laissez-faire attitude has evolved over time and is not characteristic of the past. A social and geographical attachment of the town's inhabitants to their urban environment involved an annual marking of its boundaries by all citizens, beginning in the 14th to 15th centuries in medieval Europe. This formal ceremony and associated pageantry reinforced their value system, expressing a sense of place, civic duty, and personal responsibility to maintain and secure their community.

Town fitness and preparedness was a shared responsibility. There was not a clear distinction between "public and private," which subsequently evolved. The concept of communitas, or community, prevailed. Everyone working together for the common good, in a spirit of community. Initially, police protection, as an example, did not involve the hiring of a professional police force supported by funds collected by city government. Instead it was a community effort, with all able-bodied men taking their respective turns in assuring protection. Even when laws became enacted by cities to maintain streets, as transportation evolved in the 18th to 19th centuries, adjacent residents still had their civic responsibilities to be involved, as well as to maintain the entrances and sidewalks to their property and domicile. This is a vestige that often remains today, requiring home owners to maintain the sidewalk and property adjacent to a road right-of-way where they reside.

A resilient society is essential for national protection, responsiveness, and preparedness, expressed by Steve Flynn (2007) in his book, *The Edge of Disaster: Building a Resilient Nation.* He contends that we must remain responsive as a society, including taking actions on our own when necessary, which ties back to volunteerism. Heroism is certainly alive and well in responding to catastrophic events, such as 9/11. Emergency responders at the World Trade Center (WTC) representing a blending of police, fire, and medical professionals with untold numbers of volunteers gave selflessly of their time, irrespective of the danger to their health and well-being. Clearly the passengers on United Airlines Flight 93 gave their ultimate commitment to the cause of protecting this nation by sacrificing their lives, thereby avoiding further destruction to our infrastructure and loss of human life.

Volunteerism and personal commitment remain alive, as reflected in many organizations. It is interesting to note that the great majority of fire departments in this country is composed primarily of volunteers who respond to emergencies when requested. In those instances they drop what they are doing and respond to an emergency in their community. The Civil Air Patrol, Coast Guard Auxiliary, and many nonprofits can survive only with a reliable pool of citizens, who assist this country in national preparedness.

Sustainability and Security: Conflicting Goals of Infrastructure Risk?

Although infrastructure supporting this nation's mobility-based economy is imperative, and there is consensus that it needs to be upgraded and improved, a conflict exists as to what actions we should take. Neglecting this nation's infrastructure makes the United States, among many other nations of the world, increasingly vulnerable to natural and man-made disasters and safety and security issues. The recent collapse of the Minneapolis Bridge (I-35 West bridge) across the Mississippi River is indicative of the design, safety, and operation concerns of roads, tunnels, and bridges, among other surface transportation systems.

However, there appear to be two conflicting positions. One is that of the sustainability movement group, which desires environmental protection and resource efficiency as their priority. They are often referred to as the "Green team." The second is that of the homeland and national security group, responding to threats or disasters concerning our infrastructure; they are referred to as the "Blue team." Both represent philosophical differences and approaches to infrastructure requirements. A resolution of this issue has become a subject area of interest by the Public Entity Risk Institute (PERI). Confronting government, business, and civic leaders are these profound questions: Can we afford to make our communities safe? Can we afford not to?

The elimination of this dichotomy is preferred; however, both sides bring substantive positions to the table. For example, regarding nuclear energy (i.e., expanding nuclear power plants as a primary energy source for cities), the Green group is increasingly supportive of this form of energy to minimize the carbon footprint, therefore reducing our dependency on fossil-based fuels and foreign petroleum providers. In contrast, the Blue group considers nuclear energy a vulnerable energy source from both a terrorism attack perspective as well as an accidental dispersal standpoint. Other issues involve short-term and long-term risks, benefits, and costs.

It appears that either position is self-defeating and what is needed is a synergistic- and resolution-based approach that is based not on conflict but rather on consensus. Debating differences along with reinforcing oppositional positions only defers any possibility of introducing and agreeing on a solution. For example, border-crossing solutions, particularly with Mexico, are now being contested environmentally, especially with the possible installation of a physical wall separating Mexico and the United States, which might be located in an environmentally sensitive area.

First of all the priorities among the parties should be resolution, not conflict and unrelenting opposition. Using the conflict model, the end goal is delay and compounded costs, with no solution. A resolution model must be agreed upon at the outset. Secondly, there is promising technology involving virtual communication and surveillance that could be applicable, well beyond use of traditional camera and the utilization of neoteric sensor technology (e.g., 5-D real-time spatial sensory geo-fencing). This means that a physical wall could be supplemented by a virtual wall for selected distances and locations, resulting in a combined joint-use solution. This represents a dual benefit model that needs to be the theme of any Blue–Green discussion, rather than irrefutable oppositional positions.

It appears that this Blue versus Green dichotomy is fundamentally flawed for several reasons, including the following:

1. The lack of resolution provides no inherent benefits to any parties because time is of the essence.
2. Vision-based solutions and not irresolvable conflict should be the premise since "Where there is no vision, the people will perish" (Proverbs 29:18).
3. Oppositional positions reflect a convenient and traditional approach to dialog; however, a dual benefit model should be introduced as the accepted operational strategy for all parties.
4. Systems integration, which incorporates disparate disciplines that merge ideas (e.g., fusion of science and technology), should be utilized, since synergy can produce savings and advances solutions.
5. Blue and Green alone have strengths and weaknesses, but a combination (creating turquoise or teal) is preferred, since a parallel or dual course of action is not affordable.
6. New methods and technologies need to be introduced, including use of simulation and modeling to query possible analytical solutions that can immerse all role players (avatar-based decision making).

Transportation System Elements

In order to plan and develop an effective transportation-based approach to both natural and man-made disasters, it is necessary to systematically address all modes that constitute mobility of people, goods, and information. Figure 3.6 identifies the organizational

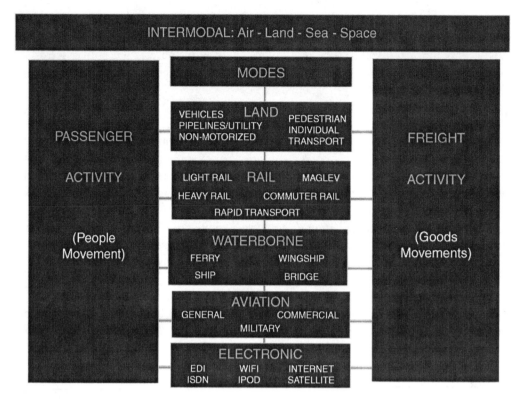

FIGURE 3.6 Transportation Modes

framework for incorporating all relevant transport modes as part of the inventory for a people and goods system of movement. This can be referred to as an *ecumenical* or *total modal* approach to transportation.

Organizational Framework: Definition

The first consideration is the definition of *transportation mobility*. Physical movement describes the manner in which a transportation mode navigates. Vehicles utilize a system of both public and private rights-of-way primarily involving a surface-based road network, while aircraft navigate through designated airspace managed by the FAA and involving both private- and public-use airports. Rail and transit involve both publicly and privately operated rail rights-of-way associated with moving freight and passengers. Maritime activity includes a range from recreational boating to commercial and military craft that navigates the waterways. In all instances, whether a car, motorcycle, bus, plane, train, boat, or ship, there is physical movement through space (i.e., air, land, or sea).

This represents only half the picture, since transportation includes the safe, secure, sustainable movement of people, goods, and information by air, land, and sea, utilizing both physical and electronic means of transport. Transporting information electronically represents the other half of the transportation system. It may include satellite, fiber optic, or other means of distributed electronic communication utilizing airways, including electronic data interchange, increasingly digital in character. Logistically, a physical system of movement includes people (i.e., operators and passengers) as well as goods and resources (i.e., natural and man-made). Electronic-based transport uses airways, frequencies, and networks that are transmitted through a secured pathway, utility, or communications information system network.

The important merging of electronic and physical movement can be described by the term *transcommunication,* which was adopted at the United Nations–sponsored Habitat II conference as an effective means of depicting this important union of all transport elements applicable to the urban habitat. Methods of both physical and electronic transport are vital components of a transportation system that is needed to respond to natural and man-made disasters. Intermodalism is a critical element, since it is associated with the linkage of all methods of movement to ensure effective management and logistical connectivity of the transportation system. Failure to include any one of these components as part of an integrated network may result in an operational dysfunction. This has been a problem identified in assessing the effectiveness of many responses to natural and terrorist events.

Transportation-Based Modal Elements

There are five basic modal categories that constitute in the aggregate all transportation. Within each of these modal categories there are various means of transport that must be examined comprehensively when a national preparedness strategy is being developed or executed. Failure to consider all these modes could seriously impair the effectiveness in response or recovery from a natural disaster or terrorist event.

Land Mode

Land-based modes of movement include surface-based transportation ranging from both motorized transport (vehicles: cars, trucks, buses, motorcycles, etc.) and nonmotorized transport (bicycling, walking, etc.). The magnitude of infrastructure needed to support these vehicles' operations in the United States is significant. Our present roadway network exceeds 4 million miles. Utilities also are classified as part of the land-based modal network, including both surface and subsurface assets (i.e., electronic transmission lines, microwave relay, pipelines, etc.) For our purpose here, land mode covers three dimensions, since it spatially deals with what is above the ground or surface, as well as what is below the ground, or subterranean space.

Rail Mode

Rail-based modes are associated with public and private movement of people and goods. This includes all freight and passenger trains and their 150,000 miles of track, generally classified as heavy rail (i.e., Class 1: CSX, BNSF, Norfolk and Southern, Union Pacific, Amtrak, etc., along with regional and local railroads). *Light rail* or *mass transit* generally refers to public transit systems operating within metropolitan areas (Long Island Railroad, Chicago L, BART, MARTA, Metro, DART, St. Louis Metrolink, etc.) within a defined rail right-of-way. Public transit buses that use a roadway network of service to the public would be part of the motorized land-based transport system. Other categories of "rail" include magnetic levitated (i.e., MagLev) transit systems, which are being explored worldwide, ranging from metropolitan commuting to large-scale regional transportation system networks (e.g., China, Japan, and Germany).

Maritime Mode

Maritime activity supporting passengers and freight represents a major portion of our economy. U.S. marine ports annually receive 9 million cargo containers and 200 million passengers. Approximately 95% of our nation's trade, in the form of cargo, enters or leaves our 361 seaports each year. Passenger vessels utilize U.S. ports as part of domestic and international cruise line activity and are the predominant operators in the world, especially those located in Florida. The Port of Miami, Port Everglades, and Port Canaveral (in that order) represent the three largest cruise ship ports in the world. Ferry operators and water taxis represent a major carrier of daily passengers and can play a strategic role during disasters (e.g., New York Waterways during the 9/11 World Trade Center crisis), acting as a critical alternative to transit and vehicle street access). The last category of maritime activity includes pleasure or recreational boats that on a volunteer basis consistently support emergency response efforts.

Aviation Mode

There are 19,847 airports operating within the United States, according to the Federal Aviation Administration's National Plan of Integrated Airport Systems (NPIAS). Of the total, 5,261 airports are open to the public, or 26.5%. The majority of airports in this country (73.4%) are privately owned. Airports are defined as landing areas developed for conventional fixed-wing aircraft, helicopters, and seaplanes. In terms of actual passengers, the large-, medium-, and small-hub airports handle over 60%, while general aviation airports support 19%. Military or joint-use (commercial–military, such as Scott Air Force Base) airports are important logistically since they can be reliever airports during emergencies, as well as handle critical missions. Both logistical issues of passenger and cargo requirements are important to coordinate. Therefore, private-use airports should not be neglected in developing national preparedness scenarios, even though the majority are not part of the FAA's NPIAS. Airports frequently are used as staging, telecommunications, and weather centers for logistical purposes.

Electronic Mode

Communication involving electronic transport, by utilizing secured electronic data interchange (EDI) between business and government, is essential as part of national preparedness. Interoperability has been one of the most serious deficiencies in communicating, particularly in the civilian sector (inter- and intra-communication involving local, state, and federal government). Command and control centers and their communication networks must be seamless and continuously responsive. It is interesting to note that the technology exists to eliminate this problem of integrated communication; however, stovepiping remains among participants (e.g., selected methods of communication between police and fire departments) and the financial support and priorities these centers receive so they can effectively accomplish their missions. Electronic communication will represent an ever-increasing part of the modal network necessary for having a successful intermodal system.

Transportation Systems Network

A second element is defining and inventorying the actual transportation network so operationally the entities involved in command and control have full knowledge of the resources available. Figure 3.7 provides a systems design example of the physical transport

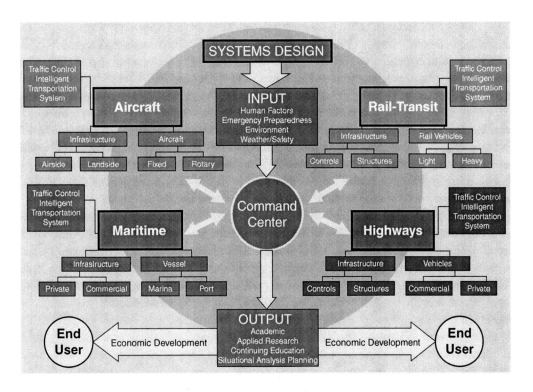

FIGURE 3.7 Integrated Transportation Network

network, identifying the primary modes of aviation, rail, maritime, and highway transport. Within each of these modes there are component parts, including infrastructure (i.e., airports, seaports, stations, highways, bridges, tunnels, etc.) and modal inventory (i.e., vessels, aircraft, vehicles, rail engines and cars, etc.), along with energy resources and maintenance elements, among others. Traffic control and intelligent transportation systems in place for monitoring, surveillance, and logistical movement of transportation modes individually and collectively are vital to both the inventory and management.

Command and control centers involving military situational awareness system technology are beginning to be evaluated for all transportation environments. A more detailed discussion of this important approach is found in Chapter 10. Civilian and military centers differ significantly, including their purpose, operational scope, scale, level of security, sophistication of equipment, and technology. As a result of inadequate responses, particularly to natural disasters, emergency response centers operated by local and state governments now routinely interface with the military, so there is greater fusion of civilian and military cooperation. Usually this involves, as a minimum, coordination with their respective state National Guard bureaus. With request for assistance from a local and state jurisdiction to the Federal authorities, the Nation Guard can provide assistance in a short response time.

Input and output systems design should include all categories of providers and responders, ranging from government, military, nonprofits, and business. The greatest deficiency is incorporating businesses into the command and control network to optimize their contribution. In many instances the private sector is the entity that is the first to effectively respond, especially in terms of support items that are logistically positioned and needed. Nonprofits (e.g., Salvation Army, Red Cross, etc.) today are active participants and typically are early arrivals on the disaster scene.

New partnerships are beginning to be structured and initiated between businesses (e.g., Home Depot, Lowe's, Wal-Mart, etc.) and nonprofit organizations. Chapter 15 discusses model nonprofit–business teaming arrangements that, because of their responsiveness and the relative freedom from red tape–bureaucracy, are beginning to offer just-in-time logistical drops to geographic areas that encounter natural and man-made disasters anywhere in the world. The number of responsible private sector providers in humanitarian disaster relief logistics is growing. For example, Gordon Atlantic through Juno Health Systems can provide emergency pneumatic shelters that can be erected in minutes for emergency and first response purposes addressing local population health needs. This responsive health-based technology is now being examined by UNESCO, in support of their Millenium Development Goals to improve the human condition by 2015.

There are also critical allied medical resources that the military can make available in a time-responsive manner, but frequently this does not occur due to a breakdown in communication (military–civilian) and general lack of knowledge as to resources. In late August of 2005, Hurricane Katrina began a path of unparalleled destruction throughout the Gulf states, leaving tens of thousands of people homeless and without hope of shelter, food, water, electricity, or even basic medical care. Little did anyone know that, within a day's journey, two massive military convoys lay waiting at the Naval Expeditionary Medical Support Command (NEMSCOM), a military medical warehouse facility located at Cheatham Annex, Virginia, ready to render aid. The convoys, outfitted with fresh water tanker trucks, generators, telescoping light stations, air conditioners, medical supplies, diesel fuel, food, blankets, toiletries, shelters, portable bathroom facilities, mobile communication centers, ambulances, fire engines, cranes, and even bulldozers were equipped to provide care for thousands of people with follow-up supplies available for tens of thousands more. The order to deploy was never initiated because relatively few entities outside the military knew what the command's capabilities were, including FEMA. This important source of assistance was never utilized. Consequently, the necessary level of responsiveness fell below what could have been employed in a time-sensitive manner.

Modeling and Simulation

Responding to real-time situations is the ultimate purpose for having an integrated transport network as part of a command and control activity. The planning, preparation, and training for an effective response management system include modeling and simulation. This is where scenarios, tabletop exercises, and what-ifs can be explored and assessed in advance of an actual response, saving considerable time and resources. Historically, simulation technology was the domain of the entertainment industry and the military. More recently this application of multimedia technology using real-time simulation software and facilities has become more widely applied by more role players (i.e., universities, architects, planners, and businesses). Playing out the role of transportation in planning its logistical role in an emergency response or disaster recovery exercise can be previewed through simulation at a significant cost saving. It also allows variations or options that need to be considered in scenario building and finalizing a plan of action, or training, before they are needed in a real-world situation. This represents electronic rehearsal, virtual real time simulation.

The world's first intermodal transportation simulation system (ITSS) was established at the National Aviation and Transportation (NAT) Center adjacent to Brookhaven Calabro Airport and the Brookhaven National Laboratory, Long Island, New York. This

$3.15 million laboratory operated by Dowling College was established for conducting applied intermodal research, consulting, and teaching. Developed and invented by Dr. Clifford Bragdon (U.S. Patent Number 5,863,203, January, 1999).

In 2000 Newsday ranked the ITSS invention as the ninth most significant new invention for the 21st Century. The ITSS lab consisted of four interconnected transportation modal rooms with simulator-based workstations (see Figure 3.8). Each room contained a transport mode involving a vehicle (front seat automobile cab), aircraft (desktop aircraft configuration with a force-feedback flight yoke, throttle and rudder pedals, multisensory connected to an airport scene), ship (desktop standing marine bridge applied to a harbor scene), and rail (magnetic levitation, or maglev, rail passenger cabin with out-the-window displays along a maglev corridor). An individual could employ 3-D simulation using an open architecture multiprocessing server (Silicon Graphics–Multi-Gen engine) to navigate a roadway, airport, rail, and maritime network. Each mode had a multichannel image generation, open GL-based high-resolution screen projection arrays, infrastructure person-in-the-lop driver–crew stations with reconfigurable vehicle consoles and controls. These modal elements were then seamlessly integrated into the adjacent situation room, where planning and transportation problems and solutions could then be merged and analyzed in a comprehensive manner.

Besides simulating virtual visual reality in three dimensions, two other senses were addressed in this ITSS laboratory at the NAT Center. Acoustically, surround sound was employed, using a domed system where the individual could be immersed in omni-directional sound connected to the community acoustical environment (e.g., roadway, aircraft, rail, ship activity superimposed on an ambient background). Acoustical instrumentation was employed to obtain accurate sound-level data that could then be incorporated into the actual scenes of interest. Numerous environmental impact assessments, noise

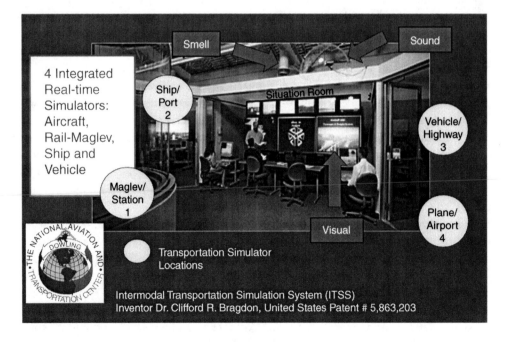

FIGURE 3.8 Intermodal Transportation Simulation System (ITSS)

abatement and control plans, and geo-sonic studies were conducted to address the acoustical issues as part of the multisensory experience.

The third sensory component was the olfactory sense, related to aroma. Smell is an important human sense since it has the longest memory recall of all five senses (i.e., vision, sound, smell, touch, and taste). Working with the olfactory company Aromasys, the ITSS facility had an aroma system installed to the air-handling system serving this laboratory area. Through a computer-based metering system, selected aromas (i.e., essence of oil compounds) were introduced into the laboratory at various thresholds. This offered a more environmentally realistic setting to conduct exercises, scientific analysis, and assessments. Working with the Aromasys president, Mark Peltier, fragrances were developed to enhance alertness and vehicle safety, particularly among drivers. One Aromasys product for drivers is called Pacific Coast Highway. Dr. Bragdon initiated human factor studies with student pilots and observed that their level of flight-training proficiency was enhanced when selected aromas were introduced into their flight simulation.

Incorporating multiple senses, besides just visual, in the simulation process improves the level of realism. By using multisensory simulation, the exercises conducted are enhanced, since it more accurately portrays the real world. This in turn improves the learning process and the outcome of the exercise or scenario. There are multiple simulation modeling tools that can effectively portray other factors related to time of day or weather conditions, along with other physical and social attributes that provide greater realism. More recent advances have been made to immerse individuals into scenes using avatar modeling, derived from the computer gaming industry. In these scenarios, multiple role players are introduced into a scene where they can operate individually and collectively, thereby influencing the outcomes. This newest interactive simulation-based technology is receiving strong interest from parties responsible for disaster management and recovery (National Guard, FEMA, and related agencies), especially in the training area.

Dr. Bragdon further developed the ITSS by raising computer simulation in transportation to a higher level with the establishment of the University Consortium for Intermodal Transportation Safety and Security, supported by The State University System of Florida as a statewide center (Type III). UCITSS was designed to aggressively address the potential threat of terrorism to the public health, safety, and general economic welfare, including Florida's population and infrastructure. The mission was to focus on strategically protecting the collective modes of transportation: airport, seaports, highways, bridges, and rail and transit facilities that compose the intermodal system of transportation by air, land, and sea. It was also established to create a road map that could serve the entire United States. Initially the consortium members were four core universities: Florida Atlantic University (FAU), University of Central Florida (UCF), University of South Florida (USF), and Florida International University (FIU). FAU was selected as the headquarters, and this author was appointed as executive director.

Federal funding was received to support this consortium, enacted upon by the Joint House–Senate Committee during the first session of the 108th Congress in 2003. President Bush signed this request into law in 2004 (Public Law 108-199) with an appropriation of $7 million for the consortium. This was the largest funding grant ever awarded to a university by the U.S. Department of Transportation, and Secretary Norman Mineta declared, "All of our Nation's interconnected transportation systems, regardless of mode, deserve to benefit from the Center's work." Another $3 million was appropriated to support the consortium's mission in Fiscal Year 2005–2006.

Two tasks initiated under the DOT federal grant to the consortium focused on developing a computer-based visualization of selected transportation modes. Developing a virtual database for a port as well as a public transit system was selected in order to develop simulation scenarios addressing possible threats of terrorism and natural disasters. A primary purpose for creating these simulations was to assist transportation providers with tools that could help them in planning, engineering, and human resource training. Port Everglades, located in Fort Lauderdale, and the Washington Metro transit system in Washington D.C. represented the two transportation modes selected for the work supported by this federal grant.

Port Everglades, Fort Lauderdale, Florida

Total cooperation was received from the executive director of the port authority, his staff, and the Broward County Port Security Task Force. This port represents the second largest cruise ship port in the world (5,700 ships call here annually), and it has diverse maritime operations. In addition to the cruise industry (i.e., Carnaval, Celebrity, Cunnard, Crystal, Disney, Holland America, Princess, and Royal Caribibean), the port maintains a large containerized cargo business (ranked 12th in tonnage at 23 million tons) besides being a major petroleum storage and distribution hub, as well as a popular U.S. Navy liberty port destination.

Initial steps were taken to establish an inventory of vehicle types and vessels serving the port, along with creating a baseline of the geographical boundaries and port infrastructure. Port Canaveral consists of 2,190 acres, of which 1,742 acres (80%) are upland, with the balance submerged. As a first priority, the director of the port requested that the UCITTS team create a simulation of the proposed port entrance (which handles approximately 20,000 vehicles per day) by examining the engineering specifications and preparing a three-dimensional model showing how it would be configured once constructed. Figures 3.9–3.11 present four computer-based 3-D renderings extracted from the final real-time database created for the port. Figure 3.9 illustrates, through computer simulation, how vehicles would enter this new gateway once it became operational, using the engineering blueprints and capturing onsite views of this corridor and the vehicle types processed through the present port entrance gate. Efforts were made to examine underground features of this gateway security system when it was necessary to restrict movement. Other gateway features and procedures were able to be examined and understood once this 3-D visualization was created. Based on the results of this gateway simulation, the port director requested and received authorization for this new gateway to be built. Today it is in place and fully operational.

The navigation, berthing, and positioning of the cruise ships in the port's inner harbor were also an important task. Having up to five major cruise ships in Port Everglades at one time was critical to the harbormaster and employing effective logistical protocol. New ships became ports of call, including the Queen Mary 2 (QM2); consequently, a detailed rendering had to be created for port planning, logistics, and security purposes. Highly skilled 3-D virtual simulation specialists were assigned to this challenging task, including Yossi Kaner and his assistant, Matt Hyner, both employees of the Florida Atlantic University UCITSS headquarters team. Figure 3.10 accurately illustrates the rendered simulation of the QM2, where passengers disembarked during one of its ports of call. With new regulations issued by the U.S. Coast Guard through the Department

FIGURE 3.9 Proposed Port Everglades Entrance Rendered

FIGURE 3.10 Queen Mary 2 (QM2) in Port

of Homeland Security, this required examining certain port security issues that could be more effectively understood by creating a real-time fly-through of the port. By simply navigating with a computer mouse through the simulated port, one can observe multiple eye points and perspectives that could not be achieved by examining a two-dimensional seaport photograph or blueprint drawing on paper.

Landside features of the port were accurately rendered into a 3-D scene of the port. Since there were 448 acres of submerged area in the harbor where ships navigated 24/7, it was important to profile the underwater views of Port Everglades for safety and security purposes. The actual underwater topography was created incorporating lidar (light detection and ranging) technology throughout this inner harbor and basin. Scenes were then created superimposing ships navigating through this area at various depths. A Maersk container ship navigating through the port harbor is illustrated in Figure 3.11. Underwater surveillance is one of the more important security issues being addressed by a few of the 347 deepwater ports operating in the United States, including Port Everglades. Certain lines of site, lighting patterns, and sensors must be capable of continuous observation irrespective of weather or time of day. Underwater surveillance, along with surface movements (land and sea) and aircraft over-flights to Fort Lauderdale International Airport represent the composite condition that is important to Port Everglades.

The overall view of Port Everglades during the day under sunny conditions is shown in Figure 3.12. This port aerial visualization can be modified by time of day and location in order to explore other venues, locations, and heuristic and climatic events, including the introduction of a variety of weather conditions (e.g., fog density and coverage,

FIGURE 3.11 Underwater Scene of Container Ship Navigating Inner Harbor

FIGURE 3.12 Aerial Visualization of Port Everglades

precipitation, etc.). In order to prepare an effective security-based plan, a comprehensive set of maritime operations reflecting all the port assets and infrastructure needs to be incorporated into the scene generation, both land- and water-based.

Adverse weather affects all intermodal facilities, including seaports. Landside operations must be concerned about access and egress under all meteorological conditions from a traffic circulation perspective. Since hurricanes are part of the Florida landscape, all port officials as well as emergency management command and control centers continually observe prevailing and forecasted weather conditions. To assist the port, an assessment was made regarding the various hurricane categories and the related water levels on the roadway network within the property boundaries of Port Everglades. Figure 3.13 shows the extent to which the port roadway network could become potentially impassable due to heavy flooding conditions, when there is a category-4 hurricane in this geographical area of Broward County. As this 3-D simulation depicts, except for the interstate system adjacent to the port, the seaport property is forecast to be under a serious flood condition, restricting use of the roadway network coming into or leaving Port Everglades.

As a result of this modeling and simulation work performed for Port Everglades by the university consortium led by Florida Atlantic University, this port has some unique advantages in preparing a more effective port security plan with realistic scenarios, exercises, and subsequent training. The real-time virtual database has now been turned over to the port authority so they can visually understand the issues and incorporate this 3-D visualization simulation technology into their planning, engineering, and management process. What was created as a prototype for Port Everglades under a federal DOT grant will become the standard for all U.S. ports.

FIGURE 3.13 Simulated Flooding Conditions Associated with a Category-4 Hurricane at Port Everglades

Washington, D.C., Metro

The Federal Transit Administration (FTA) was the federal grant coordinator on behalf of the U.S. DOT during the first contract phase ($7 million) of this congressional award to the UCITSS. An agreed scope of work was initiated to develop a transit security simulation for the Washington Metropolitan Area Transit Authority (WMATA). The focus of this work for the metro authority was replicating one of its Metro Center underground transit stations in a real-time three-dimensional scene.

Although the most significant mass transit incidents have occurred overseas, the United States is at risk. The Transit Security Administration (TSA) in an unclassified report recently stated "transit and passenger rail systems are vulnerable to terrorist attacks because they are accessible to large numbers of the public and are notoriously difficult to secure" (Unclassified Mass Transit System Threat Assessment, Transportation Security Administration, Office of Intelligence, 29 February 2008). The five largest transit attacks occurred yearly, between December 2003 and February 2007. In 2003 the first occurred in Yessentuki, Russia; followed by Madrid, Spain, in 2004; and then London in 2005. India has been the site of the most recent public transit rail attacks in Mumbai (2006) and Kashmiri (2007). Figure 3.14 summarizes these five terrorist attacks that killed 539 passengers and injured 3,363.

Multiple improvised explosive devices (IEDs) and improvised incendiary devices (IIDs) are the most common weapons used in attacks on mass transit targets, and were the devices used in these five major public transit attacks. The Transportation Security Administration (TSA) suggests that other terrorist methods are possible (i.e., chemical and biological), based on the 1995 Tokyo subway incident where plastic bags containing saran had been left on subway car floors. Within the United States suspicious activity was reported in 171 cases in 2007, involving 29 Amtrak trains, 45 heavy-rail transit, and 97

Date	Location	Fatalities	Injuries
December 2003	Yessentuki, Russia	41	150
March 2004	Madine, Spain	191	1,800
July 2005	London, England	52	700
July 2006	Mumbai, India	187	700
February 2007	Kashmiri, India	68	13
TOTALS		539	3,363

FIGURE 3.14 Major Public Transit Attacks, 2003-2007

buses. The report does state, "There is no evidence to date linking suspicious mass transit incidents (in the U.S.) to terrorism."

The primary purpose of developing a series of metro virtual simulations was to inform the metro authority personnel on what methods of activity possibly could be deployed at an underground transit station site, using computer-generated visualization (replicating different scenarios). Figures 3.15–3.17 present three scenarios that potentially could occur at various locations in a metro transit station. The use of a backpack for carrying explosives has been a significant source of concern for transit security personnel worldwide (Figure 3.15). An unattended bag left on the platform (next to or under a passenger bench) is in a highly vulnerable location, since it is close to the passenger epicenter as well as in close proximity to the track bed and the metro transit station. Mugging incidents generally are not terrorist based, unless a prominent individual (e.g., senior military officer, government official, or prominent businessperson) is the possible subject of interest (Figure 3.16). An individual using an improvised detonator on his person was created in a

FIGURE 3.15 Unattended Backpack on Passenger Platform

FIGURE 3.16 Mugging Incident with a Predetermined Target of Interest

virtual visualization based on the development of a series of storyboards for WMATA (Figure 3.17). These were just some of the possibilities identified for educational and training purposes. Recognizing that these were motion-based 3-D simulations, situational awareness conditions could be modified injecting a certain degree of realism and variation into the visualization.

When these virtual 3-D–generated scenes were completed by the OCITSS at FAU, the databases and files were turned over to the metro authority for their potential use.

FIGURE 3.17 Improvised Explosive Device Detonator on a Person

The experiences of using computer simulation applied to transportation issues (i.e., maritime and mass transit) related to emergency management and response have expanded the horizons of transportation providers. There is now a growing interest to see how this tool can have further benefits to all modal operators when they are involved in planning, engineering, education, and training for national preparedness and emergency response.

Forensic Transportation Logistics Analysis

Whenever there are national or man-made incidents investigations occur, hearings or meetings take place, analysis is conducted, reports are prepared, recommendations are made, and sometimes actions are taken. In time, documentation of incidents is filed away in some repository for record-keeping purposes. There are two problems with this process. First, incidents are generally investigated on an individual basis and comparative analysis generally is missing. Furthermore, there are few lessons learned from results, based on a compendium of incidents that are comprehensively reviewed. We generally fail from using a lessons-learned approach. Typically, we reinvent the wheel by starting over, with little regard for historical events.

The second problem with this process is the type of analysis utilized. Forensic transportation analysis that deals with logistical causality is not consistently pursued. Using lessons learned by examining the evidence forensically is gaining popularity (particularly in television series) but not necessarily in transportation logistics. Within 2 days of the 9/11 incident at the World Trade Center, Dr. Bragdon created a virtual simulation of the two American Airlines Boeing 767 aircraft impacting the Twin Towers (Figure 3.18).

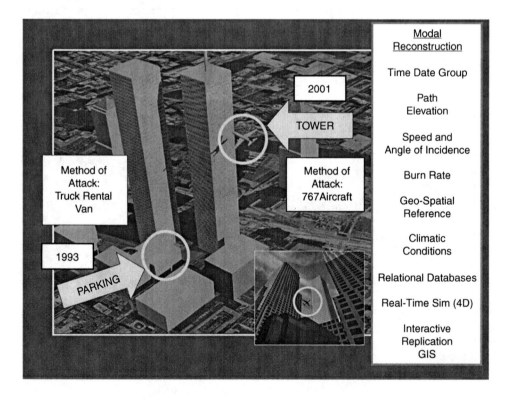

FIGURE 3.18 Safety and Security: Twin Tower Forensic Transportation Analysis

This received considerable attention from the cable television media, particularly in the New York Metropolitan Area. The motion-based visualization showed the path of the two aircraft from multiple eye points (i.e., plain view, oblique view, angle of incidence) coming into contact with the building envelope of the Twin Towers.

This modal reconstruction then provided the opportunity to analyze the time sequence, flight path and elevation, speed and angle of incidence, burn rate, geo-spatial reference, climatic conditions, and relationship databases, adapted to a Geographical Information System (GIS) platform. A Word Trade Center aircraft footprint was created comparing the percentage of the floor area for a Boeing 757 to that of a Boeing 767, if they had "landed" on one of the WTC floors (Figure 3.19). The density of the total floor area was greater with the Boeing 767, constituting 85% of a single floor within the WTC, while the Boeing 757 footprint was smaller, at 60%. The selection by the terrorists of incorporating a Boeing 767 in their planned attack by penetrating the WTC's building envelope at high speed inflicted greater damage due to the plane's mass.

Another comparison was made with the dimensions of the two aircraft, including their height (tail section elevation) and their respective wingspans. Again the Boeing 767 constituted a larger profile than the smaller Boeing 757, in terms of both vertical and horizontal dimensions. What made the difference more significant was the angle of attack the planes had when they came into contact with the steel-framed building envelope of the twin towers. The aircraft's attack angle was not parallel to the ground plain but rather at a significant angle, which meant that there was more physical damage to the building by reaching more floors (Figure 3.20). No doubt the terrorists also knew the difference in jet fuel capacity, which made the 767 a more lethal weapon once the fuel ignited on impact with the twin office buildings. In terms of fuel, the Boeing 767's capacity was nearly double (47.8% more) the capacity of the Boeing 757; that is, 23,980

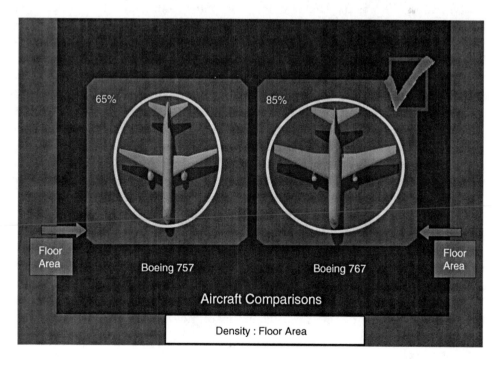

FIGURE 3.19 World Trade Center Footprint: Boeing Aircraft Comparisons

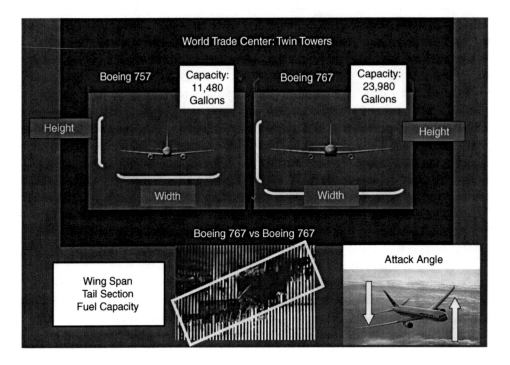

FIGURE 3.20 World Trade Center: Aircraft Penetration Profile and Fuel Capacity Comparisons

gallons compared with 11,480 gallons. This meant that there was twice the amount of combustible fuel, once the plane ignited and started burning. The secondary impact of both fire and smoke was the largest single contributor to the ultimate destruction and loss of human life to the building occupants.

All these factors influenced the strategic plan developed and adopted by the terrorists to inflict maximum damage on the World Trade Center. (It has been estimated that total cost incurred by the terrorists in planning and executing 9/11 was approximately $400,000–$500,000.) Architectural and engineering plans and drawings were found depicting these buildings in conjunction with the Bin Laden Construction Company, one of the largest construction contractors in the Middle East. It also is interesting to note that Osama bin Laden allegedly received a bachelor's degree in civil engineering from King Abulaziz University.

Following the collision of the two American Airlines planes with the WTC towers, there was a change in mobility patterns, especially in lower Manhattan. Accessibility to this geographic area was curtailed, primarily limited to emergency support and rescue vehicles. Since 30% of the New York area is composed of water, alternative methods of transport, not using ground transportation, were introduced for more effective connectivity. Figure 3.21 summarizes the logistical impact inflicted on Manhattan because of 9/11 and how the levels of services changed. Basically, transit, commercial air traffic, and local surface roads were heavily restricted.

Phone service was limited along with building access near the center of impact. Water transport became the most reliable means of movement in this area of the city. New York Waterways, which provided over 75% of the water transit, assisted pedestrians, emergency response teams, police, and volunteers during this long-term recovery period. Water access to and from Manhattan was a primary means of entry. Any effective

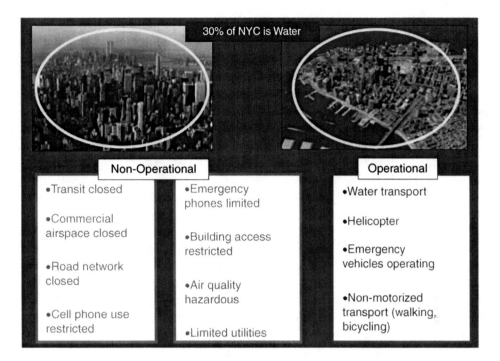

FIGURE 3.21 Man-Made Disaster Transportation Logistic Impacts: WTC Environs

approach to assessing the impact and developing a successful response requires all modes of transport, both physical and electronic. Consequently, an integrated transportation systems preparedness plan is absolutely essential for effectively responding to any type of disaster. Total modal is the best systems approach, and no mode of transport should be left out of this important national preparedness and response equation for any city, county, or state.

Summary

A systems approach to transportation that integrates all modes of movement (i.e., both physical and electronic) is absolutely essential in developing an effective national preparedness plan. *Transcommunication* was the term developed and recommended by a United Nations Habitat II task force for addressing this subject area. For the readers of this book, *transportation* refers to the safe, secure, sustainable, and efficient movement of people, goods, and information by air, land, and sea.

Transportation has been the backbone of our civilization throughout human history. Consequently, mobility is directly related to human and economic survival, and it is becoming evident that it may have genetic roots affecting our well-being. There are many impediments interfering with a total modal and seamless system of transport. These impediments include a modal bias where the motor vehicle has prevailed, resulting in the construction of over 4 million miles of roads, creating a drive-thru society way of life and an insatiable appetite for nonrenewable resources, in particular, petroleum. From drive-through divorce facilities, drive-through funeral homes, and automobile weddings to burial in a favorite car, all give an improper direction to energy independence and sustainability. Three-car garages, more personal vehicles than drivers, up to 26-lane high-ways with 400-foot right-of-ways, and increases in the size of our homes all point to spa-tial excess. We have supersized all our assets, including our bodies and the cities where we live, and it is time to develop a diet-cities initiative.

Gridlock and sprawl are now impairing not only our mobility and responsiveness but also the world's gross domestic product (GDP). If both direct and indirect costs are calcu-lated, there is an estimated global loss of $2.5-$4.0 trillion annually. Extended commuting patterns, delayed delivery of goods and services, impacted responsiveness to emergencies, decline in the quality of the biosphere and in the earth's sustainability, and increasing natu-ral and man-made disasters require new approaches and solutions. Security and threat-level conditions are taking their toll, with over 40 countries now being affected as our transport systems, so vital to a healthy economy, become the target of increased terrorism.

This chapter outlines the need for an integrated approach to transport movement and what modal elements need to be addressed. Clearly the failure to properly integrate all assets associated with air, land, sea, and space logistically has taken its toll in compre-hensively planning and effectively responding to both natural and man-made disasters. The administration and management systems in place often have been parochial, result-ing in competitive bureaucratic approaches. Envisioning the future through modeling and simulation offers an important tool to develop and create plans and initiatives that will address the type of solutions necessary for improving the human condition and national preparedness.

References

Ascher, Kate. *The Works: Anatomy of a City.* New York: Penguin Press, 2005.

Associated Press. Amtrak to step up security measures. March 14, 2008.

Berkowitz, Carl, and Clifford Bragdon. Advanced simulation technology applied to port safety and security. *Conference Proceedings, The American Society of Civil Engineers,* June 2006.

Bragdon, Clifford R. Aromacology and the built environment. Scent Society: Yesterday, Today and Tomorrow, Olfactory Research Fund, New York, 1995.

Bragdon, Clifford R. Homeland security: Developing a national center for preparedness. Brevard Community College, Melbourne, Florida, January 22, 2008.

Bragdon, Clifford R. Integrated mobility-based transportation system for logistical optimization. *Logistics Spectrum, 40*(3), July–September 2006.

Bragdon, Clifford R. Kansai: The harmonic balance of the human senses: Developing a sensory master planning system (S^5). *Proceedings, Summit 2000: The Globalization of the Senses,* New York, October 2, 1996.

Bragdon, Clifford R. Southampton intermodal transportation study. Town of Southampton, June 2003, 184 pages.

Bragdon, Clifford R. *Urban cultural evolution: An historical perspective.* Master's thesis, Michigan State University, 1966.

Brown, Patricia Leigh. The chroming of the front yard. *New York Times,* June 13, 2002.

Cairo, Robert A. *The Power Broker: Robert Moses and the Fall of New York.* New York: Random House, 1974.

Center for Transportation Excellence. A nonpartisan policy research center created to serve needs of communities and transportation organizations nationwide. The purpose of the center and its Web site is to provide research materials, strategies, and other forms of support on the benefits of public transportation.

Cohen, Charles, and Eric Werker. The political economy of natural disasters. Harvard Business School, January 16, 2008.

Commuting can kill. *Forbes*, November 14, 2007.

Dantzig, George B., and Thomas L. Saaty. *Compact City: A Plan for a Livable Urban Environment*. San Francisco: W. H. Freeman and Company, 1973.

Driving devices make scents. *New York Daily News*, March 19, 1998.

Fairlie, John. *Municipal Administration*. New York: Macmillan Company, 1908.

FAA. *National plan of integrated airport systems (NPIAS) 2007-2011*. Report to Congress, Washington, D.C., 2007.

Farazmand, Ali. Learning from the Katrina crisis: A global and international perspective with implications for future crisis management. *Public Administration Review*, Special Issue, pp. 148–158, 2007.

First Annual Conference on National Preparedness. *Proceedings, Global Center for Preparedness*, Florida Tech, December 14, 2007.

Flynn, Stephen. *The Edge of Disaster: Building a Resilient Nation*. New York: Random House, 2007.

Flynn, Stephen, and Daniel B. Prieto. Neglected defense: Mobilizing the private sector to support homeland security. *Conference on Foreign Relations*, Washington, D.C., 2008.

Freedman, Mitchell. Scouting for options less traveled Southampton. *Newsday*, July 15, 2001.

Goodman, Paul and Percival Goodman Communitas New York: Columbia University Press, 1990.

Hickey, K. The national aviation transportation center (NATC) introduces intermodal transportation simulation system that is clearly not a game. *Traffic World*, 259(7), August 16, 1999.

Hubler, Eric. The fittest and fattest cities in America. *Men's Fitness*, March 2008.

Kay, Jane Holtz. *Asphalt Nation*. New York: Crown Publishers, Inc., 1997.

Luft, Gal. "Oil dependency is America's financial ruin." *The Cutting Edge*, March 31, 2008.

Mass transit system threat analysis (unclassified). Transportation Security Administration, Office of Intelligence, Washington, D.C., February 29, 2008.

Navarro, Peter, and Aron Spencer "September 11, 2001: Assessing the cost of terrorism," Miliken Institute Review, Fourth Quarter, 2001.

Perelman, Lewis J. Infrastructure risk and renewal: The class of blue and green. Public Entity Risk Institute (PERI) Internet Symposium, Washington, D.C., January 2008.

Public Law 108-199. U.S. Congress, Joint House–Senate Committee, Omnibus Bill enacted January 2004.

Safdie, Moshe. *The City After the Automobile*. Toronto: Stoddart Publishing Company, Ltd., 1997.

Snell, Bradford C. American ground transport. *Third Rail Press*, 1974.

Strugatch, Warren A dreamer's vision vs. reality on road and rail. *New York Times*, October 29, 2000.

Texas Transportation Institute (TTI). The mission is to solve transportation problems through research, to transfer technology, and to develop diverse human resources to meet the transportation challenges of tomorrow. Texas A&M University.

Transportation for tomorrow. Report of National Surface Transportation Policy and Revenue Study Commission, Washington, D.C., January 2008.

Transportation reform and smart growth: A nation at the tipping point. Funder's Network for Smart Growth and Livable Communities, July 16, 2001.

Tri-State Transportation Campaign. Mobilizing the Region: Transportation Reform in the New York/New Jersey/Connecticut Metropolitan Region.

Williams, Stephen W. 21 Inventions for the next century. *Newsday,* June 13, 1999.

Younes, Bassem, and Carl Berkowitz. Guidelines for intermodal connectivity and the movement of goods for Dubai. *Logistics Spectrum, 40*(3), July–September 2006.

4

Mobility Security and Human Behavior

Michael Workman, Ph.D.

Objectives of This Chapter:

- Provide an overview of the complex issues involved in mobility and security
- Introduce a theory framework to help explain how governmental policies translate into actions taken in commercial enterprises and describe some of the psychosocial and behavioral outcomes
- Discuss some of the techniques and uses of information collected about people
- Provide an introduction to some of the governmental and commercial initiatives now and on the horizon in mobility and security
- Discuss the implications of surveillance and monitoring of individuals and the impacts these have on psychosocial functioning
- Introduce some ramifications of chronic vigilance and fear on individuals and societies

> *"Man is born free, but everywhere he is in chains."*
> —Jean Jacques Rousseau, *The Social Contract*, 1762

The bedrock of governing a free society is the social contract, which simultaneously seeks to preserve individual freedom and yet maintain social order. The social contract underpins democracy and asserts that a legitimate state authority is derived from the consent of the governed. However, one of the greatest challenges in a free society is found in the tension between the preservation of freedom and the control mechanisms designed to protect individuals from harm. People want freedom and also safety, but there is a danger that one may come to love the chains and consider them armor or decoration worn by choice. Dostoyevsky wrote, "Man is a pliant animal, a being who grows accustomed to anything."

Introduction

Governments gather information to identify persons of interest and ensure that only authorized persons have access to those locations and resources to which they are authorized. In the past, most democratic governments such as the United States placed relatively few restrictions on access to public resources. More recently, there has been a migration toward broader information gathering by governmental and civil authorities beyond the identification of persons, and greater emphasis is being placed on controls that restrict access, along with increasing regulation of and legislation for processes, procedures, and requirements for permits. Mobility security in the United States has come to encompass a vast landscape that spans legislation such as the U.S. Patriot Act and the McCain–Kennedy immigration bill (Secure America and Orderly Immigration Act) to a broad range of security measures from border control to terrorism, and to technologies such as surveillance devices, airport screening devices, and sensors.

Free societies model their government's policies on their commercial enterprises. Governmental laws and regulations establish the boundaries and constraints under which capitalism operates and help shape the forms of commerce, including how businesses conduct themselves as well as their strategic opportunities. Since the 9/11 terrorist attacks in America, an entire industry has been born that might be coined "fear commerce" (Greenemeier, 2006). The pervasive effects of these governmental and commercial instrumentalities profoundly shape the psychology and behavior of individuals. For example, if threats are perceived as severe and imminent, people will often trade their freedom for even the illusion of safety (Boyle, et al., 2006; Davis and Silver, 2004; Joslyn and Haider-Markel, 2007).

Rather than extolling the virtues of mobility security technologies, which by most objective measures are important and significant, this chapter takes a different tack and suggests that people in democratic societies should consider the implications of the increasing implementations of mobility security technologies and practices. For the purposes of this chapter, the coverage is limited to a few key topics related to sociopolitical and sociobehavioral security issues. Drawing from deterrence theory, terror management theory, and protection motivation theory, which together connect the sociopolitical with the sociobehavioral dimensions of security, this chapter explains how governmental policies translate into actions taken in commercial enterprises and describes some of the psychosocial and behavioral outcomes. To illustrate, it reviews some of the research literature on the behavioral effects of surveillance. It concludes with individual and societal implications of fear commerce and considers that societies might better balance their needs for safety with those of other important interests.

Mobility and Security Theory

A demonstration of the tension between mobility and security is the common experience of air travelers as they undergo U.S. airport screening at checkpoints operated by transportation security officers of the Transportation Security Administration. In recent years the list of *prohibited items* in airplanes has grown, along with types of screenings performed. The latest in thermal sensor technology, while less visible and obtrusive than the gigantic machinery currently located at airport gateways, is more intrusive. These newer screening devices are so sensitive that they are able to render images of everything under one's clothing and are designed so that they can be embedded in the walls of airport hallways. Soon, travelers and their baggage may undergo a covert screening for suspicious materials as they walk toward their gates. Anything that triggers an alarm would result in an alert message, along with photo images of the suspicious traveler, to personal digital assistants (PDAs) held by roving security officers who would detain the suspect (Meckler, 2006).

Mobility is the essential ingredient in commerce, but as observed in the airport example, in an awkward paradox mobility relies on practical levels of safety, and yet the mechanisms required to mitigate threats by their nature may constrain mobility. That is, mobility is constrained by controlling the vehicles used for transportation as well as the transactions that people perform—thus mobility security technology controls the things a person does either as physical action or as captured in data, along movement itself (Castells et al., 2007; D'Urso, 2006). The government of a democratic society walks a tightrope between working to alert its populace against threats and striving to maintain their reasonable expectations of safety through its security control mechanisms. The tightrope is strung tenuously among principles defined by three theories: deterrence theory, terror management theory, and protection motivation theory (see Figure 4.1).

Together, these theories describe the network of interrelated principles that set in motion a population's systemic behavioral responses. Each theory puts forth that unless a population is mobilized against a threat, its members become complacent. For example, people often ignore or disable security measures when they are perceived as intrusive or ineffective, unless those people are constantly placed in a psychological state of hyper-vigilance about severe impending threats (Langenderfer and Linnhoff, 2005). This is because people operate day-to-day on the basis of assumptions and personal beliefs that allow them to set goals, plan activities, and order their behavior. Their conceptual systems are developed over time and provide them with expectations regarding their environment.

While people operate on the basis of these conceptual systems, they tend not to be aware of their central postulates, among which is the belief in personal invulnerability (Dorn and Brown, 2003; Hochhauser, 2004). For instance, people may recognize that crimes are common; however, they simultaneously believe that "it can't happen to me" (Roe-Berning and Straker, 1997). Lejunene and Alex (1973) found in their research on

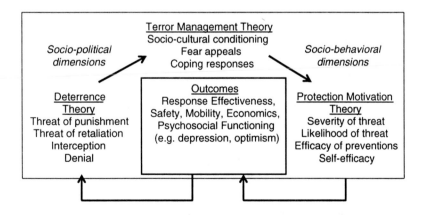

FIGURE 4.1 Theory Framework

crime that mugging victims first defined the event with disbelief and in a nonthreatening way, such as a practical joke. People operate on the basis of this "illusion of invulnerability" to support their need to view the world as orderly, stable, and meaningful, underestimating the probability of their own misfortunes and overestimating the probability of misfortunes to others (Hochhauser 2004; Roe-Berning and Straker, 1997). When people operate from this assumption, they are less likely to take security precautions and will even try to avoid or disable them completely if they intrude on some other important consideration such as their convenience or if they impinge on effective performance of their duties (Acquisti and Grossklags, 2003; Dorn and Brown, 2003; Ettredge and Richardson, 2003; Leyden, 2004; Ryan, 2004; Sasse et al., 2004).

Nevertheless, events such as criminal acts, accidents, disasters, and disease force people to recognize and make objective their basic assumptions about the environment and the world (Janoff-Bulman and Frieze, 1983). It has been found, for example, that those who have been subjected to a crime such as burglary tend to assess their chances of falling victim to future burglaries as higher than those who have never been a victim (Lejeune and Alex 1973). In general, the illusion of invulnerability is an immunizing stratagem from the fear, stress, and anxiety associated with the perceived threat of misfortune; but once victimized, even if it is induced only vicariously, it becomes easier to see oneself again in the role of victim (Janoff-Bulman and Frieze, 1983; Roe-Berning and Straker, 1997).

Deterrence Theory

Deterrence theory (Jervis, 1989; Kennan in Miscamble, 1992) stemmed from the military logic that imposing greater costs on a potential enemy than the prospective gains or benefits might neutralize the motivation for an act of aggression (Berejikin, 2002). It seeks to prevent attacks by the threat of punishment, the threat of retaliation, and proactive denial and interception of resources that could be used to perpetrate an attack.

In its original application, deterrence theory concentrated more on threats of punishment and retaliation. The starkest example of deterrence theory is found in the philosophy of mutually assured destruction (MAD) that grew out of the cold war era. Its premise is based on a rational model of human behavior, which breaks down under the conditions where a geographic enemy is unidentifiable (i.e., a country) or a rational response cannot be counted on. This is because the threat of retaliation works as a deterrent only if the enemy fears annihilation. Since 9/11, Western military and intelligence strategists have come to recognize the lethality of technology in the hands of nontraditional foes; that is, those who are neither geographically identified nor conform to traditional rational choice models of decision making.

The law enforcement, military, and intelligence communities then have leaned increasingly on a wider interdiction stance and have begun developing technologies that would assist in proactive denial of resources and interception of actors. "Project Green Dragon" evolved from an effort by the U.S. intelligence community to develop stochastic models to try to predict terrorist attacks once actors had been identified, but developing accurate predictive models proved difficult in attempting to account for nonrational motives such as ideology or brainwashing.

Because of the difficulty in predicting nonrational human behavior, most Western governments have come to rely more on a deterrence strategy of denial whereby defense and intelligence systems are aimed at thwarting attacks with information gathered extensively, and through data mining and the identification of patterns that emerge from the collection of behavioral data from entire populations. This has meant primarily the development of technologies that enable both pervasive and invasive data collection and storage of an information cache unparalleled in history.

Terror Management Theory

In order for the government of a democratic society to implement such pervasive and invasive controls from the massive collection of information about people—approve spending and gain the compliance of its populace—it must mobilize its citizens against purported threats so that they will relinquish their privacy and certain of their freedoms. Terror management theory (Pyszczynski et al., 1997) asserts that in the event that preventative (deterrent) measures fail to preclude acts of aggression, governments seek to mobilize citizens by making fear appeals based on some actual or potential threat. The perception of threat is the anticipation of a psychological (e.g., assault), physical (e.g., battery), or sociological (e.g., theft) violation or harm to oneself or others, which may be induced vicariously (Lazarus, 1991). Therefore the theory explains the processes by which governments utilize fear propaganda for mobilization and to make vigilant a population by exploitation of the human availability bias, making even remote threats seem severe and imminent but mitigated under due care of the government's watchful eye.

During the cold war era, a systematic institutionalization of fear appeals surfaced in the United States. American schoolchildren were shown movies of a nuclear explosion and then taught to hide under their desks according to a song they learned, *Duck-and-Cover.* In more recent times, the American government has attempted to enlist the mass media, which in its constant quest for sensational stories, acts as conduit for fear appeals to the public (Greenberg, et al., 1997; Huddy, et al., 2005). This has contributed to the general acceptance of invasive security measures ranging from pat downs at airports to pervasive surveillance cameras at street corners, along with acquiescence to unprecedented levels of military spending, nearly $650 billion in 2008 compared with $360 billion in 1998 (GAO, 2007), which amounts to 41% of the taxable income of Americans. Also at extraordinary levels are the emergence of private industries targeting the government sector and commercial security operations, on which spending has more than doubled since 9/11 to over 8% of a $2.16 trillion global information technology market (Bartels, 2006).

Terror management theory further purports that people exhibit greater exogenous coping behavior toward severe threats over which they perceive they have more control. However, when the perceived threat becomes so severe that it involves one's own mortality, people retreat into endogenous coping mechanisms and form more fatalistic attitudes such as clinging to religion or patriotism because these act as anxiety buffers (Rosenblatt et al., 1989). As such, given the government's need to mobilize its constituents, fear appeals must re-energize the populace once the effects of a fear appeal begins to deplete. The government must then take precautions to balance fear appeals designed to mobilize its citizenry at the risk of "crying wolf" once too often, by instilling a sense of security that enables people to conduct their lives—because these psychological processes regulate the degrees of risk that people are willing to assume, for example, whether one will fly on an airplane during a code yellow (elevated) versus code orange (high) versus code red (severe) threat level.

Protection Motivation Theory

When a threat is perceived, people behave according to the amount of risk they are willing to accept, known as *risk homeostasis*. Risk homeostasis results from the perceived severity of the potential damage, such as financial costs of repairs (Grothmann and Reusswig, 2006). Therefore, people tend to adjust their behavior in response to the extent of the damage the threat may cause (Pyszczynski et al., 1997). The perceived severity of threat and the associated acceptance of risk behavior are based on the premises that (1) people place a certain intangible value on "life," "liberty," and "property"; (2) they have a threshold level of risk they will accept, tolerate, prefer, desire, or choose; (3) this "target level" of risk they will accept before acting depends on the perceived advantages or benefits versus the disadvantages or costs of safe or unsafe behavior alternatives; and (4) this will determine the degree to which people will expose themselves to a threat or hazard before taking precautions or trying to avoid a threat altogether (Wilde, 2001).

Protection motivation theory (Rogers, 1975) further refines these principles into individual behavioral responses. It posits that how people behave will be conditioned upon their cognitive assessment of the probability or likelihood that a severe threat event will happen to them, the perceived efficacy of their available preventative measures, and their self-efficacy for implementing those preventative measures. Protection motivation theory thus translates the more global sociopolitical constructs of deterrence and terror management theories into individual-level constructs for predicting how people will behave. When fear appeals are made, individuals assess the appeals for personal impact.

Extending from this line of reasoning, Aldoory and Van Dyke (2006) tied the situational theory of publics (Grunig, 1978) and the health-belief model (Rosenstock, 1974) to issues related to bioterrorism threats. The situational theory of publics asserts that a populace may be segmented based on the activeness or passiveness of communication behavior (Aldoory and Van Dyke, 2006). The factors purported by this theory are problem recognition, level of active involvement, and constraint recognition (Grunig, 1978).

Problem recognition reflects the extent to which an individual recognizes a problem as relevant to him or her; that is, how likely a threat is perceived to impact the person. The level of active involvement results from a perception of how emotionally the problem is felt, such as the perceived severity of damage to the person that is posed by the threat. Constraint recognition reflects the degree to which people perceive their behaviors as limited by factors beyond their own control. According to Grunig (1997), if these three factors accurately depict external conditions then the environment must change before a person will respond, but if they are merely perceived (internal), they may be changed by persuasive communication, hence persuasiveness of fear appeals is a key element in whether and how people respond to messages about a threat (Petty and Cacioppo, 1986).

Mobility Security Initiatives

The focal stance of democratic governments in relation to security has been to deter people from threatening acts through fear of punishment or retaliation, and to intercept them if they try to carry out a threat (Jones, 2007). If deterrents fail, governments must have in place the facilities to deny the resources to those who might conduct an attack. These facilities comprise identification and authentication processes. Consequently, mobility security initiatives have generally involved three major elements: (1) credentialing for the purpose of allocating access rights to physical and/or virtual locations and resources; (2) the collection and distribution of data about people, their demographics, physical characteristics (biometrics), travels, and actions; and (3) surveillance, which is the physical or electronic observation of someone's activities and behavior. These three elements are broadly focused on identifying persons of interest and ensuring that only authorized persons have access to those locations and resources to which they are authorized.

The question of how narrowly or widely granted the authorization to access locations and resources determines the degree that mobility is constrained. The issue translates further to an implicit assumption of whether people inherently have rights of access, unless they commit crimes, or whether people generally have rights only to the access specifically granted them by governmental or civil authorities, which determines their degree of freedom (Jones, 2007). The range of resources that require credentials for access, the ways in which data are gathered and used, the breadth and depth of the data collection, and the breadth of surveillance in a society reflect this basic assumption.

Mobility security and freedom are inextricably linked concepts along a continuum in which each absolute resides on a polar end of the spectrum. In recent years there has been an acceleration of the slide along the scale toward governmental control. Prior to 9/11, most of the mobility security initiatives in the United States and Europe targeted immigration and national security tracking problems, or what might be called *human flow control*; namely, the pre-entry, post-entry, and post-exit processes related to foreign nationals. The purpose of the information collection was to enable governments to identify people as they entered the country and determine where they were coming from, where they went while in the country, and where they went after they left. An example system used for this purpose in the United States is the Student and Exchange Visitor Information System (SEVIS), which is a Web-enabled system for maintaining information on international students and exchange visitors administered by the Student and Exchange Visitor Program (SEVP) under the supervision of the U.S. Immigration and Customs Enforcement (ICE)—the largest investigative arm of the Department of Homeland Security (DHS).

With the signing of the North American Free Trade Agreement (NAFTA), along with attempts to better define international trade laws, many transnational commercial barriers were removed. The intent of these initiatives was to expand capitalistic opportunities amid the global economy. Simultaneously, new physical and technological barriers have been erected to try to channel the flow of people and goods through controlled checkpoints, and along with these flow management systems has come the development of an array of "smart border" technologies to credential, monitor, and track individuals and collect data about them, whether nationals or foreigners. Similar events have taken shape in the European Union (EU).

There is little doubt that the prevalence of transborder crime causes border instability. Strengthening border controls and reducing transborder crime can help solve problems related to apprehending terrorists and their weapons as they attempt to enter illegally between the ports of entry, deter illegal entries through improved enforcement, and aid in detecting, apprehending, and deterring smugglers of humans, drugs, and other contraband. The leverage of smart border technology enables law enforcement agencies to multiply the effect of enforcement personnel and reduce crime in border communities and consequently improve quality of life and economic vitality of targeted areas. However, in spite of fences, increased staffing, Web-based information systems, and the latest in thermal sensor and night vision camera technologies, it has been widely recognized that the United States has a defeated border with Mexico.

Yet, perhaps an even greater threat comes to the United States from the border that it shares with Canada since there are fewer fear appeals directed at it along with an American presumption of greater safety associated with Canada (Cordesman, 2002). Many of that nation's residents are hostile to the United States, and in recent years the Federal Bureau of Investigation (FBI) has foiled several high-profile terrorist plots, including a planned bombing of the Seattle Space Needle and bridges in San Francisco (Jones, 2007).

As a reaction, the solution to these problems by many democratic governments including the United States has been to cast a wider information collection net. As a wider net is cast in terms of security initiatives, there are many that are being enacted that affect resident citizens in democratic societies throughout the world as well and have spawned an immense commercial enterprise for a variety of mobility security technologies, including technology for the collection and dissemination of information related to human biometric characteristics (called *templates*), radio frequency identification (RFID) and implants, and nanotechnology used in passive and active surveillance devices that are so small they are called *sensor dust* (Bajcsy, 2005).

Expansion of Institutional Initiatives

Technologies are the enablers of security policy enactments, and increased capabilities lead to the inflation of implementations and security initiatives. For instance, enabled by new technologies in biometrics, database systems, and networking, in 2004 the DHS initiated the US-VISIT system, in which most people with nonimmigrant visas entering the country are fingerprinted and photographed at visa application time. The biometric characteristics are checked against a database of known criminals and suspected terrorists. When the visitor arrives at a designated port of entry, biometric characteristics are used to verify the person at the port of entry and determine if that person is the same one who received the visa.

Responding to calls by the U.S. Congress for a better integration of departmental information systems by federal, state, and local law enforcement communities regarding processes for identifying criminals and retrieving histories of aliens apprehended attempting to enter the United States, information about transborder flow control activity and foreigners is merging with information systems about domestic populations.

Hence, the US-VISIT system is now being interfaced with the Integrated Automated Fingerprint Identification System (IAFIS), a national fingerprint and criminal history system maintained by the Federal Bureau of Investigation Criminal Justice Information Services (CJIS) division. IAFIS contains the largest biometric database in the world with the fingerprints and corresponding criminal histories for more than 47 million people collected from state, local, and federal law enforcement agencies. Concurrently, the IAFIS system is also being interfaced with the biometric database (IDENT) used by the U.S. Department of Justice (DOJ). This will create a confederation of information systems about the domestic population and foreign nationals on an unprecedented scale. In Europe, there is a similar movement to consolidate European Union member states' immigration and biometric data within the Eurodac database (Van der Ploeg, 2005).

Where systems such as US-VISIT have maintained information about each visitor, and where IAFIS has been primarily a database of criminal activity, there is a broadening of the information collection about people in general that will eventually become part of a confederation of national databases. One such initiative in mobility and security taking shape is the implementation of machine-readable travel documents (MRTD).

In 1980 the International Civil Aviation Organization (ICAO) released what is known as Document 9303, which has formed the basis for the U.S. INS and DHS initiative for biometric or e-Passports. Document 9303 (and its subsequent annexes) contains specifications for machine-readable biometric passports, and through several revisions, 188 member states have adopted these specifications universally.

A press release issued by the DHS in 2005 announced that the Visa Waiver Program (VWP) would involve countries whose citizens do not require a visa to enter the United States, including countries in the European Union, Japan, Australia, and New Zealand, but that VWP countries would be required to implement MRTD containing digital photographs and other biometric information and RFID-integrated circuitry for anyone entering or leaving the United States. This announcement relates to the Enhanced Border Security and Visa Entry Reform Act of 2002 requirement that any passport issued and used for VWP travel to the United States must include a biometric identifier based on applicable standards established by the ICAO.

Along with concerns about the expansion of the data collection about general populations, critics (cf. Davis and Silver, 2004) have questioned the purpose and nature of the initiative by raising concerns about their actual security, claming that these proposed methods to strengthen security might be akin to the cliché about a locked door keeping honest people honest. They charge that extant systems have virtually ignored the creativity with which people find ways around any given system, citing that the ability of attackers to contravene security defenses continues to elude our ability to defend against them (Bresz, 2004; Sasse et al., 2004).

As a case in point, while significant financial investments are being made globally in the implementations of MRTD such as e-Passports, these may be gotten by first obtaining, with apparent relative ease, a fraudulent birth certificate from any number of sources in any number of countries. According to a recent press conference held by the U.S. Department of Justice, birth and death certificate fraud has created a serious national problem, but the extent of the problem is not known because of the underground nature of the crime. In addition to the security concerns this raises, a report from a key border state, New Mexico, documented that birth certificate fraud was estimated to have cost that state $813 million in 2005 for welfare, food stamps, and other benefits (New Mexico Department of Health, 2007). Similar reports from the governors of Arizona, Texas, and California estimated upward exponents of the New Mexico figure (Brown, 2000).

As a result the answer by many democratic governments has been to expand further the reconnaissance and data collection efforts on international, national, and local levels. In the United States, for example, there is consideration of an even more pervasive and stringent tracking and monitoring system based on a national identification system. Other countries have already moved in that direction, along with the requirement that credentialing be required for all transactions on both the sale and purchase sides. The Philippine and Malaysian national identification programs are two such instances. The Philippine National ID (PNID) program will integrate the Philippine Social Security System (SSS), the Government Services Insurance System (GSIS), driver licenses database, immigration records, and the national statistics databases (Encinas-Franco, 2005).

Moreover, the Association of Southeast Asian Nations (ASEAN) Asian Multilateral Identification Database (MID) is a repository of data established by the Southeast Asian countries of Philippines, Malaysia, Hong Kong, Indonesia, Thailand, and Singapore. The MID is an extension of the cooperation inherent in ASEAN, reinforced by recent pacts and by the extent of various threats now facing the region. It will encompass multireference personal identification on citizens of, resident aliens in, and visitors to participating countries along with data feeds of all significant print and media (Encinas-Franco, 2005).

Where the standard intelligence collection efforts tend to assemble information on individuals only after they have been identified as "persons of interest," the MID, and information repositories like it, are designed to supplement international law enforcement efforts by means of nondiscriminatory pervasive data collection. The argument for such a collection is to enable earlier detection of criminality by developing full identity profiles much earlier than possible under the current identification scheme. For instance, the MID will provide instant information to appropriate agencies in initiating and performing their own investigative activities, such as automatic notification to immigration authorities when an alien has overstayed his or her visa, or when a citizen of the home country has overstayed his or her visa in another country. However, the shared data will extend well beyond human flow control and include background information of citizens in each country such as names, dates and places of birth, identification numbers, addresses, fingerprints and biometric data, criminal and civil records, and other "unspecified" communications and transnational movement of individuals and selected materials.

While these technologies and their data confederations may assist governments in the execution of securing their populations, they also carry with them significant security risks to individuals along with other profound implications. Among them are the increased privatizations and outsourcing of human intelligence data collection and storage, along with the combining of institutional uses with commercial uses of these data (O'Brien, 2004). It has become common practice in the United States, for example, for many tax processors to outsource tax preparations to workers in foreign countries, and for immigration data used in tracking human flow control to be shared with state visitor bureaus and travel agencies for advertising purposes (Cranor and Wildman, 2003; Jørgensen, 2006).

Information gathering is potentially ubiquitous, and any kind of media or activity can become raw material for production of an information commodity (Arvidsson, 2004). In this context, both governments and private actors often use information gathered about people to maximize economic opportunities (Green and Smith, 2004). "Meanwhile government administration and law enforcement agencies following a desire to extend the traditional state surveillance and investigative power to new digital domains, consider [these data] as merely the translation of existing powers to new mediums. The surveillance extends the logic of power into the mobile arena, but also links up with commercial imperative and characters the communication economy" (Castells et al., 2007, pp. 121–122).

These practices are accompanied by a troubled history of both intentional and unintentional information leakage (Davis, 2001). To highlight this, some have pointed to fairly recent security breaches of information repositories, such as LexisNexis having 310,000 personal records stolen in 59 separate incidents, and ChoicePoint having more than 17 billion records about people that it sells to federal, state, and local law enforcement clients and businesses, some of which have been discovered to be fronts for criminal operations (Spangler, 2006).

The high-profile compromises of LexisNexis and of ChoicePoint in 2005, and many others, illustrate that having central repositories containing such sensitive information increases individual vulnerability and fuels public concerns over identity theft, of which the U.S. Department of Justice estimates that one in three people will become victims at some point in their lifetimes (U.S. Department of Justice, 2004).

The obvious technological and procedural issues to be resolved include how to protect the integrity of the devices and data used in the information capture and verification process (e.g., attending to tamper-proof materials and maintaining accurate databases) and also how to secure the information as it is stored and accessed. Beyond these technical issues is cause for alarm over the increased incidence of type-1 errors, where the innocent are singled out with potentially devastating and long-term (potentially life-long) consequences, including arrest records that remain with victims of identify theft by criminals, even if victims are later vindicated during adjudication.

The concern widens with the consideration that one's sensitive biometric information will be kept in repositories in various countries one visits. Leakage of personal information from these central information repositories continues to exacerbate both the potential and severity of the damage to individuals and industry. In spite of this an expansion of the blending of immigration, intelligence, and commercial data collection and dissemination appears to be on the horizon. The MRTD proposal in the Bahamas, as an example, contained provisions for the capture of e-Passport RFID as people visit stores, hotels, and restaurants. The proposal called for the information to be utilized by government law enforcement and intelligence communities, and also fed to information repositories for marketing campaigns and other commercial activities.

Expansion of Commercial Initiatives

As governments have stepped up their data collection about citizens, commercial initiatives have led to development of new technologies that create amazingly flexible ways of gathering and sharing information. For instance, it is widely recognized that the technologies in the World Wide Web such as search engines and the use of Hypertext Markup Language (HTML) and eXtensible Markup Language (XML) have revolutionized information advertisement and discovery, and these technologies are evolving to support an even more active system. Components that are being developed as part of the World Wide Web Consortium (W3C) standards effort toward the next-generation Semantic Web are increasingly fashioned into commercial applications to address some aspect of efficient information collection and dissemination capabilities.

The Resource Description Framework (RDF) is a technology that creates relational linkages among Web documents scattered among disparate systems. RDF can be elaborated with a markup language such as the Defense Advanced Research Projects Agency (DARPA) Agent Markup Language (DAML) and Web Ontology Language (OWL) to enable software similar to search engines, called *agents*, to seek out information encoded with this markup and act on instructions, seek goals, and make evaluations as they traverse networked computers. Software agents enable governments and companies to conduct electronic surveillance, gather intelligence, and even carry out espionage in ways that previously have been imagined only in science fiction (Kowalczyk and Alem, 2003).

These Web-based technological advancements are salient for another reason in addition to their information collection capabilities. Ngwenyama and Lee (1997) coined the term "distanciation" to describe *virtual life* as personal information is increasingly electronically disseminated. The traditional concept of physicality (physical location) has been augmented by a virtual one. Thus the concept of mobility involves not only physicality but "virtuality" as well, and this brings about a variety of electronic personal and private information, along with much easier access to these electronic resources.

The privacy concerns expressed by critics in the United States are not simply matters of constitutional debate over whether there is a right to privacy. That debate is largely irrelevant. While many have acquiesced that the concept of privacy is somewhat an illusion amid societal interests and the common good (Johnston and Cheng, 2002), especially since technologies contribute to making those boundaries amorphous, one reason why this is important is because electronic records remove some of the barriers that legislators in the United States and other democratic societies have relied upon to maintain reasonable expectations of privacy. Simply because something is in the public record does not mean that people would concede their privacy. The creation of records-access statutes and regulations have been based on the premise that people who would expend the effort to visit a courthouse and file paperwork to gain access to public records would likely have a legitimate cause, and that offers a reasonable expectation of personal privacy (Keck, 2005). However, in many places in the United States and other countries, various types of court and legal records are now available online and accessible 24 hours a day and 7 days a week by anonymous casual observers and even criminal elements. That personal information is owned not by oneself but by corporations and governments places people in precarious situations.

To address some of the concerns this might raise, there are established regulatory boundaries. For instance, the major compliance standards and regulations such as the Health Insurance Portability and Accountability Act (HIPAA), the Public Company Accounting Reform and Investor Protection Act of 2002 (Sarbanes–Oxley), and the Payment Care Industry Data Security Standard (PCI DSS) place strong restrictions on many of the software applications that provide for a range of business systems including dynamic Web sites, service-oriented architecture (SOA), electronic commerce, and business process automation and administration. While these regulations and legislations have been designed to constrain virtual mobility in an endeavor to help retain individual privacy, they have morphed into efforts toward greater governmental control of the personal information of populations in general, and simultaneously have expanded the amount and types of information that corporations must now collect and provide (Lyon, 2004).

As an example, in the United States there are federal and state records requirements (and many that are being proposed) for Internet Service Providers (ISPs) and companies such as Google to track and retain information in search logs, which have raised concerns among privacy advocates. Proposed federal mandates would require ISPs to maintain logs detailing what Internet Protocol (IP) addresses have been assigned to whom at any given time, allowing law enforcement agencies to tie IP address to a residence or place of business. Most ISPs utilize technologies and protocols that share and reuse IP addresses, thus the data that would need to be retained would become immense, and failure to retain such records is likely to carry stiff criminal liabilities.

When it comes to such data retention requirements, ISP search engines and other applications such as social network sites or e-mail, Web forums, and chat rooms, the logging retention requirements would be significant both in terms of cost and in terms of their accessibility. In the meantime, it is unclear what such requirements might mean for ISP subscribers. Although ISP subscribers such as colleges and universities and retail businesses that provide wireless access to customers typically maintain logs for billing or fraud detection, they usually purge them after a few months. Placing retention requirements on ISP users might drive many away from such provisioning services, which would dampen virtual mobility. Measures still taking shape in the United States have already been approved in the European Union by a directive requiring member states to adopt data-retention rules for ISPs, telephony (landline, mobile, and voice over IP [VoIP]) and e-mail (Mueller, 2007).

The collection of information about people in general by corporations is also frequently opportunistic, as another commercial initiative gaining ground has been corporate surveillance of employees. This has been accompanied by the development of a dizzying assortment of technologies that have begun to appear to aid in the surveillance arsenal, including context-aware perceptual interfaces used in expert systems and artificial intelligence. Global positioning satellite (GPS)–enabled context-aware software executing in PDA and mobile telephones can "recognize" proximal services and infrastructure such as hospitals and police stations as people travel from point to point, along with anticipating their movements. From a security point of view, while such software can be an immense benefit in cases of emergencies, it carries with it the implicit ability for various parties to track and monitor people on the move (Canny, 2006).

Again, one of the central issues in terms of surveillance initiatives concerns worker *rights to their privacy* versus employer *rights to know* along with their responsibilities to report illegal activities. Laws have tended to support the rights of corporations to inspect and monitor work and workers, which arise from needs related to business emergencies and the corporation's rights over its assets and to protect its interests (Borrull and Oppenheim, 2004). Employees may use e-mail or the Web inappropriately for purposes such as disseminating harmful information, infringing on copyrighted or patented materials, harassment, corporate espionage or insider trading, or acquiring child pornography. In these instances, the courts have leaned on the principle of discovery to allow inspection, granting any party or potential party to a suit including those outside the company access to certain information such as e-mail and backups of databases, even to reconstruction of deleted files (Bureau of Transportation Statistics, 1992). There are protections in place to mitigate unreasonable search and seizure, such at the Electronic Communications Privacy Act, which defines an electronic communication as the following:

> *Any transfer of signs, signals, writing, images, sounds, data, or intelligence of any nature transmitted in whole or in part by a wire, radio, electromagnetic, photoelectronic or photooptical system that affects interstate or foreign commerce, but it does not include the radio portion of a cordless telephone communication that is transmitted between the cordless telephone handset and base unit, wireless networks, any wire or oral or video communication, any communication made through a tone-only paging device or any communication from a tracking device.*
> —Losey, 1998

The original intent of the law was to protect one from the interception of communications not otherwise subject to interception, but the U.S. Patriot Act (HR 3162) has vastly expanded the powers of government and enforcement agencies to conduct surveillance and communications interception pertaining to criminal activity (Borrull and Oppenheim, 2004). The fairly recent disclosure of the National Security Agency's monitoring of telephone conversations highlights the concern, especially if wrongdoing is admitted on the part of a government but the behavior goes unpunished.

The majority of work surveillance to date has consisted primarily of electronic observation of Web-surfing activity, monitoring e-mails, telephone-call monitoring for office workers, and GPS tracking of mobile workers. Two recent studies found that in 1999, 67% of employers surveyed electronically monitored their employees, but by 2001 that number had grown to 78%, and 92% of employers indicated the use of electronic monitoring by 2003. Most of the employers surveyed monitored employee Web surfing, more than half reviewed e-mail messages and examined employees' computer files, and roughly one-third tracked content, keystrokes, and time spent at the computer (Harvey, 2007).

Increasingly, employers are adding video surveillance to their monitoring repertoire (Fairweather, 1999). Of companies surveyed, only 18% of the companies used video surveillance in 2001, but by 2005, that number had climbed to 51%, and 10% of the respondents indicated that cameras were installed specifically to track job performance (Harvey, 2007). These uses of video surveillance to monitor work activity may be supported by an analogy to the "plain view" doctrine, which allows the rights of a company to monitor employees in plain sight, and is strengthened if the company takes overt action to notify its employees of the practice (Scholz, 1997).

From a legal perspective, the extent of an employee's expectation of privacy often turns on the nature of an intended intrusion. With the exception of personal containers, there is little reasonable expectation of privacy that attends to the work area. That is, a worker may have reasonable expectation of privacy in personal possessions such as a handbag in the office, but the law holds that employers possess a legitimate interest in the efficient operation of the workplace and one aspect of this interest is that supervisors may monitor at will that which is in plain view within an open work area, even within a home if a home is used for teleworking (Braithwaite and Drahos, 2000).

The legal protections afforded employers combined with increasing technological sophistication and decreasing costs are allowing companies to monitor the actions of workers more closely than ever before (Akdeniz et al., 2000; Fairweather, 1999). Software that covertly monitors computer activities and surveillance devices are being fashioned to blend into the environment by hiding them in pens, clocks, or bookends, which has a dampening affect on employee surveillance awareness regardless of employer notifications. This creates a psychological bind on employees who know they are being monitored, but the unobtrusiveness of the technology and devices make them seem innocuous (Akdeniz et al., 2000). Moreover, the concepts of privacy and security are not simply legal matters. Privacy is an element of the psychological state of security. When people's privacy is reduced combined with an increase in exposure to fear appeals, there can be deleterious consequences for human psychosocial functioning.

Commercial Fear

An important question in the dialog about mobility and security is how does a democratic society come to increasingly relinquish mobility and freedoms in exchange for security controls (Huddy et al., 2005)? People become accustomed to their environment, but a radical change in an environment is made through systematic threats of imminent and severe harm. The mass media, often a conduit for fear appeals (Boyle et al., 2006; Davis and Silver, 2004; Joslyn and Haider-Markel, 2007), is a focal point in this dialog because it "has whipped up sentiment that has taken countries into disastrous wars, and suddenly, has whipped up sentiment to evacuate them disastrously" (Jones, 2007). For example, reports about activity in a given geography in Iraq during the U.S.-led war have not been referred to by the media according to its location, landmarks, or coordinates but rather by the term *triangle of death*.

Studies by Chatsworth (1996) and Vasterman et al. (2005) of media reporting about disasters chronicled the 1995 bombing of the Alfred P. Murrah Federal Building in Oklahoma City, which showed that the media blitz lasted more than 6 months, from April to October of that year.[1] Following the media frenzy, a Jacksonville, Florida, television news reporter ran a story on someone buying large amounts of chemicals and some sodium nitrate at a local ACE Hardware store. Replete with vivid eyewitness testimonies, the reporter warned of impending doom and that the person was *acting like a terrorist*—noting the make and model of the vehicle the shady character was driving along with telecasting the license plate number to any vigilante who might rid the community of the "evil doer," who turned out upon further police investigation to be a completely innocent person buying chemicals for a swimming pool and yard.

The media fury over incidents such as Oklahoma City and the first World Trade Center bombings whipped up public panic and reinvigorated earlier calls for federal regulation and controls on all precursors to explosives such as ammonium and sodium nitrate (ATF, 2007; NAC Report, 1998). What followed was the development of an array of technologies that enabled the documentation and tracking of precursor chemicals and of persons purchasing them. "The findings to date support the use of chemical markers to improve detection of plastic and sheet explosives... One key strategy is prohibiting the retail sale of the types of packaged ammonium nitrate fertilizers that can be detonated, unless consumers produce identification and retailers keep accurate records of transactions" (Pippenger and Nye, 1998, p.1).

Fear appeals call to action societies and their governments, and while countermeasures such as chemical marking help to reduce threats from that given source, little can be done to stop a determined attacker. As sodium nitrate has become more difficult to purchase in large quantities without falling under the watchful eye of law enforcement, terrorists in a sort of cat and mouse game simply resort to another catalyst, such as liquid peroxide-based precursors. While screeners at airports relieve travelers of their shampoos and toothpastes, the fear appeals germinate a reactionary development of a host of new technologies that can detect such explosives, which by the time they are widely deployed will be largely obsolete (Castells et al., 2004; Huddy et al., 2005).

In a similar vein, the 2007 attack perpetrated by a student at Virginia Tech killing more than 30 people drew intense media saturation coverage, along with broadcasts of the perpetrator's gun-waving rants. The press naturally linked the incident with the Columbine High School shootings 8 years earlier, as parallel stories were run in both print and television media. As Brian Williams reported on *NBC News*, "Jitters from schools all across the country, 28 states had bomb threats and lockdowns"; the shootings

---

[1] See for instance: http://www.cnn.com/US/OKC/bombing.html.

then were tied to the shootings at Virginia Tech and to a hostage and shooting incident at the Johnson Space Center in Texas, followed by a report on the carnage in Iraq where "hundreds have been killed and hundreds more wounded," stating further that other news such as six deaths from flooding in the Northeast had taken a backseat to the events in Blacksburg (*NBC Nightline,* April 20, 2007).

A democratic capitalistic society states its priorities by its spending habits. Despite the fact that more school-aged children die as a result of traffic accidents in 1 year (nearly 400,000) than have been killed in all the recorded school shootings combined, there is unprecedented spending on surveillance equipment and metal detectors in schools, while driver education courses have been eliminated from nearly all school curricula (Albright, 2007). Clearly the shooting tragedies were horrific, but the spending on screening devices in schools is not proportionate relative to other more lethal concerns, and the devices may not be particularly effective in preventing the kinds of crimes for which they are targeted (Vasterman et al., 2005).

What is conclusive, however, is that the intensification of surveillance and monitoring and data collection of large populations has fueled a technological renaissance, and the numbers of public and private companies in the security business as well as the price tags have soared. For example, the Department of Homeland Security initiated the United States Visitor and Immigrant Status Indication Technology (US-VISIT) program that produced nearly $10 billion in government contracts with private companies, and sales estimates for biometric technologies grew from $400 million in year 2000 to $1.9 billion by year 2005 (Monahan, 2006).

Along with advancing pervasive technological countermeasures, companies are increasingly institutionalizing security policies, processes, and practices. No less than 100 well-recognized security certifications have appeared in the vernacular, filling an acronym dictionary, including CEH, CISM, CISSP, CINSec, CSP, ECSA, GIAC, LPT, PCIP, SSCP, STA, SCSE, SCSP, Security+, ABCP, CBCP, MBCP, CCSA, CISA, CISM, and so on. Security technology and process infrastructure, as an overhead cost of doing business, have reached more than 8% of a company's average budget (Bartels, 2006). In 2003, private industry spending on information security in the United States was over $1 billion, and over $6.5 billion for the U.S. government. According to the Information Security Oversight Office (ISOO) of the U.S. National Archives and Records Administration:

> *The [2003] cost estimate on information security for the US government indicated a 14 percent increase over the cost estimate reported for FY 2002. For the second year in a row, industry reported an increase in its cost estimate. The total cost estimate for Government and industry for 2003 is $7.5 billion, $1 billion more than the total cost estimate for Government and industry in 2002. In particular, physical security cost estimates went up by 47 percent. All other categories noted increases: Personnel Security (1%); Professional Education, Training and Awareness (18%); Security Management, Oversight and Planning (16%); Unique Items (8%); Information Security/Classification Management (19%); and Information Technology (17%).*
> —ISOO, 2004, p.1

Perhaps among the most ambitious of the initiatives are the technologies being developed to help secure maritime ports. Since seaports are gateways for commerce, significant attention has been given to their vulnerability, and port authorities have been ordered to comply with mandates and legislation such as the Maritime Security Act of 2002, Safety of Life at Sea (SOLAS), International Ship and Port Facility Security (ISPS) code, Container Security Initiative (CSI), Customs Trade Partnership Against Terrorism (C-TPAT), the Safe Port Act, and the Green Lane Maritime Cargo Security Act. Although these legislations and the implementation of technologies have contributed tremendous deterrent and interdiction potential, such protections come at a substantial price tag, and while there may be little doubt that improvements have been made in the actual security of citizens in democratic societies, it seems that enemies of these societies have taken a page out of the U.S. cold war manual to spend them into oblivion (Jones, 2007). It has been conceded that the current spending on security and the war on terror is unsustainable both financially (GAO, 2007) and psychologically (Albright, 2007; Huddy et al., 2005). Moreover, the perpetual state of fear in which citizens such as many Americans find themselves creates a codependent relationship and facilitates opportunities rife with massive wasteful spending by their governments on commercial projects. Questionable selections of particular ports chosen for security upgrades is an example, which was criticized by John McCain for congressional special interest appropriations that go toward protecting seaports without focusing on the ports that are most vulnerable (McCain, 2006).

Fear Appeals and Human Coping Behavior

The theories related to mobility and security presented in this chapter have been used to study coping behaviors in different contexts such as those related to taking information security precautions or in health and traffic safety–related behaviors, particularly relative to dislodging inertial tendencies. However, frequently overlooked are the effects of persistent fear stimuli and protracted hyper-vigilance of the individuals themselves, especially in terms of their psychological states, such as depression and other frames of mind, or the effects of systemic monitoring on prosocial behaviors; this despite a large body of research literature on the effects of corporate monitoring and surveillance. Areas of particular concern are the effects of surveillance on the psychology and behavior of the law abiding (Calluzzo and Cante, 2004; Proctor et al., 2002). Thus there are two principal aspects relating to mobility security, fear appeals, and human behavior, addressed in this chapter. The first relates to the effects of continuous fear appeals on people in general, especially as it relates to a state of chronic fear and anxiety, and the second relates to the specific effects of monitoring and surveillance on individuals.

Fear Appeals and Psychosocial Functioning

Behavior may be framed in terms of psychosocial functioning. That is, people make cognitive assessments concerning threats in relation to their social settings. These relationships are triadic: The ways people perceive, conceptualize, and cognitively evaluate information interacts with their social environments, and this determines their behavior, which is either positively or negatively reinforced, or is punished (Bandura, 1978). Therefore, fear appeals lead to cognitive appraisal of the threat, which results in a physiological response.

People have optimal levels of stimulation that they seek from their environments, and they try to adjust the amount of stimulation they receive from their environments so that it closely approximates their own optimums (Eysenck, 1994). When stimulation is excessive, people try to decrease it using coping mechanisms in order to return to their optimal levels. Such actions produce defensive posturing or avoidance reactions. Therefore while hyper-stimulation becomes averse to everyone, there are differences among people in terms of their optimal levels of stimulation, depending on their cognitive makeup and the environments to which they have become adapted; for example, where thrill seekers might choose skydiving or bungee jumping as a stimulating pastime, others find these activities terrifying. Regardless, over time people become desensitized to the extant stimulation. Thus when fear appeals are made, people have different reactions to them, and eventually, the impact of these fear appeals attenuates such that those making them constantly ratchet up the espoused severity and frequency of the threats in their rhetoric (Albright, 2007).

One of the most stressful conditions people experience is when they anticipate a threat of physical or psychological harm to themselves or others that they feel helpless to avoid. A form of perceived helplessness derives from media-based fear appeals about pervasive crimes or impending doom from terrorist attacks (Huddy et al., 2005). The state of fear concomitant with perceived helplessness may be induced by seeing or hearing about an accident or attack upon another. Since most people have an empathetic response to a threat affecting another person, they experience the episode vicariously.

To illustrate this, Lazarus (1991) showed films about accidents in a woodshop and genital operations performed with crude stone instruments as part of initiation rites in primitive tribes. The fear arousal increased sharply during the critical scenes in which operations were performed or accidents occurred, and decreased during neutral scenes. The psychological reaction from the vicarious threat in the film was similar to a direct threat induced by telling subjects that they would receive a painful electric shock. Lazarus also showed that using loaded terms in narration of the threat event exacerbated the physiological responsiveness when he accompanied the movies of accidents with different sound tracks. Subjects who were shown the movie accompanied by an "intellectualized" version where the narration consisted only of the technical facts about the incidents had lower fear arousal than those who heard a "traumatized" version in which loaded terms such as *horror, suffering, pain,* and *sorrow* were used.

People develop defensive coping behaviors toward chronic fear appeals to buffer or neutralize their anxieties. Anxiety is the consciously perceived feelings of tension and apprehension associated with chronic fear arousal and persistent stress (Spielberger, 1972). The most stressful situation is one in which a person is unable to produce an appropriate response or to avoid or escape a threat (Lazarus, 1991), thus when people perceive little or no control over a threat they report greater anxiety (and more fear arousal) than those who expect to be able to avoid it. People cope differently to prolonged anxiety. Some will attempt to avoid the anxiety by suppressing the associated thoughts, which results in depression (Hansen, 2002), while for others it leads to hostility and aggression (Houston, 1972) and even to rage (Janis, 1958).

Because according to the theory of terror management people must be kept in a state of arousal toward threats, the quantity and intensity of fear appeals are continuously elevated, since repeated exposure to threat stimuli becomes less arousing (Chatsworth, 1996). Unless a population is mobilized against a threat, its members tend to become complacent, and people will often ignore or disable security measures perceived as intrusive or ineffective (Workman and Gathegi, 2005). Thus people must be constantly placed in a psychological state of hyper-vigilance about severe impending threats (Vasterman et al., 2005). Democratic governments therefore often use the media as a conduit to make fear appeals along with other stimuli and cues laced throughout societies, such as using color-coded threat levels to characterize states of emergency, or weekly publications of terror lists by the U.S. State Department, or the institutionalization of coping responses, also called "preparedness drills." Government spending on fear commerce creates its own perpetual marketing campaigns as businesses advertise to seize opportunities to generate revenue from government contracts and to commercialize the security-related technologies in private industry (Greenemeier, 2006).

Situational studies (e.g., Albright, 2007; Vasterman et al., 2005) have shown that major network TV news stations used fearful stories with loaded terms 63% of the time on average in 2005, up from 41% in 2000. Meanwhile, research (Chatsworth, 1996; Delisi, 2006) shows that populations in America and northern Europe have sought increasingly intense stimuli from virtual media such as video games and television. In the short term, fear appeals tend to elevate vigilance and facilitate broader situational attention, but being in a constant state of fear has long-term negative consequences to human psychology, physiology, and behavior (Houston, 1972). Hence, where fear appeals facilitate and heighten arousal and vigilance, if chronic, they may lead to unrealistic and maladaptive responses and even to carrying out violent acts. In this sense, chronic fear appeals create a vicious cycle where fear breeds violence that breeds fear (Janis, 1958).

Surveillance and Psychosocial Functioning

Escalation of fear leads to further monitoring and surveillance. Surveillance is the physical or electronic observation of someone's activities and behavior (Ball and Webster, 2003). As technologies become cheaper, less obtrusive, and more sophisticated, governmental and civil authorities increasingly perform both overt and covert surveillance (Bennett and Regan, 2004). In a suburb of Tampa, Florida, in the United States, for example, cameras are used to scan faces of people as they walk down the streets and compare them with images in a database of wanted criminals. Tampa was among the first of what appears to be a trend in America where police use the facial-recognition technology for routine surveillance. The unblinking eye potentially records all movements, subjecting them to scrutiny, captured, stored, manipulated, and used subsequently for various purposes (Bennett and Regan, 2004; Rotfeld, 2006).

It is unequivocal that security measures can improve actual security, and that surveillance, while not a preventative measure, improves prosecutorial ability. Hence, institutional practices carry over into the commercial sphere as democratic societies model their government's policies on their commercial enterprises, and "The current climate supporting surveillance is a potential signal to many organizations that surveillance of employees continues to be tolerated at unprecedented levels" (D'Urso, 2006, p. 282). The range and intensity of monitoring and surveillance has expanded along two axes: first to sort activities and characteristics for profiling purposes in order to more effectively manipulate people (Bennett and Regan, 2004), and to reduce risks associated with potential harm and/or liability to individuals and companies (Ball and Webster, 2003; Norris and Armstrong, 1999).

As many as 80% of organizations now routinely employ some form of electronic surveillance of employees (D'Urso, 2006). Vehicles, telephones, and computers have become tools for monitoring people and their activities (Bennett and Regan, 2004). In the workplace, employers often use video, audio, and electronic surveillance; perform physical and psychological testing including pre-employment testing, drug testing, collecting DNA data, and conducting searches of employees and their property; and collect, use, and disclose workers' personal information including biometrics (Johnston and Cheng, 2002; Langenderfer and Linnhoff, 2005).

The use of security measures and surveillance contributes to the public's general perceptions of security, but there is a point at which security measures and surveillance psychologically undermine the perceptions of security. This can affect behaviors in unintended ways (Bennett and Regan, 2004). At a minimum, people may repress and internalize the emotional impact of the simultaneous effects of feeling under constant threat and scrutiny. Studies (e.g., Hansen, 2002; Whitbeck et al., 2000) show, for example, that such persistent stress conditions may lead to as much as half of all clinical diagnoses of depression, and ongoing research suggests that when people are placed in a continuous fearful state, this can permanently alter the brain's neurology that controls emotions, exaggerating later responses to stress.

The trajectory of the security stance in organizations is toward treating anomalies as the rule (Jones, 2007). A significant body of research has produced a wide spectrum of negative outcomes from pervasive surveillance in the workplace, beginning with the implied presumption that everyone is potentially guilty (cf. Langenderfer and Linnhoff, 2005). A study by Fairweather (1999) examined the use of surveillance of teleworkers (people who work from home offices) and found an association between monitoring and low employee morale. A study by Holman et al. (2002) of surveillance and monitoring in call centers found an inverse relationship between monitoring intensity and employee feelings of well-being. Lee and Kleiner (2003) found that while monitoring provided certain objective advantages in aiding firms and workers, it coincided with feelings that employees' privacy had been invaded by employers' constant monitoring.

In their meta-analysis of electronic surveillance in the workplace, Johnston and Cheng (2002) identified an association with health problems experienced by employees who had their performance technologically monitored, including hypertension, headaches, extreme anxiety, depression, anger, severe fatigue, and skeletal-muscular problems, and that these health problems resulted in absenteeism, turnover, and decreases in productivity. They also found that monitored employees described their environments as toxic and that it undermined employee morale and created divisions between employees and management. Furthermore, in cases where management used electronic monitoring as a means to improve consistency in employee performance evaluations, employees frequently questioned the fairness of their employer's use of electronic monitoring in the reviews, and the perceived lack of procedural fairness in monitoring sometimes led to negative employee reactions such as withdrawal, sabotage, resignation, or some other form of diminished organizational citizenship.

Summary and Conclusions

The research and popular literature involving mobility and security has tended to focus on technical rather than behavioral perspectives. This technical perspective has identified techniques and technologies that lead to better defenses against security vulnerabilities (Dhillon and Backhouse, 2001), such as performing risk analyses followed by the appropriate application of technological countermeasures (e.g., Duh et al., 2002). Doubtless these technological advances have helped tactically in securing populations against very real threats, but the systematic barrages of dire warnings and fear appeals have become part of the extensive commercial enterprise that we might call *fear commerce* (Greenemeier, 2006).

Mobility security and freedom are inextricably linked concepts along a continuum in which each absolute resides on a polar end of the spectrum. In recent years there has been an acceleration of the slide along the scale toward greater control by governmental and civil authorities. People in democratic societies must act for more centrist positions because while the abundant fear-appeal landscape and the surrounding technological and procedural advances have tactically improved the security of people living in these societies, they have come at a strategic cost to both societies and individuals (cf. Rotfeld, 2006). This chapter will conclude with a brief discussion of behavioral objectification and the implications.

Human beings are social creatures but they have strong tendencies to group themselves along political and ideational lines because armies can defeat the individual. Social identity theory explains how people form groups (ingroups) and designate outsiders (outgroups), and how the psychological and physical boundaries between them are defined (Chen et al., 1998; Heald et al., 1998). The theory is concerned with three components: categorization, identification, and comparison (Tajfel and Turner, 1979), which are comprised in the processes used to form attachments with certain others that create perceptions of belonging to a collective, and detachments from those who fall outside the psychologically and socially defined boundaries (Scott, 1997). Patriotism is perhaps the most common and broad manifestation of this process.

An element of social identity is defined by the epistemological weighting hypothesis (Gass and Seiter, 1999), in which behavioral conformance depends on how closely an individual's norms match those of a group to which he or she belongs. It postulates that people are socialized as both active trial and error and passive observation, and through the socialization processes of communication and observation, their behaviors are shaped into conformity with the dominant logic of the group as they come to identify with and become de-individuated by the group. De-individuation into a group results in the reduction of an individual's identity as one is subsumed into the identity of the larger collective (Zajonc, 1969). Terror management theory explains further, for example, that people who violate the norms that circumscribe a social group are devalued and ostracized by the group, who then assert if harm subsequently befalls the "deviates" and "evil doers," that they got what they deserved (Rosenblatt et al., 1989).

Epistemological weighting is regulated by the enculturation processes in a society (Straus and McGrath, 1994; Zajonc, 1969). Therefore the degree to which people comply with the demands of their society depends on the weighting they place on their personal views and behaviors versus the social norms acquired through the socialization processes (Gass and Seiter, 1999; Zajonc, 1969). Those who are cordoned off by the societal boundaries, to varying degrees, are increasingly subjected to objectification (Deutch and Gerard, 1955) as the society amplifies their comparative dissimilarity (Shaw and Barrett-Power, 1998; Tajfel, 1982). As such, conflict increases between societies as comparative dissimilarity and objectification increase (Heald et al., 1998; Shaw and Barrett-Power, 1998).

In the solutions to the inevitable conflicts, a society may facilitate social interaction and social solutions, or they may seek to motivate ingroup norm-congruent behavior, which impedes social interaction with other societies as they subsequently act more uni-laterally—which further intensifies social isolation, engenders deeper social divisions, and exacerbates further intra-group or societal conflict (Gass and Seiter, 1999; Janis and Mann, 1979; Moscovici, 1984; Shaw and Barrett-Power, 1998; Straus and McGrath, 1994; Zajonc, 1968). This is because the less people interact socially, the more they tend to objectify and stereotype others categorically. For example, stereotyping leads people to make predisposed (typically inaccurate) judgments about others based on the representa-tive heuristic where one individual is surmised to be representative of an entire group, but this effect can be reduced as people interact more with the stereotyped others (Chen et al., 1998; Heald et al., 1998).

The contact hypothesis (Sherif, 1966) states that as people interact more with each other, information is exchanged, which dilutes the reliance on the representative heuristic (Shaw and Barrett-Power, 1998; Suzuki, 1997), and the more people interact with others outside their social organizations, the more likely they are to come to iden-tify with them (Zajonc, 1969). The contact hypothesis suggests that even inveterate enemies reduce their objectification of each other as social interaction increases, which results in lower levels of conflict (Gass and Seiter, 1999; Suzuki, 1997; Tajfel, 1982; Zajonc, 1969).

Thus the solution to the problem of objectification is intuitive but repulsive in the climate of fear. The cliché that the best defense is a good offense may work well in a game such as American (or even European) football, where the rules are clear and everyone is playing the same game, but it is also the case that violence begets more violence and oppression fosters rebellion. It has been established in the research literature that an esca-lation in fear rhetoric will ultimately lead to a systemic doctrine of comparative dissimi-larity, which increases violence as people objectify others along cultural, religious, political, and ethnic lines. There can be long-lasting changes in societal structure and function as a result of objectification: People act with a lack compassion, feel and express indifference, and exhibit open hostility, which intensifies and spreads the causes of secu-rity threats as people try to impose even a society's will upon another through the barrel of a gun.

References

Akdeniz, Y., Walker C., and D. Wall. *The Internet, Law, and Society.* Harlow, UK: Pearson Pub-lishing, 2000.

Albright, A. R. Media rhetoric and public outrage: The antiphonal imperative. *Journal of Mass Media*, 14, 233–252, 2007.

Bajcsy, R. Information technology in service to humanities and social sciences. *Proceedings of the World Conference on Security Management and Applied Computing*, Las Vegas, NV, 241–247, June 2005.

Bandura, A. The self-system in reciprocal determinism. *American Psychologist*, 33, 344–358, 1978.

Bartels, A. Global IT spending and investment forecast, 2006 to 2007. *Forrester Research.* November 4–31, 2006.

Borrull, A. L., and C. Oppenheim. Legal aspects of the Web. In Blaise Cronin (Ed.), *Annual Review of Information Science and Technology, 38,* 483–548. Medford, NJ: Information Today, 2004.

Boyle, M. P., Schmierbach M., Armstrong C. L., Cho J., McCluskey M., McLeod, D. M., and D. V. Shah. Expressive responses to news stories about extremist groups: A framing experiment. *Journal of Communication, 56,* 271–288, 2006.

Braithwaite, J., and P. Drahos. *Global Business Regulation.* Cambridge, UK: Cambridge University Press, 2000.

Brown, J. G. *Birth certificate fraud.* Washington, DC: Department of Health and Human Services, Office of the Inspector General. Document 0EI-07-99-00570, 2000.

Bureau of Transportation Statistics. *Transportation implications of telecommuting.* Washington, D.C.: National Transportation Association, 1992. Accessed online April 9, 2007, at http://ntl.bts.gov/DOCS/telecommute.html.

Canny, J. The future of human-computer interaction. *ACM Queue, 4,* 24–32, 2006.

Cranor, L. F., and S. S. Wildman. *Rethinking Rights and Regulations: Institutional Responses to New Communications Technologies.* Cambridge, MA: MIT Press, 2003.

Davis, D. W., and B. D. Silver. Civil liberties vs. security: Public opinion in the context of the terrorist attacks on America. *American Journal of Political Science, 48,* 28–46, 2004.

D'Urso, S. C. Who's watching us at work? Toward a structural-perceptual model of electronic monitoring and surveillance in organizations. *Communication Theory, 16,* 281–303, 2006.

Duwadi, S. R., and S. B. Chase. *U.S. Department of Transportation multiyear plan for bridge and tunnel security research, development, and deployment.* Washington DC: USDOT, Publication FHWA-HRT-06-072, 2006.

Fairweather, B. N. Surveillance in employment: The case of teleworking. *Journal of Business Ethics, 22,* 39–49, 1999.

GAO. *The nation's long-term fiscal outlook.* Washington, DC: The United States Government Accountability Office: GAO-07-510R, 2007.

Gass, R. H., and J. S. Seiter. *Persuasion, Social Influence, and Compliance Gaining.* Needham Heights, MA: Allyn and Bacon, 1999.

Greenemeier, L. The fear industry. *Information Week,* 35–49, April 2006.

Hansen, K. L. Anxiety in the workplace post-September 11, 2001. *The Public Manager, 31,* 133–151, 2002.

Harvey, C. The boss has new technology to spy on you. *Datamation,* 1–5, April 2007.

Holman, D., Chissick C., and P. Totterdell. The effects of performance monitoring on emotional labor and well-being in call centers. *Motivation and Emotion, 26,* 57–81, 2002.

Huddy, L., Feldman S., Taber C., and G. Lahav. Threat, anxiety, and support of antiterrorism policies. *American Journal of Political Science, 49,* 593–608, 2005.

ISOO. *The report on cost estimates for security classification activities for 2003 from the Information Security Oversight Office (ISOO).* Washington, DC: National Archives and Records Administration, 2004.

Janoff-Bulman, R., and I. H. Frieze. A theoretical perspective for understanding reactions to victimization. *Journal of Social Issues, 39,* 1–17, 1983.

Jervis, R. Rational deterrence: Theory and evidence. *World Politics, 41,* 183–207, 1989.

Jørgensen, R. F. *Human Rights in the Global Information Society.* Cambridge MA: MIT University Press, 2006.

Joslyn, M. R., and D. P. Haider-Markel. Sociotropic concerns and support for counterterrorism policies. *Social Science Quarterly, 88,* 306–319, 2007.

Keck, R. *Disruptive Technologies and the Evolution of the Law.* Atlanta, GA: Hi Motion Publications, 2005.

Kowalczyk, R., and L. Alem. *Supporting Mobility and Negotiation in Agent-Based E-Commerce: Managing E-Commerce and Mobile Computing Technologies.* Hershey, PA: IGI Publishing, 2003.

Langenderfer, J., and S. Linnnhoff. The emergence of biometrics and its effect on consumers. *Journal of Consumer Affairs*, 39, 314–338, 2005.

Lee, S., and B. H. Kleiner. Electronic surveillance in the workplace. *Management Research News*, 26, 72–81, 2003.

Losey, R. C. The electronic communications privacy act: United States code. Orlando, FL: The Information Law Web, 1998. Accessed May 10, 2007, at: http://floridalawfirm.com/privacy.html.

Meckler, L. Technology promises to improve airport screening process. *The Wall Street Journal*, 3–4, April 2006.

Miscamble, W. D. *George F. Kennan and the Making of American Foreign Policy, 1947-1950.* Princeton, N.J: Princeton University Press, 1992.

Mueller, P. R. Will the feds run your log servers? *Network Computing*, 16–18, April 2007.

New Mexico Department of Health. Vital records and statistics, 2007. Accessed May 14, 2007, online at: http://dohewbs2.health.state.nm.us/VitalRec/.

Ngwenyama, O., and A. Lee. Communication richness in electronic mail: Critical social theory and the contextuality of meaning. *MIS Quarterly*, 21, 145–167, 1997.

Perry, B., Southwick S., and E. Giller. Adrenergic receptor regulation in post-traumatic stress disorders in advances in psychiatry: Biological assessment and treatment of post-traumatic stress disorder. Washington, D.C: American Psychiatric Press, 1990.

Pippenger, E. B., and K. Nye. Government should focus on controlling access to explosives and using existing detection technologies to prevent illegal bombings. *National Academies News*, 1–7, March 1998.

Pyszczynski, T., Greenberg J., and W. Solomon. Why do we need what we need? A terror management perspective on the roots of human social motivation. *Psychological Inquiry*, 8, 1–20, 1997.

Roe-Berning, S., and G. Straker. The association between illusions of invulnerability and exposure to trauma. *Journal of Traumatic Stress*, 10, 319–327, 1997.

Rogers, R. W. A protection motivation theory of fear appeals and attitude change. *Journal of Psychology*, 91, 93–114, 1975.

Rosenblatt, A., Greenberg J., Soloman S., Pyszczynski T., and D. Lyon. Evidence for terror management theory: The effects of mortality salience on reactions to those who violate or uphold cultural values. *Journal of Personality and Social Psychology*, 57, 681–690, 1989.

Rotfeld, H. J. Depending on the kindness of strangers. *Journal of Consumer Affairs*, 40, 407–410, 2006.

Scholz, J. T. Enforcement policy and corporate misconduct: The changing perspective of deterrence theory. *Law and Contemporary Problems*, 60, 153–268, 1997.

Scott, C. R. Identification with multiple targets in a geographically dispersed organization. *Management Communication Quarterly*, 10, 491–522, 1997.

Tajfel, H., and J. C. Turner. An integrative theory of inter-group conflict. In W.G. Austin and S. Worchel (Eds.), *The Social Psychology of Intergroup Relations* (pp. 433–455). Monterey, CA: Brooks-Cole, 1979.

Tajfel, H., and J. C. Turner. The social identity theory of inter-group behavior. In S. Worchel and L. W. Austin (Eds.), *Psychology of Intergroup Relations* (pp. 129–140). Chicago: Nelson-Hall, 1986.

U.S. Department of Justice. *Violation of the 18 U.S.C. 1030(a)(5)(B): Gaining unauthorized access.* Information Security Report of the Government Accounting Office, *17*, 188–219, 2004.

Vasterman, P., Yzermans J., and A. J. E. Dirkzwager. The role of the media and media hypes in the aftermath of disasters. *Epidemiologic Reviews*, *27*, 107–114, 2005.

Whitbeck, L. B., Hoyt D. R., and W. Bao. Depressive symptoms and co-occurring depressive symptoms, substance abuse, and conduct problems among runaway and homeless adolescents. *Child Development*, *71*, 721–732, 2000.

Workman, M., and J. Gathegi. Observance and contravention of information security measures. In *Proceedings of the World Conference on Security Management and Applied Computing, SAM*, 241–247, Las Vegas, NV, July 2005.

Wu, Y., Stanton B. F., Li X., Galbraith J., and M. L. Cole. Protection motivation theory and adolescent drug trafficking: Relationship between health motivation and longitudinal risk involvement. *Journal of Pediatric Psychology*, *30*, 122–137, 2005.

Zajonc, R. B. Attitudinal effects of mere exposure [Monograph]. *Journal of Personality and Social Psychology*, *9*, Supplement 2-2, 1968.

Modal Aspects of Transportation Security

Road Transportation and Infrastructure Security

L. David Shen, Ph.D.

Objectives of This Chapter:

- Provide an overview of road transportation and infrastructure security
- Introduce the National Highway System and the critical role it plays
- Discuss security issues and threats against the road transportation system
- Provide an introduction to some of the challenges for state transportation officials
- Discuss the potential countermeasures and estimated cost

> *"The terrorist only has to be right once, we have to be right all the time."*
> —Tom Ridge, former secretary of the Department of Homeland Security

Introduction

The road transportation mode is unique in that it consists of privately owned vehicles traveling on publicly maintained roads. With more than 3,995,635 miles of roads, America has the largest and most comprehensive road transportation system in the world (BTS, 2007). Road transportation and infrastructure security were not a top priority for federal and state transportation officials in the past. However, the experience of September 11, 2001, has shown that terrorists can successfully attack any targets in the United States, including our extensive road transportation system. In the United States, road transportation infrastructure includes the following (TSA, 2006; U.S. DOT, 2007):

- 46,747 miles of interstate highway (U.S. DOT, 2007)
- 115,500 miles of other National Highway System roads (U.S. DOT, 2007)
- 3,849,259 miles of other roads (U.S. DOT, 2007)
- 582,000 bridges over 20 feet of span (AASHTO, 2002)
- 54 tunnels over 500 meters in length (AASHTO, 2002)

- More than 50 freeway traffic operations centers (U.S. DOT, 2002)
- More than 300 municipal traffic operations centers (U.S. DOT, 2002)

Note that AASHTO refers to the American Association of State Highway and Transportation Officials and that U.S. DOT refers to the U.S. Department of Transportation.

The road transportation system is a crucial component of the U.S. public infrastructure and plays a vital role in maintaining the vigor of the nation's economy. In 2001, the Bureau of Transportation Statistics (BTS) reported that the 50 states spent $104 billion to build and maintain highway infrastructure that supported some 2.7 trillion vehicle miles of travel (TSA, 2006). The use of private automobiles on America's extensive road transportation network provides Americans with an unprecedented degree of personal mobility and freedom, continuing to allow people to travel where and with whom they want. In 2001, 87% of daily trips involved use of personal vehicles on the road transportation network.

The U.S. vehicle fleet includes 7.9 million trucks, 750,000 buses, 137 million cars, 4.9 million motorcycles, and 84 million other two-axle vehicles (TSA, 2006). Roads are also a key conduit for freight and cargo movement in the United States. Trucks carried 60% of total freight shipments by weight and 70% by value (not including shipments moved by truck in combination with another transportation mode). Trucks are playing an increasingly important role as businesses turn to just-in-time delivery systems to minimize logistics costs (e.g., warehousing and storage).

Road transportation has a central role to play in the continued health and growth of the nation's economy. Americans expect cars to travel and goods to be delivered door-to-door to all corners of the continental United States, quickly, economically, efficiently, and on time. Often road transportation is the only answer to the demand for such high levels of personal mobility and flexibility—a situation that will remain despite increasing investment in other modes, especially in the major urban areas. The road transportation sector itself already contributes significantly to the nation's economy.

Without an efficient, vibrant road transportation system, other modes cannot function properly, as most freight and passenger journeys begin and end with a trip on the road. Only road transportation can provide door-to-door service. Road transportation, therefore, also plays a key role in development of America's integrated transportation networks and intermodal transportation solutions.

The nation's economy is totally dependent on this critical infrastructure. It includes many historically and culturally significant structures that are easily accessible to vehicles of all kinds without screening or inspection. Some of these structures also have high economic value and could easily be targeted for attack or sabotage. Trucks routinely carry hazardous materials that could be used to attack targets that are part of, or are adjacent to, the road transportation system. This was sadly demonstrated with a truck bomb at the Murrah Federal Building in Oklahoma City in April 1995 (TSA, 2006).

In light of the tragic events of September 11, 2001 (9/11), enhancing the security of our road transportation system is expected to be one of the highest priorities of local, state, and federal transportation agencies. Tom Ridge, former secretary of the Department of Homeland Security, has often said, "The terrorist only has to be right once, we have to be right all the time." These few words perfectly explain the difficulty and challenge in protecting our nation's road transportation infrastructure against terrorist attacks 24/7.

The Transportation Research Board (TRB) and the National Academies have generated extensive information on this issue in recent years. In addition, the Federal

Highway Administration (FHWA) and AASHTO contributed significantly to this effort. This chapter attempts to brings together much of this information. Also included are links to other related Web sites that contain discussions of these issues, actions that can be taken, and guidance and training opportunities on road transportation and infrastructure security.

The National Highway System

The National Highway System (NHS) includes the interstate highway system as well as other roads important to the nation's economy, defense, and mobility (U.S. DOT, 2007). The NHS was developed by the U.S. DOT in cooperation with states, local officials, and metropolitan planning organizations.

The NHS is approximately 160,000 miles (256,000 kilometers) of roadway important to the nation's economy, defense, and mobility. It includes the following subsystems of roadways (note that a specific highway route may be on more than one subsystem):

- Interstate: The interstate system of highways retains its separate identity within the NHS.
- Other principal arterials: These are highways in rural and urban areas that provide access between an arterial and a major port, airport, public transportation facility, or other intermodal transportation facility.
- Strategic Highway Network (STRAHNET): This is a network of highways that are important to the United States' strategic defense policy and that provide defense access, continuity, and emergency capabilities. The official STRAHNET Web site is located at http://www.tea.army.mil/pubs_res/strahnet/strahnet.htm.
- Major Strategic Highway Network connectors: These are highways that provide access between major military installations and highways that are part of the Strategic Highway Network.
- Intermodal connectors: These highways provide access between major intermodal facilities and the other four subsystems making up the National Highway System. A listing of all official NHS intermodal connectors can be found at http://www.fhwa.dot.gov/hep10/nhs/intermodalconnectors/index.html.

The NHS includes 87.5% of urban and other freeways and expressways, 35.9% of urban and other principal arterials, and 83.8% of rural and other principal arterials. While the NHS makes up only 4.1% of total U.S. mileage, it carries 44.8% of total travel. The National Highway System is shown in Figure 5.1 (U.S. DOT, 2007).

In 2006, the National System of Interstate and Defense Highways, commonly known as the interstate system, turned 50 years old. The 46,747 miles of interstate highways serve as the backbone of transportation and commerce in the United States (U.S. DOT, 2007). About 67.1% of this 2004 mileage was in rural areas, 4.5% was in small urban areas, and 28.3% was in urbanized areas. In 2004, Americans traveled approximately 267 billion vehicle miles on rural interstates, 26 billion on small urban interstates, and 434 billion on urbanized interstates. Taken together, this represents approximately 24.5% of all U.S. travel in 2004.

In 2004 the interstate system included 55,315 bridges, 27,648 in rural areas and 27,667 in urban areas. About 15.9% of rural interstate bridges were considered to be

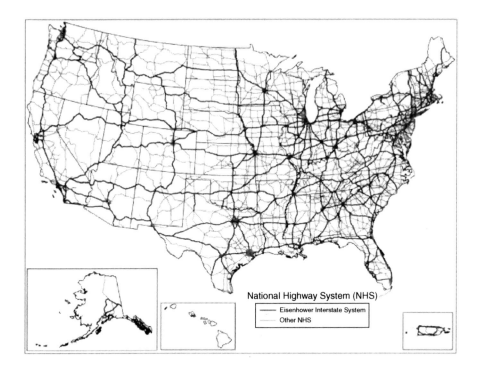

FIGURE 5.1 The National Highway System (FHWA, 2007)

deficient, including 4.2% classified as structurally deficient and 11.7% classified as functionally obsolete. Among urban interstate bridges, about 26.5% were considered to be deficient, including 5.1% classified as structurally deficient and 20.5% classified as functionally obsolete. Structurally deficient bridges are more vulnerable to attacks. Additionally, the U.S. transportation system includes 337 highway tunnels and 211 transit tunnels; many are located beneath bodies of water, and many have limited alternative routes due to geographic constraints (U.S. DOT, 2007), thus making them tempting targets for terrorists.

The mobility needs of the American people were served by a network of 4 million miles of public roads in 2004. About 75.1% of this mileage was located in rural areas (those with populations less than 5,000). In contrast, urban interstate highways made up only 0.4% of total mileage but carried 15.5% of total vehicle miles traveled (VMT). Percentages of highway miles, bridges, and vehicle miles traveled by functional system in 2004 are shown in Table 5.1 (U.S. DOT, 2007).

Security Issues

Twenty-four hours a day, seven days a week (24/7), road transportation touches the lives of nearly every citizen of the United States, including everyone who commutes to a job site, rides a bus, or hauls freight. The highways involve even more. Virtually every item in a person's house or place of employment and in shopping malls, department stores, or supermarkets spent time in a truck and traveled into that person's life via one of our nation's highway systems. Our road transportation system serves to unify America and sustain the American way of life, and without it, the world stops.

Table 5.1 Percentage of Highway Miles, Bridges, and Vehicle Miles Traveled (VMT) by Functional System, 2004

Functional System	Miles	Bridges	VMT
Rural Areas			
Interstate	0.8%	4.7%	9.0%
Other principal arterials	2.4%	6.1%	8.1%
Minor arterials	3.4%	6.8%	5.7%
Major collector	10.5%	15.8%	6.7%
Minor collector	6.7%	8.3%	2.0%
Local	51.3%	35.1%	4.4%
Subtotal rural	75.1%	76.8%	35.9%
Urban Areas			
Interstate	0.4%	4.7%	15.4%
Other principal arterials	0.3%	2.9%	7.0%
Minor arterials	1.5%	4.1%	15.2%
Major collector	2.5%	4.2%	12.3%
Minor collector	2.6%	2.6%	5.5%
Local	17.7%	4.7%	8.6%
Subtotal urban	24.9%	23.2%	64.1%
Total	100.0%	100.0%	100.0%

Source: U.S. DOT, 2007.

The success and safety of our extensive road transportation system, combined with the perceived number of parallel routes, can lead to the conclusion that the road transportation system is so robust that it is not susceptible to significant disruption by terrorist attack or sabotage. Unfortunately, this conclusion is misleading and incorrect. In many parts of the country, the road transportation system is straining to keep up with the current demands of society and the economy. Many urban roads are already operating close to or above their design capacity. Bridges and tunnels are especially vulnerable. The actions of terrorists can impose critical damage to some bridges, and with explosive forces, exert loads that exceed those for which components are currently being designed. Worse yet, in some cases the loads can be in the opposite direction of the conventional design loads and cause significant damages (AASHTO, 2003).

Tunnels are an increasingly important part of the critical infrastructure network that keeps traffic moving nonstop all day long. Due to a tunnel's enclosed environment, it is a challenge for incident prevention and management, fire protection, and security against terrorist attacks. The recent fiery truck pileup on a Los Angeles–area freeway tunnel that shut down the north–south traffic for nearly 2 days clearly shows the vulnerability of our roadways. Three people died and at least 10 were injured in a fiery 15-truck pileup on Interstate 5 in a Los Angeles–area freeway tunnel (Daily News, 2007). The accident blocked major traffic arteries near Santa Clarita that feed into the highway through which the tunnel runs, leaving many drivers unable to move on the roads for more than 7 hours.

It is clear that with so many miles of roads, bridges, and tunnels, our road transportation system is vulnerable to terrorist threats. The nature of a terrorist attack is sudden and unexpected. In reality, no road transportation system can claim to be 100% secure 100% of the time because there is just no such thing. Even if the government has some information on a possible attack, it will generally not know exactly where, when, or how an attack will occur. Without specific information, the most effective strategy is to plan in advance to prevent and mitigate, where possible, as well as to respond, when necessary, with flexibility, coordination, and speed (Stovall and Turner, 2004).

The security principles for road transportation and infrastructure security can be summarized by the following four simple steps (Maunsell, 2006):

1. Deterrence: Keep the bad guys out; make it easier for them to go elsewhere.
2. Detection: If they do get in, make sure you know about it.
3. Assessment: Once something happens, know what is unfolding.
4. Response: Be able to respond appropriately and manage the result.

Deterrence is generally the main responsibility of the federal government. The departments of Defense, Intelligence, National Security, Homeland Security, and the Federal Bureau of Investigation (FBI) all play a role in this. What state and local transportation officials can do about this is limit access to key components of road transportation infrastructure such as tunnels and bridges. In addition, use of security guards, security dogs, surveillance cameras, and prominent warning signs can make it more likely for potential attackers to go elsewhere.

Detection is very important because it provides advance warning on threats about to happen. State and local transportation officials should deploy motion detectors, sensors, and surveillance cameras around key bridges and tunnels so that if attackers do get in, officials will quickly learn about this and respond. The information can be fed into a traffic operations center or security office manned by trained professionals.

Assessment can be only as successful as the person doing the assessment. Proper training and education are critical if the assessment is to be done in a prompt and correct manner. Depending on the size of the facility, it can be either a traffic operations center or a security office that will conduct the assessment. Depending on the severity of the threats, upper administration and other law enforcement officials may have to be contacted immediately.

Response is the key to guarantee that normal traffic operations would continue and minimum disturbance would result from an incident. The staff and/or security officials who staff the office must be trained to respond appropriately and manage the result. In the case of a major attack or serious security threat, upper administration and other law enforcement officials must be contacted immediately. Proper training and routine exercises are keys to make the four-step security principles work.

With more than 4 million miles of roads and nearly 600,000 bridges in the United States, it is nearly impossible to make the system 100% secure 100% of the time. However, priorities can and should be established and critical areas identified to facilitate the process. The basic principle of cost and benefit applies to road transportation and infrastructure security. Terrorists would attempt to maximize the damages, and we must carefully apply our resources to minimize the adverse impact. The 46,747 miles of interstate highway are the most important part of our road transportation system. The 115,500 miles of other National Highway System roads are the next most important. The 3,849,259 miles of other roads would be considered when resources are available (U.S. DOT, 2007).

In addition to highway classification, urban and rural areas are also used to determine the priority of the list. Since nearly 70% of the population live in urban areas, major roadways in urban areas would be ranked higher on our priority list. Major urban areas such as New York City, Boston, Chicago, Los Angeles, and Washington, D.C., are more important than rural areas with much less population and development.

Tunnels, bridges, and interchanges are the most important components of the road transportation system due to their vulnerability to possible sabotage and the disruption they would create once out of service. Interstate highway tunnels in major urban areas thus would be on the top of the priority list of any state or federal transportation agency's security consideration. For instance, in the United States there are 20 road tunnels that are longer than 1,700 meters; 9 of those top 20 tunnels are located in New York City and the Boston area. The list of the 9 key tunnels is shown in Table 5.2 (Merzagora, 2005).

The 1.5-mile (2.4 km) Lincoln Tunnel (with three separate tubes–tunnels) is one of the most important tunnels in the nation (Wikipedia, 2007). The Lincoln Tunnel is part of a transportation network that the Port Authority of New York and New Jersey (PANYNJ) maintains to support the community with vital links throughout the region. This famous Hudson River tunnel, a three-tube underwater vehicular facility, provides a vital connection between midtown Manhattan and central New Jersey, and forms part of New Jersey Route 495 (PANYNJ, 2004).

The tunnel's three tubes provide paramount flexibility in traffic handling and urban transportation management. They carry six traffic lanes in total. By converting the center tube to a two-way operation, it has the ability to change the six lanes to four lanes in one direction, or three lanes in each direction. The tunnel carries about 120,000 vehicles per day, making it one of the busiest vehicular tunnels in the world (NYSDOT, 2006). Each day, approximately 1,700 buses carrying more than 62,000 commuters use the Exclusive Bus Lane (XBL). Annually, this amounts to approximately 419,000 buses carrying over 15 million passengers (PANYNJ, 2004). The profile of Lincoln Tunnel is shown in Figure 5.2. For bridge crossings serving Manhattan, New

Table 5.2 Selected Tunnels in the United States Longer Than 1,700 m

Tunnel	Length (m)	Date of Opening	State	Comment	Road
Ted Williams/I90 Extension	4 200	15.12.1995 17.01.2003	MA	Ted Williams (2600 m, also named 3rd Harbor) + I90 extension (1600 m, opened in 2003). Immersed tunnels. City of Boston.	I-90
Brooklyn Battery	2 779	15.05.1950	NY	Second tube: 2779 m. East River. NYC.	I-478
Holland	2 608	13.11.1927	NY-NJ	Shortest tube: 2551 m. Hudson River. NYC.	I-78
Ted Williams	2 600	15.12.1995	MA	Two tubes. Immersed tunnel. Boston Harbor.	I-90
Lincoln Center	2 504	22.12.1937	NY	Hudson River. NYC.	495
Lincoln South	2 440	25.05.1957	NY	Hudson River. NYC.	495
Lincoln North	2 281	01.02.1945	NY	Hudson River. NYC.	495
Queens Midtown	1 955	15.11.1940	NY	Second tube: 1912 m. East River. NYC.	495
Sumner	1 725	1934	MA	Immersed tunnel. Boston. Opposite to Callahan.	1A

Source: Merzagora, 2005.

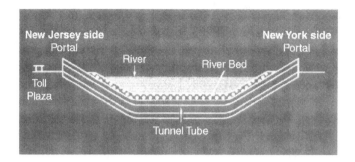

FIGURE 5.2 Profile of Lincoln Tunnel, New York City (PATH, 2007)

York City, the George Washington Bridge carried more traffic than any other river crossing serving Manhattan. Some 309,300 daily vehicles used this bridge in 2001, 2.6% fewer than the average volume of 317,600 in 2000 (NYCDOT, 2002). The impact of 9/11 is significant.

Statistically speaking, the probability of occurrence of terrorism on our road transportation system is much lower than traffic collisions and vandalism. However, the impact of occurrence of terrorism is much greater than accidents or vandalism as evidenced in the 1995 Murrah Federal Building bombing in Oklahoma City. The possible occurrence of transportation system threats is shown in Figure 5.3. While it is impossible to make our road transportation system 100% secure 24 hours a day, 7 days a week (24/7), we can definitely focus our attention and resources on key areas and minimize the negative impact of attacks.

Tunnel systems, in their design, have a safe environmental order and are capable of withstanding the assaults "normally" presented by everyday use. For example, below-grade tunnels are watertight, with proper water evacuation capability and safety systems to move air into and out of the tubes. The nature of the closed environment makes a tunnel system vulnerable to power failure or attacks on its power system. Without lighting and ventilation, a tunnel system could easily become a death trap. Without an emergency pumping system, a tunnel system could become a swimming pool and thereby a death trap. The tunnel structure is designed and built to exist within the soil or seabed that it occupies. Despite these and other safety features, however, damage or disruption to a tunnel, its operations, and/or occupants can result from the impact of hazards or threats (NCHRP, 2006). The longer the tunnel, the more susceptible it is to attack or sabotage because the incident could be magnified due to the tunnel's limited-access nature.

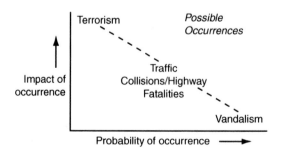

FIGURE 5.3 Transportation System Threats (AASHTO, 2007)

Table 5.3 Bridge and Tunnel Security Issues

Key Topics in Infrastructure Security	Specific Issues
1. Foundations for policy	• Criteria establishing investment priorities
	• Institutional continuity
2. Planning, design, and engineering	• Design review for secure structures
	• Research and development (R&D) needed to support "Design for Security"
	• Design criteria
	• Design specifications
3. Management and operational practices	• Best practices
	• Practice review
	• Institutional relationships
	• Preparedness
	• Personnel and vehicle security
	• Communication–outreach
4. Information security	• Procurement practices
	• Information security
5. Mobilization ("notice") and response ("trans-event")	• Threat warning
	• Early response
	• Initial response
6. Recovery (post-event)	• Damage assessment
	• Functional continuity

Source: AASHTO, 2003.

Bridge and tunnel security, like security for any infrastructure asset, includes a broad range of issues that must be addressed to ensure that adequate measures are taken to protect the asset and the people and goods that utilize the asset. Table 5.3 shows the bridge and tunnel security issues considered by the AASHTO Blue Ribbon Panel (AASHTO, 2003). Interchanges are basically bridges over a roadway in a different direction. As such, security issues and concerns that apply to bridges can also be applied to interchanges. Therefore, the discussion of bridge security in this chapter also applies to interchanges.

Several of the topics and related issues are of general interest and apply to all transportation infrastructure (including traffic operations centers); others relate more directly to bridges and tunnels. For example, the "management and operational practices" issues apply to most infrastructure assets, as do "information security," "mobilization and response," and "recovery" issues. However, issues that fall within the "planning, design, and engineering" area may be unique to bridges and tunnels and require special solutions that go beyond what might be needed to reduce the vulnerability and improve the security of other infrastructure assets.

Security Threats

A list of security threats that may adversely impact the normal operation of a tunnel and associated infrastructure is shown in Table 5.4. The transportation tunnel and associated infrastructure include all electrical and mechanical operations within the tunnel environment, such as lighting, surveillance, ventilation, and fire suppression. Threats to the

Table 5.4 Major Threats to Transportation Tunnels and Associated Features

Security Threat	Vulnerable Tunnel Feature									
	Tunnel Construction and Engineering Feature							Tunnel System Feature		
	Immersed tube	Cut-and-cover	Bored or mined	Vent shaft	Portal	Station	Distribution channel	Control center	Substation	Utility building
Introduction of Small IEDs	✓	✓	✓	✓		✓	✓	✓	✓	✓
Introduction of Medium-Sized IEDs	✓	✓	✓	✓	✓	✓	✓	✓	✓	✓
Introduction of Large IEDs	✓	✓	✓	✓	✓	✓	✓	✓	✓	✓
Introduction of Chemical Agents				✓		✓				
Introduction of Biological Agents				✓		✓				
Introduction of Radiological Agents	✓	✓	✓		✓	✓				
Cyber Attack	✓							✓		
Maritime Incident	✓	✓	✓		✓					
Fire (Arson)	✓	✓	✓	✓		✓	✓	✓	✓	✓
Sabotage of MEC Systems	✓	✓	✓	✓		✓	✓	✓	✓	✓

Note: IED = improvised explosive device; MEC = mechanical, electrical, and communications.

Source: NCHRP, 2006.

tunnel environment also include actual or perceived physical threats affecting users of the transportation system (NCHRP, 2006).

For a tunnel, due to its enclosed nature, the primary criterion used for analysis of security threats was the level of impact that a major threat would have on the tunnel system. All threats considered in depth are capable of closing a tunnel for an extended period of time (i.e., lasting more than 25 hours). These threats encompass potential incidents that have not been routinely encountered or planned for by a tunnel operator. Fire caused by accident is the most common event that shuts down a tunnel, as experienced in the recent L.A. tunnel fire (Daily News, 2007). A deliberate attack or sabotage may create even more damage to tunnels because the attackers would focus the attack on the most vulnerable part of the facility to maximize impact.

The damage potential of threat scenarios—often a sequence of physical events such as fire, explosion, or flooding, and their secondary impacts such as injuries, fatalities, or loss of function—determines the key characteristics of countermeasures that can mitigate the impact of hazards and threats, if not prevent them. All types of damage, including fire, explosion, and radiation, may be mitigated. Certain damages may be more difficult to recover than the others. Possible damage includes the following (Better Roads, 2003):

- Fire/smoke: Any active conflagration or postconflagration condition of smoke and harmful vapors.
- Flooding: The condition of excessive water inflow to a tunnel area exceeding the pumping capacity of the tunnel systems and causing a hazard or threat to people and property.
- Structural integrity loss: Any decrease in the fitness of the tunnel to carry passengers or freight that requires inspection by the tunnel owner and major repair prior to its reopening for beneficial use by the public.
- Contamination: The condition of being unfit for normal habitation due to the presence of radiation, biological agents, harmful chemicals, hazardous airborne particles, or sewage sufficient to require professional remediation.
- Utility disruption: Loss of power, air, steam, water, or communication service for more than 25 hours.
- Extended loss of asset use: Loss of the ability to safely move passengers or allow vehicular traffic for more than 25 hours.
- Extended public health issue: Actual or potential ability to cause illness in a significant portion of the population sufficient to overwhelm the medical treatment capacity of the area.

Similarly, bridges and interchanges are designed to withstand assaults normally presented by everyday use. However, when extreme force is presented, all bets are off. The recent San Francisco freeway elevated section that was hit by a gasoline tanker truck and collapsed is a good example of a bridge's vulnerability. Nearly 75,000 vehicles used the damaged portion of the road in San Francisco every day, but because the accident occurred where three highways converge, authorities said it could cause problems for hundreds of thousands of commuters (Associated Press, 2007). Elevated freeway sections acted similarly to a bridge and are equally vulnerable to accident and attack. A diagram showing the critical components that are vulnerable to possible attacks is shown in Figure 5.4. Bridges over shipping channels are vulnerable to attack from both top and

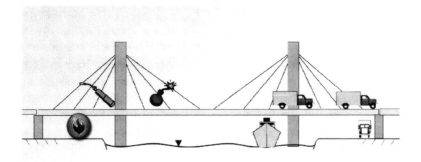

FIGURE 5.4 Critical Bridge Components That Are Vulnerable to Possible Attacks (FHWA, 2007)

bottom. A ship loaded with explosives can easily collapse a major bridge support and create havoc.

There are many types of bridges. For longer bridges over major waterways, there are suspension and cable-stayed bridges, truss bridges, and arch bridges. Multigirder–freeway overpass bridges can be found at highway interchanges and grade separations. Critical components of various bridge structures are shown in Table 5.5 (AASHTO, 2003).

For bridges and interchanges–overpasses, due to their open-air nature, the impact of chemical, biological, and radiological agents is not as serious a problem as compared with tunnels. However, the April 2007 San Francisco elevated freeway section collapse due to the fiery inferno caused by a gasoline tanker truck exploding clearly shows the vulnerability of the system to fire and explosion. Subsequently, improvised explosive devices (IEDs) mounted on vehicles–vessels present the most serious challenge to bridge structures. In addition, for major bridges with elaborate traffic control measures, cyber-attack on a traffic operations center is also a serious threat to normal operations. Critical asset factors and values for road and bridge infrastructure are presented in Table 5.6.

Table 5.5 Critical Bridge Components

Suspension and Cable-Stayed Bridges	Truss Bridges	Arch Bridges
• Suspender ropes, stay cables	• Suspended span hangers	• Tension-ties
• Tower legs	• Continuous and cantilever hold-down anchorages	• Connections
• Main cables	• Compression chords or diagonals	• Decks
• Orthotropic steel decks	• Connections	• Piers
• Reinforced and prestressed bridge decks	• Decks	
• Cable saddles	• Piers	
• Approach structures		**Multigirder–Freeway Overpass Bridges**
• Connections		• Decks
• Anchorages		• Connections
• Piers		• Piers

Source: AASHTO, 2003.

Table 5.6 Critical Asset Factors and Values for Road and Bridge Infrastructure

Critical Asset Factor	Value	Description
Casualty risk	5	Is there a possibility of serious injury or loss of life resulting from an attack on the asset?
Emergency response function	5	Does the asset serve an emergency response function and will the action or activity of response be affected?
Government continuity	5	Is the asset necessary to maintain government continuity?
Military importance	5	Is the asset important to military functions?
Economic impact	5	Will damage to the asset have an effect on the means of living, or the resources and wealth of a region or state?
Available alternate	4	Is this the only asset that can perform its primary function?
Replacement cost	3	Will significant replacement cost be incurred if the asset is attacked?
Replacement–downtime	3	Will an attack on the asset cause significant replacement–downtime?
Relative vulnerability to attack	2	Is the asset relatively vulnerable to an attack?
Functional importance	2	Is there an overall value of the asset performing or staying operational?
Ability to provide protection	1	Does the asset lack a system of measures for protection?
Environmental impact	1	Will an attack on the asset have an ecological impact of altering the environment?
Communication dependency	1	Is communication dependent on the asset?
Symbolic importance	1	Does the asset have symbolic importance?

Note: Value ratings range from 1 to 5, with 5 being the most critical and 1 being the least critical. This ranking will help agencies decide which assets need the most immediate protection.

Source: Better Roads, 2003.

Casualty risk, emergency response function, government continuity, military importance, and economic impact are the five most important critical asset factors, and all have the highest ranking value of 5. Casualty risk deals directly with possibility of serious injury or loss of life resulting from an attack on the asset and is a key factor in our consideration. Emergency response function has a direct impact on our relief action when emergency events strike, and it is critical that we keep these facilities open at all times. Government continuity is vital in the administration of emergency relief actions once attacks take place. All assets necessary to maintain government continuity will be high on our priority list. Assets important to military functions are critical as well because of the military's role in national security and recovery operations. Economic impact is crucial in consideration of asset factors because we want to minimize an attack's negative impact on the resource and wealth of a region and state.

There are several potential threats that exist for bridges and tunnels. The first and most serious is a precision demolition attack using high impact explosives. An attack can be launched either from a car or truck or from a vessel (for bridges). The possible scenarios are shown in Figure 5.4. If carried out, this attack will destroy or seriously damage the bridge or tunnel. Therefore, this threat must be mitigated so that it will not happen. Other threats to consider are conventional explosives, collision, and fire. Their potential magnitude is presented in Table 5.7.

Table 5.7 Magnitude of Threats

Threat Type	Largest Possible	Highest Probability
Conventional explosives	Truck: 20,000 lbs*	Car bomb: 500 lbs*
	Barge: 40,000 lbs	
Collision to structure (i.e., the size of a vehicle that could collide with a structure)	Truck: 100,000 lbs GVW	Truck: H-15
	Water Vessel: depends on waterway	Water Vessel: (see AASHTO spec. LRFD on vessel impact)
Fire	Largest existing fuel or propane tank	Gasoline truck (3S-2)
	Largest fuel vessel or tanker	Fuel barge
Chemical–biological HAZMAT	These threats exist; however, the panel is not qualified to quantify them. Therefore, other experts should assess these threats in this way.	

*Largest possible conventional explosive. For a truck, based on largest truck bomb ever detonated internationally by a terrorist act. For a barge, based on the assumption that it is the largest explosive that could pass by unnoticed by current security at place at major waterways.

**The size of an explosive charge that can be concealed within the trunk of an automobile without being visually detected when inspecting the automobile.

Source: AASHTO, 2003.

Challenges for State DOTs

The experience of September 11, 2001, has again shown that terrorists—armed with weapons of mass destruction (WMD) and determined to harm large numbers of the civilian population—can successfully attack targets in the United States. Major buildings were the targets of September 11 and of the Oklahoma City bombing, and nobody can say for sure what the next target will be. Road transportation infrastructure is a tempting target, and we must be prepared to confront this reality.

Although the nation's 50 highway and transportation departments (state DOTs) may be reasonably prepared for playing a key role in "normal" disasters, WMD in the hands of terrorists introduce new considerations and challenges that were unthinkable just a few years ago. September 11 changed all that, and unfortunately, for many terrorists the United States remains an attractive target. The relatively "peaceful" world we lived in changed permanently after 9/11. The new security threats introduce new challenges such as the following (PB, 2002):

- People are the intended target.
- Advance warnings are unlikely.
- Multiple simultaneous attacks are possible.
- Emergency responders may be targets.
- The weapons may introduce serious and long-lasting hazards.
- The weapons may introduce large-scale damage or contamination to critical equipment and facilities.
- Public reaction is unpredictable.

It is impossible to guard every inch of the nearly 4 million miles of roads in the nation against terrorist attack. For such a large road transportation system, 100% security 24 hours a day, 7 days a week (24/7) is definitely an unrealistic goal. Terrorists could wait patiently in the dark and pick their target systematically or randomly, while we are left to guard the road system from coast to coast day in and day out. Just as a famous Chinese proverb says, "You can be a thief for a thousand days straight, but you cannot prepare yourself against the threat of theft for a thousand days straight." A thief always has the upper hand because he or she can strike when your guard is down.

As transportation professionals, what we can do is focus our attention on the most important components of our road transportation system to discourage terrorists from launching an attack in the first place. The introduction of WMD also signals the need for modifications to the existing set of state DOT roles and responsibilities that were originally focused on natural disasters and accidents (PB, 2002):

- Law enforcement and national security agencies would play a larger role in a terrorist incident. State DOT personnel need to understand the different relationships inherent during and after a terrorist WMD incident.

- If an incident occurs on or near a highway, state DOT personnel may be first or early responders. Therefore, basic training may be needed in identifying possible signs and consequences of terrorist incidents for appropriate actions, including the consideration of the responders' own safety.

- Specific traffic control regimes may be needed to evacuate people or establish emergency access. Preplanning strategies, signage, and equipment may be appropriate, together with capitalizing on Intelligent Transportation Systems (ITS) and traveler information resources.

- Some resources may become unavailable for use if contaminated. Having procedures and equipment in place for decontamination becomes more important. Medical treatment and facilities could be overwhelmed quickly.

- Response resources may be required far beyond those originally anticipated, especially where a WMD is used that initially leaves few distinguishing marks. State DOT response resources need to be available but may also need to be protected as the consequences spread.

- Addressing public concerns is critical. Panic and uncontrolled flight are possible, and controls may need to be quickly put into effect. A comprehensive public information strategy is necessary. Where highways are concerned, state DOT personnel will be expected to provide information, e.g., through variable message signs and communication systems to motorists evacuating an area.

Most state DOTs provide support functions in the existing emergency plans of state emergency management agencies. These all-hazard statewide plans have proved to be robust tools for natural disasters. However, the 9/11 experience has indicated the need to update and modify these statewide plans, including supporting the emergency operations plans of state DOTs.

This is because of the significant difference between a natural disaster and a terrorist attack. While the probability of a terrorist attack on our road transportation infrastructure is low, the impact of such an attack can be traumatic and costly. New and continuing challenges recommended by "A Guide to Updating Highway Emergency Response Plans for Terrorist Incidents" (PB, 2002) include the following:

- Absence of interoperable and reliable communications among agencies
- Lack of familiarity with the roles and personnel of other agencies
- Need for responding to the introduction of federal security agencies and crime scene factors
- Unfamiliarity with Incident Command System practices of public safety agencies
- Need for protection of first responders from biological, chemical, and radiological hazards
- Need for specific operations regimes such as evacuation and emergency access
- Need to capitalize on Intelligent Transportation Systems technology for traffic control and communications

These challenges constitute some of the key agenda items in tailoring current state DOT emergency response plans to the new reality of terrorism. It is neither easy nor cheap to adopt these measures. However, we are living in a different world from the one prior to 9/11, with completely different paradigms. As transportation professionals, we could only have hoped this would never happen, whereas now we must prepare as if it will happen tomorrow. Due to resource limitation, we must identify our key assets and the most vulnerable components of our road system. This will allow us to maximize our resources and minimize potential damage.

Road Elements and Vulnerability

A study conducted under a contract from the AASHTO Security Task Force prepared "A Guide to Highway Vulnerability Assessment for Critical Asset Identification and Protection." The vulnerability assessment process presented in this guide is derived from careful review of information compiled from state, federal, and international agencies and their personnel (SAIC, 2002).

The guide provides six steps for conducting a vulnerability assessment of highway transportation assets. These six steps provide a straightforward method for examining critical assets and identifying cost-effective countermeasures to guard against terrorism (SAIC, 2002). The criteria used in selecting the preferred approaches include availability, accessibility, transparency, replicability, reasonableness, scalability, robustness, cost-effectiveness, and modularity.

The six-step assessment process is as follows:

1. Identify critical assets.
2. Conduct a criticality–vulnerability assessment.
3. Conduct a consequence assessment.
4. Determine countermeasures.
5. Estimate cost of countermeasures.
6. Review operational security planning.

These six steps represent an integrated and iterative approach to vulnerability assessment. The systems approach is recommended for use in conducting vulnerability assessment of road transportation assets. A systems analysis provides a clear perspective of problems and hence leads to more relevant solutions. This approach depends upon formation of a dedicated, multidisciplinary team with ready access to a range of resources—from databases to personnel—as well as a commitment from senior state DOT officials to examine critical assets carefully and identify cost-effective countermeasures to provide better protection against the threats of terrorism involving use of WMD. With solid information and well-trained personnel, vulnerability can be properly assessed and countermeasures developed (SAIC, 2002).

Even though the major security threats mentioned previously have not occurred in the United States, this does not mean that they will not happen in the future. Terrorist bombing of the London subway train–bus and Madrid commuter train clearly shows that if it can happen on a public transportation system, it can also happen on our road system. We are living in a dangerous and uncertain world and the potential for terrorist threats is real. Their capacity to close a tunnel or bridge, however briefly, is a real possibility and cannot be overlooked. A damaged bridge or tunnel in a major city has a tremendous impact on a city, region, and even the nation. Attention must be directed to mitigate their detrimental effects and provide relief.

Among surface transportation's modal systems, the nation's road transportation infrastructure is relatively robust and redundant. Nevertheless, the consequences—both direct and indirect—of an attack on critical links could be significant. There are certain contexts across the United States in which the loss of key links could have major economic and mobility impacts and result in immediate loss of life. This is especially the case for major urban areas such as New York City, Boston, Washington, D.C., and Los Angeles. Furthermore, as demonstrated on September 11, 2001, roads are essential for evacuation, as well as in the response and recovery effort.

The principal security threat against highway physical assets is explosive attacks on key links such as bridges, interchanges, and tunnels. Facilities most vulnerable to disruptions are those playing important economic and strategic roles, the loss of which would be maximally disruptive and involve greater replacement challenges. On a nationwide basis, approximately 450 bridges and 50 tunnels meet relevant criteria as critical assets (Ham and Lockwood, 2002).

Countermeasures

While full road transportation system asset protection is not feasible, reasonable program objectives include the deterrence of terrorist attacks by (1) adding new and clearly visible security features and reducing vulnerabilities, and (2) minimizing the potential for damage in the event of an attack (Ham et al., 2002). The overall practical objective of the proposed security program is, therefore, not to provide full protection but to discourage terrorist attacks through visible security and reduced vulnerability, as well as to minimize damage in the event of an attack. Typology of countermeasures is shown in Table 5.8.

Due to the sheer size of the American highway system, 100% security at all places all the time is not a realistic goal. The goal is, therefore, to discourage terrorist attack through visible security and reduced vulnerability, as well as to minimize damage in the event of an attack. Because bridges and tunnels are the most vulnerable parts of the road transportation system, most countermeasures are focused on these two key components.

Table 5.8 Typology of Countermeasures

Countermeasure	Definition
1. Prevention	Actions taken to try to ensure that harmful security and hazard events do not occur
2. Preparation	Actions taken to anticipate and minimize the harmful impacts of events and ensure that DOT reactions to events are efficient and effective
3. Mitigation	Actions taken to reduce or eliminate long-term risk harm from hazard and security events and reduce the human consequences or asset loss of an event
4. Response	Actions taken to react to an incident and events consistent with plans and procedures using all appropriate resources
5. Recovery	Actions taken to promptly return to normal

Source: AASHTO, 2007.

Eleven countermeasures adopted in Maryland to enhance bridge security are shown as follows (Better Roads, 2003):

- Built-in monitors on bridges
- Motion-detection devices below bridges
- Increased armed security
- Regular checking of truck traffic
- Application of X-ray technology
- Improved training for toll collectors and other tunnel personnel
- Enforcement of hazardous materials (HAZMAT) requirements
- Increased lighting
- Closed-circuit TV (CCTV) cameras for surveillance
- No-fly zones around bridges
- Suspension cable protection

Built-in monitors on bridges, motion-detection devices below bridges, increased armed security, regular checking of truck traffic, application of X-ray technology, increased lighting, CCTV cameras for surveillance, and no-fly zones around bridges are eight good measures to discourage and deter terrorists from launching the attack.

Enforcement of hazardous materials (HAZMAT) requirements and suspension cable protection are two measures that can reduce the vulnerability of the bridges. Improved training for toll collectors and other tunnel personnel would allow them to better identify possible attackers and know how to respond promptly.

Similarly, 11 countermeasures adopted in Texas to enhance bridge security are shown as follows (Better Roads, 2003):

- Eliminate parking areas beneath bridges
- Restrict ingress and egress routes from adjacent areas
- Provide additional lighting
- Limit–monitor access to plans of existing bridges
- Install motion sensors or other active sensors
- Install surveillance cameras
- Apprise local law enforcement officials of critical bridges

- Provide column protection
- Provide pass-through in concrete median barriers
- Install an advance warning system
- Patrol boats under and around bridges

Texas countermeasures are somewhat different from those in Maryland because Texas emphasizes the importance of denying access to potential attackers as deterrence, which makes sense and is very proactive. Two of these 11 measures do that: eliminate parking areas beneath bridges, and restrict ingress and egress routes from adjacent areas. Those are simple measures that can yield significant benefits by pushing potential car bombs farther away from the bridges, thus minimizing the explosive impact.

Provide additional lighting, install motion sensors or other active sensors, install surveillance cameras, apprise local law enforcement officials of critical bridges, install an advance warning system, and patrol boats under and around bridges are six measures used to deter potential terrorists from launching their attack. Limit–monitor access to plans of existing bridges, provide column protection, and provide pass-through in concrete median barriers are three measures used to reduce vulnerability of bridges.

Cost of various countermeasures can range from low (L) to medium (M) to high (H), depending upon the type utilized. Labor cost is typically a major part of any cost associated with bridge security. Examples of countermeasure costs for Smith Bridge are shown in Table 5.9.

The United States needs a Homeland Security Advisory System to provide comprehensive and effective means to disseminate information regarding the risk of terrorist acts to federal, state, and local authorities as well as the American people. Such a system

Table 5.9 Examples of Countermeasure Costs for Smith Bridge

Countermeasure Description	Countermeasure Function			Estimated Cost H/M/L		
	Deter	Detect	Defend	Capital	Operating	Maintenance
Increase inspection efforts aimed at identifying potential explosive devices as well as increased or suspicious potential criminal activity.	√			L	M	L
Institute full-time surveillance at the most critical assets where alternate routes are limited or have not been identified.	√	√		H	H	H
Eliminate parking under the most critical bridges. Elimination of the parking can be accomplished with concrete barriers.	√			L	L	L
Place barriers in such a way as to eliminate ease of access where a vehicle could be driven right up to the asset.	√		√	L	L	L
Improve lighting.	√	√		L	L	L

Source: Better Roads, 2003.

would provide warnings in the form of a set of graduated "threat conditions" that would increase as the risk of threat increases (White House, 2002).

The Homeland Security Advisory System shall be binding on the executive branch and suggested, although voluntary, for other levels of government and the private sector. There are five threat conditions, each identified by a description and corresponding color. From lowest to highest, the levels and colors are

- Low = Green
- Guarded = Blue
- Elevated = Yellow
- High = Orange
- Severe = Red

Threat level to bridges and additional security measures for high priority bridges are shown in Table 5.10. Threat levels are based on the Homeland Security Advisory System (White House, 2002).

The preceding countermeasures mentioned are for bridges; however, many of the measures can also apply to tunnels, with the exception that no vehicles or vessels can pass beneath a tunnel. One distinct difference between a bridge and a tunnel is that a tunnel is an enclosed environment and ventilation is vital for the safety and security of its users.

The selected countermeasures shown in Table 5.11 are available to bridge and tunnel owners and operators for use in planning and implementing more effective security practices (AASHTO, 2003). The list covers some of the most effective measures that can be adopted for security enhancement.

Table 5.10 Threat Level–Based Measures for Bridges

Threat Level to Bridges	Additional Security Measures ("High Priority": Bridges that score a high R)
Severe	• Restrict access with guards, barriers, and vehicle searches • Take all other measures listed below
High	• Increase frequency of patrols and checks • Conduct unscheduled exercises of emergency response plan • Postpone nonessential maintenance • Coordinate with the National Guard or law enforcement for possible closure and vehicle searches when Severe level is reached • Take all other measures listed below
Elevated	• Implement regularly scheduled police patrols • Take all other measures listed below
Guarded	• Review and update emergency response procedures • Increase frequency of periodic checks of cameras, fences, etc. • Take all other measures listed below
Low	• Monitor security systems in place (including periodic checks) • Disseminate threat information to personnel • Regularly refine and exercise emergency operations plan • Conduct emergency responder training • Continually update threat and vulnerability assessments

Source: AASHTO, 2003.

Table 5.11 Countermeasure Options for Bridges and Tunnels

Measures	Selected Countermeasure Options
Planning and coordination measures	• Coordinated response, responsibilities, and liaisons • List of potential areas of vulnerability • Procedures for notification and activation of crisis management teams • Evacuation and shutdown procedures • Identification of emergency evacuation routes and staging areas for response teams • Regular drills, tabletop exercises, no-notice responses, and full-scale simulations • Plans for rapid debris removal and repairs • Establishment of a security policy
Information control measures	• Reviewing and sanitizing Web sites for potential information • Establishing a common classification system for sensitive information
Site-layout measures	• Improved lighting with emergency backup • Clearing overgrown vegetation to improve lines of sight to critical areas • Elimination of access to critical areas • Elimination of parking spaces beneath bridges • Providing pass-through gates in concrete median barriers to enable rerouting of traffic and access for emergency vehicles • Review of locations of trashcans or other storage areas that could be used to conceal an explosive device, ensuring they are not near critical areas
Access control–deterrent measures	• Police patrol and surveillance • Enhanced visibility • Signs issuing warnings that property is secured and being monitored • Exterior and interior intrusion–detection systems • Boundary penetration sensors (below bridge) • CCTV placed where it cannot be easily damaged or avoided while providing coverage of critical areas • Denied–limited access to critical structural elements • Denied–limited access to inspection platforms • Physical barriers to protect piers • Rapid removal of abandoned vehicles • No-fly zones around critical bridges
Retrofit options	• Reinforcing welds and bolted connections • Using energy-absorbing bolts to strengthen connections and reduce deformations • Adding stiffeners and strengthening lateral bracing on steel members • Wrapping the lower portions of cables on cable-stayed bridges and suspension bridges with CFRP or other types of protective armor to protect against damage from blast and fragmentation • Increasing standoff distance and reducing access to critical elements • Including reinforcing steel on top and bottom faces of girders to increase resistance to uplift forces • Providing system redundancy to ensure alternate load paths exist • Strengthening the deck on curved steel trapezoidal girder bridges

Source: AASHTO, 2003.

Most of the bridges, interchanges, and tunnels were built prior to September 11, 2001. As such, security was only one of the factors considered and not to the extent as since 9/11. Consequently significant changes and modifications must be done when the bridges, interchanges, and tunnels are ready for renovation and rehabilitation. This change of priority and emphasis would take time to implement and be costly. The measures that are easy to implement and less costly should be considered first.

To protect key road transportation assets, the following countermeasures are proposed as retrofits on critical bridge, interchange, and tunnel assets (Ham et al., 2002):

- Maximize potential explosives placement standoff distance to key structural members or mechanical systems via various types of barriers.
- Deny access to locations where placement of explosives would affect points of structural integrity and vulnerability for infiltration of mechanical systems through installation of locks, caging, and various types of fencing.
- Minimize time-on-target for terrorists via installation of real-time intrusion-detection and surveillance systems.
- Selectively protect the structural integrity of key members against collapse by strengthening key substructure members and blast shielding.

Additional Resources

The Transportation Research Board (TRB) has taken an active role in hazard and security research. The TRB Committee on Critical Transportation Infrastructure Protection shares research results from all sources and identifies research needs. The American Association of State Highway and Transportation Officials (AASHTO) Special Committee on Transportation Security (SCOTS) identifies and refers research needs to the National Cooperative Highway Research Program (NCHRP). NCHRP's 20-to-59-member panel funds applied research or refers prioritized requests to federal, state, local, and nongovernmental agencies, or to the private sector (TRB, 2007; AASHTO, 2007).

Since September 11, 2001, 90 security-related projects have been authorized in the Cooperative Research Programs: 57 of these projects have been completed, 21 are in progress, and 12 have contracts pending or are currently in development (TRB, 2007). Capsule descriptions of products and links to a variety of security-related products produced by TRB, other divisions of the National Academies, and other transportation research organizations can be found on the Web page of the Transportation Research Board and National Academies' security-related products and links (www.TRB.org/NASecurityProducts):

- TRB security-related publications: TRB-published reports at TRB.org/SecurityPubs
- TRB Cooperative Research Programs Security Research Status Report: Updated monthly (in PDF)
- Transportation Security: A Summary of Transportation Research Board Activities: Updated monthly (slide show in PDF)
- TRB Transportation System Security Web Site
- Key Hazards and Security Products of The National Academies: Updated monthly (in Microsoft Word, with live links)
- Slides—Hazards and Security Activities of the National Academies: 28 MB in PowerPoint with live links

- Transportation security information contained in TRB's Transportation Research Information Services database
- Transportation Security Research in Progress
- Select non-TRB transportation security information: Material highlighted in past TRB e-newsletters

Summary

The road transportation mode is unique in that it consists of privately owned vehicles traveling on publicly maintained roads. With nearly 4 million miles of roads, America has the largest and most vibrant road transportation system in the world. Road transportation has a central role to play in the continued health and growth of the nation's economy. It is essential for mobility and commerce, and it plays a critical role in times of crisis. America's roads are indispensable for evacuation, as well as in the response and recovery effort.

This chapter attempts to provide an overview of road transportation and infrastructure security. It introduces the National Highway System and the critical role it plays. Security issues and threats against road transportation system are reviewed. Challenges for state departments of transportation are also presented. Potential countermeasures and estimated costs are discussed. Additional resources from TRB, AASHTO, and other agencies are also presented. The objective of the proposed security program is not to provide full protection but to discourage terrorist attack through visible security and reduced vulnerability, as well as to minimize damage in the event of an attack.

Bridge, interchange, and tunnel security is important because they are the most vulnerable parts of our road transportation system. The threat from terrorists is real: Attacks at choke points could be devastating for the nation, as experienced on September 11, 2001. We must focus our attention and resources on these key components and prioritize our actions. When using this chapter, one must recognize that most mitigation countermeasures fall between two extremes. One extreme is to prevent all damage at enormous cost, which is very difficult, and the other is to spend nothing and risk enormous damage, which is foolish and not advisable. Tunnel and bridge owners, operators, planners, and engineers must make balanced decisions in selecting countermeasures for their facilities, preferably to risk an acceptable level of damage at a reasonable cost. As big as the problem appears to be, it may be made manageable through prioritization and risk assessment.

Finding this balance becomes more complicated when considering possible loss of human life, which is extremely difficult if not impossible to assign a value to. Protection of human life should always receive the highest priority. Due to the sheer size of the American road transportation system, 100% security at all places all the time is not a realistic goal. The goal is, therefore, to discourage terrorist attack through visible security and reduced vulnerability, as well as to minimize damage in the event of an attack. In addition, interagency cooperation is necessary for the development of cost-effective and implementable policies to respond in times of crisis.

References

American Association of State Highway and Transportation Officials (AASHTO) Blue Ribibon Panel on Bridge and Tunnel Security. *Recommendations for bridge and tunnel security*. The American Association of State Highway and Transportation Officials (AASHTO) Transportation Security Task Force, September 2003.

American Association of State Highway and Transportation Officials (AASHTO) Special Committee on Transportation Security. *Fundamentals of effective all-hazards and security management for state DOTs*. SCOTS Annual Meeting, August 2007.

American Association of State Highway and Transportation Officials (AASHTO) Special Committee on Transportation Security (SCOTS), Washington, D.C., 2007. Accessed at: http://security.transportation.org/?siteid=65.

Associated Press. San Francisco workers get free transit after partial freeway collapse, April 30, 2007.

Better Roads. Security and our road and bridge infrastructure. April 2003.

Bureau of Transportation Statistics (BTS). *National transportation statistics—2007*. U.S. Department of Transportation, Washington, D.C., 2007.

Daily News. 3 dead in fiery truck pileup on a Los Angeles area freeway. Los Angeles, Sunday, October 14, 2007.

Federal Highway Administration (FHWA). *National highway system*. Washington, D.C., 2007. Accessed at: http://www.fhwa.dot.gov/hep10/nhs/index.html.

Federal Highway Administration (FHWA) AASHTO Special Committee on Transportation Security. *Establishing the next generation of surface transportation security*. NCHRP Project Panel, 20–59, 2007.

Ham, D. B., and Lockwood S. *(PB) and SAIC, national needs assessment for ensuring transportation infrastructure security*. The American Association of State Highway and Transportation Officials' Transportation Security Task Force, October 2002.

Lincoln Tunnel. Wikipedia, the Free Encyclopedia. 2007. Accessed at: http://en.wikipedia.org/wiki/Lincoln_Tunnel.

Maunsell, F. Security aspects in the construction and maintenance of infrastructures of the inland transport sector. Richard Harris Intelligent Transport Systems, AECOM, 2006.

Merzagora, E. A. *List of tunnels in the U.S.* Wikipedia, the free encyclopedia, 8th revision, January 2005.

National Cooperative Highway Research Program (NCHRP). Report 525, Volume 12, *Making transportation tunnels safe and secure*. Transportation Research Board (TRB), Washington, D.C, 2006.

New York State Department of Transportation (NYSDOT). *2005 NYSDOT traffic data report: AADT values for select toll facilities*. 2006.

New York City Department of Transportation (NYCDOT). *Manhattan river crossings 2001*. 2002.

Parsons Brinckerhoff—PB Farradyne. *A guide to updating highway emergency response plans for terrorist incidents*. Prepared for The American Association of State Highway and Transportation Officials' Security Task Force, May 2002.

Port Authority of New York and New Jersey (PANYNJ). *Lincoln Tunnel interesting fact*. New York, New York, October 2004.

Science Applications International Corporation (SAIC). *NCHRP Project 20-07/Task 151B, a guide to highway vulnerability assessment for critical asset identification and protection*. American Association of State Highway and Transportation Officials' Security Task Force, National Cooperative Highway Research Program, May 2002.

Stovall, M. E., and D.S. Turner. *Methodology for developing a prioritized list of critical and vulnerable local government highway infrastructure*. University of Alabama, 2004.

The White House. *Homeland security presidential directive-3*. Washington, D.C., March 2002. Accessed at: http://www.whitehouse.gov/news/releases/2002/03/20020312-5.html.

Transportation Research Board (TRB). *Cooperative research programs: Security research status report*. The National Academies, Washington, D.C, September 11, 2007.

Transportation Security Administration (TSA). *Transportation sector network management office, assessment of highway mode security: Corporate security review results*. Highway and Motor Carrier Division, Washington, D.C., May 2006.

Transportation Research Board (TRB). *Hazard and security activities of transportation research board*. The National Academies, Washington, D.C., September 11, 2007.

U.S. Department of Transportation. *2006 status of the nation's highways, bridges, and transit: Conditions and performance*. Report to Congress, Washington, D.C., 2007.

6

Aviation Security

Thomas L. Jensen

Chapter Objectives:

- Briefly outline the history of aviation security
- Define the entities responsible for aviation security and the roles they play in protecting passengers and infrastructure
- Define the separate areas of the airport and the respective associated personnel, technologies, and processes
- Describe the envisioned future of aviation security—the anticipated increase in passengers, as well as new and emerging technologies

The Evolution of Aviation Security

Aviation security began as an industry-driven, reactive course of action and evolved into a proactive, government-driven process. The most significant transformations were made following the events of 9/11. However, the history of terrorist attacks on civil aviation certainly did not start on 9/11. Up to that point, skyjacking was the primary concern for aviation security.

The public increasingly became concerned about criminal activities involving aircraft when skyjacking began to make headlines on a regular basis. Throughout history, hijacking occurred with sailing ships, trains, and stagecoaches. However, when aircraft became the target, the term *skyjacking* was coined.

Although skyjackings were reported much earlier, the first incident with a U.S.-based airliner occurred on May 1, 1961, thereby beginning a wave of skyjackings. It started when a man forced a commercial airliner en route from Miami to Key West, Florida, to detour to Cuba. This was the first of four flights diverted to Cuba that year. Skyjackings were considered criminal acts, usually involving escape from law enforcement, political terrorism, or extortion. In 1968, there were 22 attempts at escape by forcing aircraft from the United

States to Cuba; 18 of these were successful. Later, political terrorism directed at the United States and Israel resulted in the destruction of four jet aircraft.[1]

The only recorded successful extortion attempt occurred in 1971, when a skyjacker received $200,000 from Northwest Airlines and then parachuted from the plane. The story is reported as follows: When a Boeing 727 was taking off from Portland, Oregon, en route to Seattle, Washington, a man in seat 18C handed a note to the cabin attendant that said he had a bomb onboard and would blow up the plane unless he received $200,000 and four parachutes when the plane landed. He had purchased the ticket under the name "Dan Cooper," a.k.a. the infamous D. B. Cooper. After receiving the money and parachutes, he released the 36 passengers and two members of the crew. He then ordered the pilot and remaining crew to fly him to Mexico. At 10,000 feet over an area north of Portland, he parachuted from the plane's rear stairs. The case remains unsolved.

Following the four U.S. skyjackings in 1961, air piracy became punishable by 20 years in prison or death under U.S. law. In 1963, skyjacking was addressed among several nations by the International Civil Aviation Organization (ICAO) through the Tokyo Convention, which required the return of aircraft and passengers and mandated the prosecution of perpetrators under international law. Some nations that attended the convention did not change their own laws to harmonize with the recommendations. Still others simply ignored the whole matter. In 1964, the Federal Aviation Administration (FAA) stepped in and specified that cockpit doors must be locked during flights. In 1969 it authorized physical searches of passengers at the airlines' discretion. These efforts, however, did not result in significant improvement.

It is believed that the reduction of skyjackings, which dropped from an average of 72 per year worldwide during the period from 1969 to 1972 to less than 29 per year for the next decade, was the result of improved deterrent and prevention measures. It took armed sky marshals that were introduced in 1970 with the public mandate to shoot to kill, more stringent screening of passengers and luggage in 1972, and requirements that U.S. airlines inspect all hand luggage and screen all passengers with metal-detecting devices to make significant reductions in skyjackings. By 1974, 25 potential skyjacking attempts were averted, with 2,400 firearms confiscated at U.S. airports. The institution of these more stringent screening procedures resulted in a large reduction in skyjackings of U.S. aircraft.[2]

The period of relative calm that followed extended for roughly a decade; then the faith in aviation security enjoyed by U.S. citizens was shattered on December 21, 1988, when Pan Am Flight 103 was blown out of the sky over Lockerbie, Scotland. This was a U.S. carrier with 180 U.S. nationals onboard. The flight originated at London Heathrow International Airport and was headed to John F. Kennedy International Airport in New York City. All 243 passengers and 16 crew members, along with 11 people on the ground, perished in that incident.

The Lockerbie loss resulted from the detonation of a small bomb (perhaps as little as 1 pound of plastic explosive) inside a radio cassette player in a suitcase. This bomb blew a 20-inch hole in the forward cargo hold. In contrast to the 9/11 flights, where 19 terrorists committed suicide in order to carry out their missions, the bag containing the explosive device was not matched to a terrorist riding to his death on the same flight.

---

[1] *Criminal Acts Against Aviation 1988 and 1999.*

[2] *Criminal Acts Against Aviation 1988 and 1999.*

For many years, protection of the aircraft was thought to be the primary responsibility of the industry (the airlines for the aircraft itself and the airport for the landing field and terminal facilities). However, the loss of a 747 over Lockerbie, Scotland, in December 1988 called attention to the world that very large losses can occur from terrorism and other criminal acts. Governmental response included the creation of the White House Commission on Aviation Safety and Security (Gore Commission), which, among its extensive findings and recommendations stated, "The federal government should consider aviation security as a national security issue and provide substantial funding for capital improvements." In other words, attacks against U.S. air carriers were now being considered attacks against the United States' policies and culture rather than a specific airline.[3]

Aviation Security Turning Point: 9/11

As the day began at the annual convention of the Airports Council International-North America (ACI-NA) in Montreal, Canada, the convention hall was filling with delegates enjoying a continental breakfast and getting settled for the first session of the day. The meeting room's large screens displayed routine conference information. Suddenly, the screens changed to live news broadcasts from New York City, where American Airlines Flight 11, a wide-body Boeing 767, had crashed into the North Tower of the World Trade Center at 8:46 EDT, hitting floors 94 to 98.

The conference disintegrated as managers of most of the major U.S. airports tried to conduct business via their cell phones.

Their airports were actually all out of business because Department of Transportation Secretary Norman Mineta, working from a bunker with Vice President Dick Cheney, ordered that all civilian aircraft leave the skies over the United States under the edict of a 100% "ground stop." Flights approaching the United States would be diverted to other nations, and air operations at all airports in the country would be shuttered until FAA personnel could personally visit, inspect security, and approve a plan to reopen each individual airport.

Airport managers of the Port Authority of New York and New Jersey rushed home to deal with all the problems associated with getting three airports operational, while their offices, which were in the World Trade Center (WTC), were completely gone.

Contributing to the devastation that September morning, a second Boeing 767 crashed into the World Trade Center. That plane, United Airlines Flight 175, hit floors 78–85 of the South Tower at 9:02 a.m. This second hit was captured by television broadcasters and amateur photographers who had focused their cameras on the buildings after the first crash. Within a few short minutes, the United States had suffered the largest attack on its soil since Japanese planes struck Pearl Harbor. There could be no doubt that the attacks were the deliberate work of terrorists.

A third airplane, American Airlines Flight 77, a Boeing 757-200, crashed into the Pentagon in Washington, D.C., at 9:37 a.m.

One planned attack that failed to reach its target, United Airlines Flight 93, a Boeing 757, went down in a field near Shanksville, Pennsylvania. It is believed that the terrorists onboard were attempting to fly the aircraft to Washington, D.C., to crash it into the White House or Capitol Building. The plane burrowed into the ground at 10:03 a.m. Passengers onboard the flight are credited with a heroic effort to wrest control of the plane from the

---

[3] *Criminal Acts Against Aviation 1988 and 1999.*

terrorists, who had already killed cabin attendants. People on the ground learned of these happenings through cell phone calls from passengers. In turn, the passengers learned about the use of pirated aircraft that morning to attack the World Trade Center and the Pentagon. Apparently, the call to action among the passengers was a shouted command, "Let's Roll."[4]

Early responses to the 9/11 attacks have mitigated the opportunity for terrorists to use the same kind of tactics again. In addition to official responses, attitudes of passengers have greatly changed. Many passengers are apparently ready to subdue anyone who attempts to confront the flight crew.

The responses from government include the following:

- Strengthening the cockpit door and requiring that it be kept locked during flight.
- Requiring pilots to remain in the cockpit rather than going into the main cabin to negotiate with unruly passengers and potential skyjackers. In the past, the practice included going along with demands in order to keep as much order as possible.
- Increasing the number of mission air marshals.
- Arming cockpit crews. A program known as Federal Flight Deck Officer (FFDO) allows pilots to have sidearms in the cockpit only.[5]

Aviation Industry and Economic Impact

The attacks of 9/11 caused not only the greatest loss of lives; they were also the most horrendous in terms of economic cost. The total bill has been estimated to be as much as 2 trillion dollars when opportunity costs are included.[6]

To keep this in context one must recognize that the economy of the United States is the world's largest, totaling an estimated 13.2 trillion in 2006.

The civil aviation industry is a major economic contributor. In Memphis, Tennessee, an economic impact study for the Memphis–Shelby County Airport Authority by Sparks Bureau of Business and Economics and the University of Memphis estimated that one employee in four is employed directly or indirectly by airport-related activities. The Federal Express headquarters has a substantial influence on this economic effect. The economic effects in the Chicago, Illinois area are even more daunting. The city of Chicago recently sponsored an analysis of the economic benefit of O'Hare, Midway, and Meigs to the nine-county region based on 1997 data. The analysis concluded that activities related to airline travel have created 340,000 jobs in those nine counties, or 7% of the total employment for the region. Nearly 50,000 of these jobs are held by people who are employed directly by the air travel industry, which generates $13.5 billion annually in wages, benefits, and other income.[7]

Airports are recognized as economic engines in communities throughout the country. A study conducted for the Des Moines Iowa International Airport in 1998 concluded "that the Des Moines airport is a critical component of its region's infrastructure, sustaining commerce, industrial activity, and supporting household and economic growth." The study showed the total economic value for the year 1998 was $182 million and provided jobs for 2,352 individuals.

---

[4] http://www.martin.ws/html/911.

[5] *The 9/11 Commission Report.*

[6] Institute for the Analysis of Global Security, 2003.

[7] Metropolitan Planning Council for the Chicago Region.

This comparison illustrates that while big airports like Chicago's O'Hare (36 million boardings in 2004) are giant economic engines, smaller airports like Des Moines (<1 million boardings in 2004) also make very substantial contributions to local economies throughout the nation.

By the same token, when airports are shut down, sometimes even for a few hours, substantial losses are encountered.

Although terrorists thought the 9/11 attacks would destroy the U.S. economy and perhaps Western civilization, the total cost in material terms resulted in roughly 2 months' worth of our gross national product (GNP). America was able to withstand the ripple effect of that massive devastation.

U.S. Aviation Security Players

Aviation Security in the United States is composed of multiple levels and emanates from many sources. These disparate groups can be arranged into four categories: government, airport, airline, and industry. While the government category is larger by far than the others, all offer important aspects and challenges to the changing face of public protection.

Use of the word *players* here is not to suggest that the security of aviation is somehow a game. It is a means to show that a team is at work to address vulnerabilities that provide opportunities for the terrorists, criminals, and malcontents who would do harm to America and its citizens.

Government Players

The breadth of governmental involvement in aviation security can be mind boggling. Extending over two U.S. departments (Homeland Security and Transportation) and composed of multiple agencies, it is sometimes hard to know who is responsible for what. The following sections detail some of those delineations.

Federal Aviation Administration

The Federal Aviation Administration (FAA) is the modal agency within the U.S. Department of Transportation (DOT) that regulates all aspects of civil aviation. The FAA's major roles include

- Regulating civil aviation to promote safety
- Encouraging and developing civil aeronautics, including new aviation technology
- Developing and operating a system of air traffic control and navigation for both civil and military aircraft
- Researching and developing the National Airspace System and civil aeronautics
- Developing and carrying out programs to control aircraft noise and other environmental effects of civil aviation
- Regulating U.S. commercial space transportation[8]

Prior to 9/11, the FAA was also responsible for development and enforcement of rules relating to the security of aircraft, passengers, and crews. Before and immediately

---

[8] http://www.faa.gov/about/mission/activities.

following 9/11, it undertook programs that ran the gamut from laboratory development of defensive materials and equipment to the enforcement in the field by inspectors and "red teams." However, with the creation of the Department of Homeland Security (DHS), most of the security functions were transferred to the new department.

Department of Homeland Security

On November 25, 2002, President George W. Bush signed into law the Homeland Security Act, which created the cabinet-level Department of Homeland Security (DHS). DHS, under direction of Secretary Tom Ridge, entailed the largest government reorganization since the Department of Defense was created 50 years prior. It is one of the largest departments of the U.S. government, carrying the responsibility of domestic protection of United States territories.

In March 2002, the president, under Homeland Security Directive-3, established the Homeland Security Advisory System to be used by the DHS to determine, announce, and execute various levels of threat conditions. The color code system, depicted in Figure 6.1,

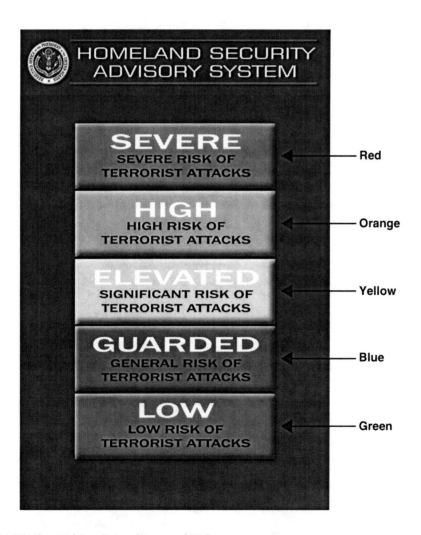

FIGURE 6.1 DHS Threat Advisory System (Courtesy of DHS)

was met with many jokes from late night comedians and politicians as being too simplistic and lacking imagination.

However, the system is actually quite effective because with it, DHS can instantly advise law enforcement agencies and first responders at all levels of government, airports, airlines, and the general public about changes in the assessment of threat level.

The system is used both for a national indicator and to allow higher or lower levels at various locations to match local conditions. For example, the airports in New York are at "High Risk of Terrorist Attacks" (orange), while most airports are at "Elevated-Significant Risk of Terrorist Attacks" (yellow).

The announcement of a change in level requires the immediate modification of many procedures in accordance with a preset script required for that level.

At airports, higher levels often require setbacks from terminal doorways restricting vehicular access. These adjustments have led to the use of barricades, both by physical constraints and with security personnel. These requirements are especially burdensome to airports because they not only disrupt traffic flows and access points but also often make areas unusable inside the restricted areas. For instance, all or a part of a parking garage may be required to shut down during the higher threat level.

Transportation Security Administration

The Transportation Security Administration (TSA) was created by the passage of the Aviation and Transportation Security Act (ATSA) on November 19, 2001. It was originally part of the DOT but was subsequently moved to DHS authority in 2003. TSA was assigned most of the responsibility for security of all modes of transportation, though it is most recognized for its role in aviation security. The majority of the agency's funding is appropriated for aviation security ($4,731,814,000 in 2007).[9]

TSA hires, trains, and deploys federal employees to screen passengers and luggage at 450 carrier airports of the 5,000 airports nationwide. While uniformed Transportation Security Officers (TSO) conduct security screening at most airports across the country, private screeners are still utilized at a handful of airports under TSA's Screening Partnership Program (SPP). The SPP allows an airport to contract its screening services to TSA-approved providers. Besides meeting security qualifications, the contractors' pay, benefits, and training must be equal to or better than those of TSA personnel. At its inception, TSA generally took over the screening functions and arranged for use of outdated screening equipment from the airlines.

TSA's Evolution   TSA was initially headed by law enforcement executives who had little experience in airport operations and related business needs, which sometimes resulted in unnecessary confrontations. TSA personnel would suggest that a prime business spot near the checkpoint was needed for some administrative function.

The airports countered that they were landlords who received much of their revenues from space rental; they asked Congress for rental fees for the space that TSA occupied. Meanwhile TSA, recognizing the high value of retail space in the concourse, began locating administrative offices in commercial buildings near airports rather than requiring so much prime space in the terminals.

---

[9] House Bill HR 5441.

The Maturing of TSA   As TSA has expanded its authority, it has been the target of major public criticism in response to a number of unpopular regulations.

One controversial action was the passenger liquid ban, instituted by Administrator Kip Hawley in August 2006. This controversial and somewhat confusing rule was introduced in response to intelligence reports and arrests of suspects, suggesting that terrorists were planning to carry IED (improvised explosive device) components onboard aircraft in the form of innocuous liquids, which would become bombs when combined. On August 10, 2006, TSA banned all liquids carried by passengers—even those purchased in the sterile area—and passengers found themselves divested of what they considered essential.

After much public derision of the decision, TSA modified its rule. On September 26, Hawley announced the 3-1-1 plan: maximum of 3-ounce quantities of liquids, gels, and aerosols contained in a 1-quart ziplock transparent bag, with one bag allowed per passenger. The passenger was to present the bag to screening personnel separate from luggage. At the same time, TSA lifted the ban on liquids purchased within the sterile area before boarding.

Hawley has also instituted a development team initially called PAX 2.0. This group, composed of both government and private individuals, is looking for technologies and procedures that will make security screening at airports more effective while increasing appeal to travelers. These efforts are giving rise to a model airport approach in which a specific airport is chosen to operationally test the technology or procedure being proposed.

Federal Air Marshals (FAMS)

The Federal Air Marshal Service (FAMS) is the primary law enforcement entity within TSA. Its official mission is to "Promote confidence in our Nation's civil aviation system through the effective deployment of Federal Air Marshals to detect, deter, and defeat hostile acts targeting U.S. air carriers, airports, passengers and crews."[10]

Federal air marshals (FAM) are discreetly deployed, armed personnel who travel aboard commercial flights. Though FAMs (originally known as sky marshals) have existed within aviation security since the 1960s, their role was not significant until the 9/11 attacks, when the reactionary legislation that created the DHS also expanded the air marshal program. The exact number of officers employed by FAMS is classified, though it is estimated to be in the thousands. In 2006, changes to the system included moving its direction from the U.S. Bureau of Customs and Immigration Enforcement to TSA, thereby aligning FAMS' function as consistent with that of DHS.

Federal Flight Deck Officers

After 9/11, security onboard aircraft was tightened to protect the aircrew of commercial aircraft. Cockpit doors were hardened to withstand mid-air blasts, and federal air marshals were recruited in greater numbers to accompany flights. As an additional layer of security, the Transportation Security Administration instituted the Federal Flight Deck Officer (FFDO) program under the Homeland Security Act in November 2002. FFDOs, classified as pilots, flight engineers, or navigators, may use deadly force to protect the

---

[10] http://www.tsa.gov/lawenforcement/people/index.shtm.

cockpit from seizure by terrorists or air pirates. Under this program, crew members are instructed on use of firearms, legal issues, defensive maneuvers, psychology of survival, and use of force policies. Although it is strictly voluntary, the program has experienced no dearth of participants. In 2003, the program was expanded to include cargo pilots and other flight crew members.[11]

Customs and Border Protection

Customs and Border Protection (CBP) is another unit of the U.S. Department of Homeland Security that has direct involvement with airports and aviation security. CBP, though its own entity since 1789, became an official agency within DHS in 2003, combining with employees from the departments of Agriculture, Immigration and Naturalization Services, and Border Patrol. Its mission, besides its original duty to collect import duties, is to regulate and facilitate international trade and enforce U.S. trade laws, including prevention of terrorists and terrorists' weapons from entering the United States.

In 2004, CBP instituted the US-VISIT (United States Visitor and Immigration Status Indicator Technology) program. Funded by DHS and the Department of State, US-VISIT makes it possible to identify the person who arrives at a port of entry as the same individual to whom the visa was issued. Photos and fingerprints of travelers to the United States are checked against a database of known criminals and suspected terrorists. The information is collected at overseas U.S. consular offices and locations where visas are issued to travel to the United States. The program started with a 2-fingerprint system, but as of November 29, 2007, has begun to be replaced by 10-finger scanners.

Airport Players

Security Guidelines for Terminal Construction

In 2002, the FAA issued a set of guidelines for security needs in terminal construction. These guidelines were the product of a government–industry committee that considered the changing security needs in airport design. The document was revised and rewritten by the Airport Security Design Working Group of the Aviation Security Advisory Committee (ASAC). "Recommended Security Guidelines for Airport Planning, Design, and Construction" was published on June 15, 2006.

A concern expressed often in deliberations over design issues is the need for flexibility. Many beautiful, soaring designs sometimes proposed by renowned architects for airport terminals have offered almost insurmountable bars to the continuous march of new and different technologies and procedures.

Many new terminals for major airports have entire floors beneath the main floor devoted to the handling and screening of checked luggage. These facilities, though solving aesthetic issues, require not only the facility for routine screening but also the facilities to handle items that appear to contain explosives. This threat possibility requires special hardened rooms and an exit plan for removing the bag from the building, among other security-based architectural concerns.

On the other hand, airport terminal buildings that predate the introduction of significant security programs have caused many challenges. Corridors and concourses were not built to accommodate screening equipment and personnel, and gates were not large

---

[11] http://www.alpa.org and http://www.tsa.gov/lawenforcement/programs/ffdo.shtm.

enough to allow additional security before passengers boarded. There also was no provision for expansion of administrative functions.

Some alternatives being considered involve screening of some luggage before it arrives at the terminal. Luggage would be screened and retained in the custody of a properly vetted handler, which would substantially reduce the load on airport screening.

Pilot programs have involved checking luggage at hotels, car rental returns, train or subway stations, and cruise line terminals. One such offsite program allows travelers leaving McCarran International Airport in Las Vegas to check bags at their hotel. For a fee, the luggage is transported to the airport and sent on to the departing plane without contact with the passenger. Other airports and prospective handlers are investigating similar plans. So far, the number of bags handled by this system at McCarran represents a very small percentage of the total.

Airport Federal Requirements

All airports in the United States are operated either directly by a government entity or by an airport authority created by the government.[12]

Though the inception of TSA greatly reduced the burden of security from the airport, several functions are still administered by airport personnel, with TSA's approval.

Federal security regulations describe in detail the requirements that must be adhered to at airports. These regulations require airports to provide for the safety and security of persons and property on an aircraft against an act of criminal violence, aircraft piracy, and the introduction of an unauthorized weapon, explosive, or incendiary device. Airports must create and maintain a written document that describes the compliance measures employed at their airport. The document is referred to as the airport security program. Some of the program requirements that must be addressed are as follows:

- Measures used to restrict access to areas that are controlled for security purposes, as well as detailed maps of the areas
- Procedures for the distribution, storage, and disposal of sensitive security information
- Procedures for responding to security incidents
- A description of the law enforcement support used to comply with federal regulations

It is important to point out that the requirements listed in this section are distinctly different and separate from those imposed at security screening checkpoints. To be clear, TSA has the responsibility for screening checkpoints that are located inside the airport. Airports bear the responsibility for securing all other areas.

Airports are still responsible for meeting federal standards for securing the perimeter of the airport, and airlines, until recently, provided an agent at the entrance of each checkpoint to ascertain if the person trying to gain entry had a government-produced picture ID and a boarding pass for his or her flight from that concourse. Through negotiation, this duty was assumed by TSA. One of the vulnerabilities being addressed was to deny access to the sterile area beyond the screening point by anyone other than ticketed passengers and personnel employed at the airport.

---

[12] The two notable exceptions are Indianapolis International Airport and Stewart International Airport.

This requirement affected goods and services offered by vendors in the airport. Post-checkpoint restaurants could no longer serve nonticketed patrons, and ticketed passengers felt uncomfortable shopping or dining before passing through security. Since airports depend heavily upon vendor revenues, new approaches were considered and tried at various locations. Some airports moved shopping into the sterile area with only newsstand-type retail before the checkpoint. Others tried to maintain full service in both areas. The heavier retail activities inside the sterile area also presented challenges to airport designers to allow service carts, vehicles, and suppliers into the sterile area. Many devised underground tunnels and elevators to provide access.

Airport operators find themselves in the middle of many controversies. Security experts suggest that tightening of the checkpoint—the front door—just adds additional vulnerability to other areas. There is the chance that someone could walk right into the baggage claim area with explosives either on his or her person or in a suitcase. There is the possibility that an intruder could gain access by hiding in vehicles arriving at the airport. The question that was asked to panel members at a recent aviation security conference was, "What keeps you up at night?"

There is growing interest in Congress to have airports provide checkpoints for arrivals of employees and contractors. However, Airports Council International-North America (ACI-NA) and American Association of Airport Executives (AAAE), two organizations representing airports, have led an effort in Congress to find alternatives to this screening.

Airline Players

Airlines are another partner in providing security for passengers and air cargo. Airlines that offer regularly scheduled service usually have long-term contracts with the airports where they will land and be serviced. Scheduled airlines in the United States had been public utilities regulated by the Civil Aeronautics Board (CAB), from its creation in 1938 until it was finally closed on January 1, 1985. CAB had approved the routes that airlines would fly, the amount of competition from other carriers, and the fares that they would charge. There has been a long period of adjustment from those days of tight regulation to the freedom to compete both in schedule and price. The result has been an air carrier system that has suffered from financial difficulties. When the added losses from the turndown after 9/11 occurred, carriers were looking for ways to provide less service for items like meals and additional luggage and for ways to reduce cost such as flying smaller aircraft, reducing the number of flights on many routes, and shutting down unprofitable routes.

Under these constraints, airlines have been going to Congress to shift costs including security to others. For example, the checker at the public side of each checkpoint previously was an airline employee or contractor. That duty has shifted to a TSA employee.

Industry Players

The aviation industry is international in scope. It includes manufacturers of aircraft; a vast array of support systems and products; and information systems, consultants, and planners; as well as the air carriers and airports.

With the market-driven business culture of the United States, these players are coming forward with new and improved products, procedures, and services at a bewildering rate.

Management of security programs must include evaluation of the potential products and services, testing them and, where appropriate, deploying them. Under such dynamic conditions in the marketplace, most systems are destined for replacement or modification almost by the time they are placed in operation.

Airport Security Areas

Checked Baggage

Congress also assigned TSA the daunting task of providing explosives screening for 100% of checked luggage by December 31, 2002. To illustrate the enormity of this requirement, although the number was not publicly acknowledged, estimates were that prior to this mandate, screening of checked baggage was performed on only a very small (single-digit) percentage of the total. Since the 2002 requirement, many processes and much equipment have been implemented in this important area. Explosives detection systems (EDS), which can be the size of a minivan and weigh as much as 2 tons, had to be accommodated in baggage-handling areas of many sizes. These systems required floor space that was already at a premium and in many instances required new structural additions to existing facilities to withstand the loads on floors.

Technologies

Explosives Trace Detection   Explosives trace detection (ETD) is a labor-intensive task of examining luggage with swabs wiped on the handles and edges of the bag. The small pieces of fabric are then screened by a trace device that allows discovery of very minute amounts of explosives.

Explosives Detection Systems   TSA uses the large EDS equipment and complex conveyor systems at many airports. These machines are designed to detect explosives based on density of the contents. EDS efforts are augmented with K-9 teams and hand searches.

Soon after the 2002 deadline passed, with TSA declaring success, airport operators and the government started designing systems that would replace or augment many of the hastily assembled systems. There is general agreement that use of in-line systems, where bags move through the process on conveyors, will be cost effective, but the capital cost of the system and space to house it require very large appropriations. Proponents for the additional facilities point out that if this were the private sector, debt financing would be used to acquire assets that would be amortized by the projected labor savings. TSA is using letters of intent to airports in some cases whereby the airport can use its credit to make the improvements with assurances from the government that the airport will be repaid.

Some efforts have also been made by smaller airports to use in-line checked baggage screening. By building an addition on the back of its terminal, Blue Grass Airport in Lexington, Kentucky, a Category II airport, was able to obtain approval for an in-line system that could be a pilot program for other airports of its size. In addition, this pilot program has enabled TSA to test the screening of air cargo on the same EDS equipment that is primarily there for handling checked baggage during idle periods. If the plan works, it will reduce the cost of both equipment and labor while offering better service.

Procedures

The Aviation and Transportation Security Act of 2001 (Public Law 107-71) mandated that all passengers be screened prior to aircraft boarding and that property be screened prior to aircraft loading as well. Currently, checked baggage can be screened by one of two ways, by using either the EDS or the ETD machines as described earlier.

Most checked baggage screening pods utilize EDS machines for primary screening and use the ETD machines only to resolve alarms that cannot or should not be resolved by on-screen alarm resolution procedures. However, with 100% EDS screening, there are times when the bag flow is in excess of EDS capacity, and those bags may be primarily screened using ETD machines.

Other pods may use ETD machines as the primary means of screening when EDS machines are absent or experience a fault.

Most passengers are aware of the items prohibited from being carried onto aircraft but are not as knowledgeable when it comes to prohibited items within checked baggage. Effective January 1, 2008, spare lithium batteries (e.g., cell phone batteries, computer batteries, AA/AAA batteries, etc.) were prohibited from being included in checked baggage. Devices that use the batteries can be checked with the battery inserted, but spare batteries must be packaged properly and included in carry-on luggage. Following are some items prohibited from checked baggage; however, this list is not all inclusive, and therefore, items considered dangerous by the discretion of TSA may also be prohibited.

- Blasting caps
- Dynamite
- Fireworks
- Flares of any form
- Gunpowder
- Hand grenades
- Plastic explosives
- Realistic replicas of explosives
- Aerosols, except for personal care or toiletries (limited quantity)
- Fuels (including cooking fuels)
- Lighter fluid
- Matches
- Turpentine and paint thinner
- Spare lithium batteries
- Items not properly packaged and/or declared at check-in (i.e., ammunition, firearms, lighters, etc.)

Most items not included in the preceding list are permitted to be transported through checked baggage. Permitted items that require special screening procedures when included in checked baggage are firearms, precious or high-value items (by request of passenger), evidentiary–grand jury and lifesaving materials, U.S. military chemical suits, animals, parachutes, and large musical instruments. Most of these special items should be declared with the airlines and tagged as such when checked to be easily identified for screening.

Lithium Batteries   As of January 2008, passengers are no longer allowed to pack spare lithium batteries in checked luggage. The new regulation, designed to reduce the risk of lithium battery fires, will allow lithium batteries in checked luggage only if they are installed in the devices they power. Lithium batteries are considered hazardous materials because they may overheat and ignite under certain conditions. Safety testing conducted by the FAA found that current aircraft cargo fire-suppression systems would not be capable of extinguishing the fire if a shipment of lithium batteries were to ignite in flight.

Passenger Checkpoint

The most visible face of aviation security for passengers flying aboard commercial airplanes is the passenger screening checkpoint. This checkpoint is where passengers and their carry-on luggage are examined for weapons and other dangerous objects under the watchful eye of the uniformed TSA officers (also known as TSOs). The basic process of screening passengers is largely the same as in years past, though a lot of the details have changed. Passengers with a valid boarding pass and government-issued photo identification are asked to walk though a metal detector while their personal items and carry-on bags pass through an X-ray. Additional screening may be required as a result of alarms at either the metal detector or the X-ray, but once all checks are passed, the passenger is free to proceed to his or her gate. Every airport and every checkpoint is different, but the general idea remains consistent.

Responsibilities

Historically, the airlines were taxed with the responsibility of meeting federal standards for operating passenger screening checkpoints with limited federal oversight, which was provided by the FAA Civil Aviation Security Division. After President Bush signed ATSA in 2002, the screening checkpoints became the responsibility of federal workers, first with DOT and later with DHS through TSA.

TSA takes responsibility for passengers from the point at which they submit their belongings for screening until the passenger exits the checkpoint. The areas before and after the checkpoint are generally the responsibility of the airport authority and the airlines, and incidents there are typically handled by law enforcement officers. TSOs may also refer passengers to law enforcement officers for various reasons, including being in possession of weapons or other dangerous items. Passengers who bring such items to the screening checkpoint may face civil penalties and criminal charges.

Technologies

X-Rays   Along with the metal detector, the traditional transmission X-ray is one of the staples of the airport security checkpoint. TSA has improved the standards for these machines with additional tools and techniques, but the idea is the same. Items placed on the X-ray belt are examined on the screen to determine if there are any items of interest inside that warrant further search. Screeners also look for areas where something could be concealed (such as behind a big piece of metal) and may have the bag searched or put through the X-ray at a different angle to be sure that nothing is there.

X-ray images can get quite cluttered when electronics or other items with a lot of parts are visible, which is the main reason why laptop computers and video game consoles must be removed from carry-on baggage and placed in a separate container. If the laptop stays in the bag, it is difficult for the screener to see things that might be behind or underneath it.

X-rays used in today's airports are typically referred to as TIP-ready X-rays (or TRXs). TIP, or Threat Image Projection, is a program used by TSA to ensure high screener performance and help with design of training programs.

Walk-Through Metal Detectors　Not all walk-through metal detectors (WTMDs, for those in the know) have been created equal, as most frequent fliers have experienced. As time has passed, newer WTMDs have improved in detecting legitimate threats while letting benign items through. Older WTMDs tend to alarm on belt buckles, watches, and other metal. Still, passengers would be advised to remove any wearable metal to avoid the hassle of re-divesting.

Explosives Trace Detection—Swabs and Portals　A relatively new entrant to the field of passenger screening is the concept of explosives trace detection. Right now, there are two primary means of doing this at the checkpoint. The more common is the swab method. A TSO takes a sample from a bag or other item with a specially treated piece of cloth or paper (it varies by machine manufacturer), which is then analyzed by a machine for trace amounts of explosives.

An explosives trace portal (ETP, or "puffer") takes samples directly from the passenger. The passenger enters a booth, which blows puffs of air onto the person, and then the machine automatically samples the air for explosives' residue. This method is deployed at only a handful of airports at the present time.

Procedures

Boarding Pass and Identification Checks　In order to pass through the security checkpoint, passengers must present a valid boarding document issued by the airport and a valid government-issued photo identification such as a passport or driver's license. TSA has moved to using specially trained TSOs to screen boarding documents and IDs for signs of tampering or falsification. ID checks at some airports may include the "lights and loupes" method, whereby the TSOs use black lights and magnifying glasses to verify the specific security features of passengers' licenses and passports. Passengers who are exposed as attempting to travel under false identification are generally referred to law enforcement personnel for further investigation.

Selectees: Passengers Selected for Additional Screening　The airlines maintain a program that selects passengers for additional screening. These passengers are typically given a more intense screening than other passengers, including use of explosives trace detection technology on their baggage and person and a screening with a handheld metal detector. Exact factors that influence selection of an individual for secondary screening are unknown to the public, but range from watch lists to randomized selection.

Shoe Removal　As of August 2006, all passengers must X-ray their shoes at the passenger checkpoint. This requirement was spurred by the potential for shoes to contain explosive materials, as in the case of Richard Reid in 2001 (who attempted to hijack a plane in this manner). X-ray screening allows for any potentially dangerous items inside shoes to be seen by TSOs. Many shoes have metallic shanks in them, which would require that they be removed for passage through a WTMD anyway. Passenger shoe removal is a standard procedure in place today.

Liquid Restrictions: The 3-1-1 Rule   In August 2006, officials in London intercepted a terrorist plot to blow up planes using liquid explosives. The incident spurred worldwide alarm as to the perceived threat of seemingly benign liquids and gels. In the United States, TSA's immediate reaction was to ban all liquids from carry-on luggage. The rule, however, was soon relaxed to merely restrict the amount of liquids a passenger could carry through the checkpoint, i.e., the 3-1-1 rule.

The 3-1-1 rule states that passengers can carry containers no larger than 3 fluid ounces, contained in a 1-quart bag, with only one bag per passenger—3 ounces, 1 quart, 1 bag. Medicines and certain other liquids, such as baby formula and breast milk, are exempt from the rule.

Electronic Devices   Because of the difficulty of analyzing their X-ray images, large electronic devices, laptop computers, game consoles, and video cameras must be removed from baggage before passing through the X-ray. The electronic devices are X-rayed as separate items to simplify the image on the X-ray screen.

Cargo

Although many new security requirements have been implemented by the government since 2001, such as enhanced background checks of employees and shippers, the majority of air cargo shipments are still not screened before being loaded onto aircraft. However, technology may soon play a role in increasing the amount of cargo screened, thanks to new equipment designed to examine large quantities of cargo efficiently. Increased cargo security is the major area of airport security that TSA is beginning to address, although belatedly. Efforts are now being made to address this shortcoming, including a full tracking surveillance system of all cargo from point of origin to delivery.

Responsibilities

If cargo is screened at all, it is the responsibility of the individual airlines. The devices used are airline owned and maintained. TSA tests equipment and procedures, as well as providing a degree of security consultation, but does not officially provide resources for cargo screening.

Technologies

Explosives Detection   Both explosives detection systems (EDS) and explosives trace detection (ETD) equipment are being used to screen various types of cargo throughout the United States. The two technologies are used in conjunction to provide a complementary screening approach.

Items that indicate potential threats are handled in accordance with airline procedures, which usually involve contacting supervision and/or the local security personnel for resolution.

Procedures

TSA uses standardized operating procedures for EDS and ETD screening. The procedures have evolved from checked baggage screening procedures and are now specific to cargo screening. In addition, threat notification procedures including alarm resolution and safety are specific to the cargo screening process.

Perimeter

In security terms, an airport's perimeter consists of the outer boundaries of airport grounds (i.e., the fence line), all structures enclosed therein, and all entryways onto airport grounds. Though great strides have been made toward improving perimeter security, hundreds of miles of perimeter boundaries at airports across the United States remain vulnerable to intrusion. Ultimately, weak perimeter security can compromise efforts that have been made in other airport sectors.

Responsibility

Traditionally, the security of the perimeter of an airport has been the responsibility of its airport authority. The personnel and technology devoted to perimeter security thus vary by airport, as do the security protocols. The protocols, however, are subject to TSA revision and approval before they may be officially implemented.

Technologies

Fence-Mounted Sensors Many of the current fence-mounted perimeter intrusion detection devices (PIDDs) use fence vibration as a means of detecting an intruder. These devices can be separated into two categories: devices that use a specially designed cable for detection, and systems that use a sensor contained within a mountable housing that detects vibration. The cable vibration systems can be further separated into two types: fiber optic and magnetic (metal) cable.

Volumetric Sensors Volumetric-type sensors include a variety of technologies, most of which use a specific radio frequency to monitor the volumetric displacement of objects within a particular location.

One specific frequency range commonly used among systems is the infrared range. Several systems use these radio frequencies either in a passive or active configuration. The passive infrared sensor typically uses only a transmitter to transmit the radio frequency and determine the volume displacement within the area. Active infrared sensors, on the other hand, use both a transmitter and a receiver to determine the volume displacement.

One technology that is well tested throughout the military is radar. Several volumetric PIDD manufactures use ground-based radar systems to detect objects within a specific area. This type of system is being used not only to detect intruders entering the perimeter but also to monitor airline–airport employees within the Air Operations Area (AOA).

A fairly recent technology being used for volumetric detection is video analytics. This term is often used to describe systems that use computer-based intelligence to analyze and access threats within a video image of a particular area.

MANPADS Man Portable Air Defense systems, have also been identified as potential major threats to commercial aircraft. Wikipedia, the online encyclopedia, describes the MANPAD weapon as follows:

> *MANPADS were originally developed in the late 1950s to provide military ground forces protection from enemy aircraft. They are receiving a great deal of attention as potential terrorist weapons that might be used against commercial airliners. These missiles, affordable and widely available through a variety of sources, have been used successfully over the past three decades both in military conflicts[1] as well as by terrorist organizations. They can be purchased on the black market anywhere from a few hundred dollars for older models to upwards of almost a quarter million dollars for newer, more capable models. Seventeen countries, including the United States, produce man-portable air defense systems.[2][3] Possession, export and trafficking in such weapons is officially tightly controlled, due to the threat they pose to civil aviation, although such efforts have not always been successful.[4][5] The missiles are about five to six feet in length, weigh about 35 to 40 pounds, depending on the model. Shoulder-fired SAMs generally have a target detection range of about 6 miles and an engagement range of about 4 miles so aircraft flying at 20,000 feet (3.8 miles) or higher are relatively safe.[6]*

In November 2003, an incident occurred involving a DHL Airbus A300 when a MANPAD was fired and struck the aircraft as it took off from Baghdad Airport. The crew was able to return the aircraft to the Baghdad Airport.

In January 2008, DHS announced that three American Airlines passenger planes that fly in and out of JFK International Airport in New York will be equipped with antimissile systems. The test program will determine if the technology deployed will be effective in preventing loss of aircraft from a missile attack.

Procedures

The procedures by which an airport's perimeter is monitored or secured are proprietary to each individual airport. The means by which the aviation security view and/or monitor each area are distinct. This could include video surveillance, manned patrol, sensors, and/or barricades.

Access Controls and Biometrics

An airport's access control system consists of maintaining accessibility of specific airport locations, such as the air ramp, administrative office areas, or sterile area. Though many improvements have been made to increase the ability to restrict airport access, weak areas such as employees allowing entrance to unauthorized persons or definitively verifying the person seeking access into the controlled area "is who he says he is" still exist in the day-to-day operation of the airport. Implementation of identity verification biometrics and systems that allow only one person to enter a secured area at a time improve security in these zones.

Responsibility

Traditionally, the security of the access-controlled areas of an airport has been the responsibility of its airport authority. The technology devoted to access security thus varies by airport and availability of airport resources to expend on improvement technology.

Technologies

Anti-Piggybacking–Anti-Tailgating Systems To prevent the entrance of more than one person at a time through a secured door, anti-piggybacking–anti-tailgating systems can be installed at that location. Piggybacking is knowingly admitting an unauthorized person into the secure area by an authorized individual. Tailgating is unknowingly admitting an unauthorized person into the secure area through devious means. Prevention technologies range from camera-based systems to laser detection systems to physical mantraps.

Biometric Identification Systems Biometric identification access control systems include a variety of methods to verify the identity of the individual seeking access into the secured areas through a series of personal physical features. The typical physical features used for reference include fingerprint identification, facial recognition, iris recognition, and hand geometry pattern.

The most popular choice of biometric identification is the use of fingerprint recognition technologies. The methods of fingerprint verification range from use of an optical photographic system or silicon chip system that captures the fingerprint surface to infrared lights and ultrasonic waves that capture the underlying information. These technologies can be seen in the use of portable recognition systems and stationary wall-mounted systems.

Facial recognition systems typically involve use of cameras to capture the face structure of the individual seeking access, either in normal or three-dimensional ranges. Iris recognition systems often use a camera and infrared light system to capture the pattern of color and veins in the user's iris. Hand geometry systems measure various points along the shape of the hand to create a matching template, or use infrared lights to capture underlying skin and vein patterns to create a matching template.

A recent testing program has been instituted by TSA to identify biometric technologies that meet specific standards for failure-to-enroll rates and user identification rates. These technologies are being placed on a list for airport reference whenever purchasing biometric access control systems.

The Future of Airport Security

Technologies

Many efforts have been made in recent years to introduce the "airport of the future." Some forward thinkers envision a transparent or passive checkpoint where passengers are screened as they pass through a tunnel or passageway or ride on an escalator. The concepts involve various sensors disguised in wall finishes and trim elements. One is reminded of the motion picture *Total Recall,* in which Arnold Schwarzenegger was shown moving along a walkway with his weapons exposed in an X-ray fashion.

Mike Golden, assistant administrator for operational process and technology–chief information officer–chief technology officer for TSA, said in a recent article in *International Airport Review,* "The fundamental challenge of protecting passengers and transportation networks against an act of terrorism is a constantly changing, unpredictable threat environment. TSA knows that terrorists seek to exploit our weaknesses. We also know that terrorists attempt to adapt to the security measures we put into place. A static fortress-like defense is not the answer."

In addressing these challenges, TSA is looking to many emerging technologies:

- **Advanced technology (AT):** This is the designation for the next generation of X-ray machines for checkpoints. They employ two-dimensional images and better penetration of dense objects. They also enhance TIP technology already in use at checkpoints.

- **Whole-body imagers:** These devices include millimeter wave imaging and back scatter technology. Backscatter has been tested in several major airports and has been somewhat controversial because of the concern for privacy. It shows outlines of the anatomical parts of the body. The developers have therefore worked at modifications that will show the hidden prohibited materials without being explicit about the body. Tests are beginning on the millimeter waver device, which uses radio frequency (RF) energy to reflect an image of the body. By using these technologies, the screener is able to detect both metal and nonmetallic weapons and explosives. The promise is that passengers can be screened for hidden threats without any physical contact.

- **Bottled liquids scanners (BLS):** Many versions of this type of device have been tested over the last several years. The current models are handheld devices that detect specific chemical vapors coming from a bottle or other container. The devices are designed to identify explosives and flammable liquids at the checkpoint.

- **Screening of cargo:** Cargo transported via aircraft creates unique screening challenges within the transportation arena. The volume of cargo along with the size and timely need for shipment are obstacles that are driving potential screening solutions. Baggage screening explosives detection systems placed in cargo facilities along with integrating cargo within checked baggage areas of airports are among potential screening solutions. In addition, hardened containers capable of withstanding blasts are under evaluation as potential methods of mitigating a catastrophic incident.

- **Credential–personal authentication and biometrics:** Ensuring that the right individual has access to secure areas or even the cockpit of an aircraft continues to be a fundamental requirement for public security. Biometric devices are used to identify and recognize humans based upon one or more intrinsic physical or behavioral traits. The various devices include but are not limited to facial recognition, fingerprints, iris scanners, hand measurements, and vascular identification. In addition, technologies that enable security professionals to verify individuals via two- and three-dimensional bar codes from licenses and passports are capable of supporting the authentication process.

- **Registered traveler program:** This program, still in its infancy, continues to seek acceptance from the traveling public as well as the airports, airlines, and government. Individuals capable of passing a background check and willing to pay a yearly fee gain an advantage through personal authentication and verification.

- **Intelligent video analytics:** The technology uses security video to provide an advanced solution that fully automates video monitoring. It uses advanced software algorithms to automatically track and identify objects, analyze motion, and extract video intelligence. This also allows integration of smart sensors into video detection and alarming into airport security-monitoring systems.
- **Perimeter intrusion technologies:** Various perimeter intrusion technologies are being assessed that could be used as a layered security approach to detect non-passenger threats to airport infrastructure. These could include fence-mounted or volumetric devices such as microwaves, IR, fiber optics, radars, vibration sensors, etc. These could also include vehicle screening technologies for monitoring service vehicles entering airport operation areas.

There are numerous technologies and systems on the horizon that may become solutions in the transportation arena in the coming months and/or years. Among the areas under evaluation are thermal imaging technologies that examine temperature changes within an individual and suspicious behavior devices or recognition methods. Other areas include technologies related to shoe screening, aircraft–fuselage hardening, intelligent aircraft monitoring systems, and the overall integration and information dissemination of security systems.

All schemes and technologies, whether deployed or under development, are each just one facet of the "layered security" espoused by TSA and its parent, DHS. This approach was recommended as far back as the White House Commission on Aviation Safety and Security report of 1997. Today, the rings or layers start at the global levels, such as intelligence reports and alerts to the aircraft itself. The idea is to have a system of systems that is changing with moving targets and keeping would-be terrorists off balance.

JPDO–NextGen

The FAA was recreated by Congress through passage of a "reauthorization bill" that, in recent years, has covered periods of 4 years at a time. One such law passed in 2003, commonly referred to as Vision 100 Century of Aviation Reauthorization Act (P.L. 108–176), endorsed the concept of the "next generation air transportation system" (NextGen). It directed creation of the Joint Planning and Development Office (JPDO) that is to spearhead the long-term transformation of the national airspace system.

The law assigned the task of establishing and operating JPDO to seven departments and agencies that were to work together under the Senior Policy Committee chaired by the Secretary of Transportation. An early task of the fledgling office was to estimate the growth of civil aviation during a 20-year period. The studies that were made estimated that there could be a threefold growth in passengers by the year 2025. Some more conservative estimates suggest doubling the present levels.

The advantage of the NextGen process is that it includes representatives of the departments and agencies of the federal government and representatives of the industries that must coordinate many disparate activities. JPDO has established working groups, made up of both government and industry representatives, who are experts in nine specific fields. The working groups are responsible for development programs in nine distinct areas: aircraft, airport, air navigation services, environment, global harmonization, net-centric operations, safety, security, and weather.

The JPDO Security Working Group is planning various projects entailing computer modeling and demonstrations of emerging technologies through pilot projects in selected airports.

Indeed, if this nation moves from approximately 50,000 flights over a 24-hour period to 100,000 or 150,000, many of today's methods for receiving passengers at airports and moving them through check-in, security, and boarding will be inadequate.[13]

PAX 2.0

The TSA has an ongoing program dubbed PAX 2.0 in which a committee of TSA staffers and industry professionals are searching out products, procedures, and ideas that could bring about a change in the way passenger checkpoints are operated in the future.

Pre-Operational and Operational Testing

Pre-operational testing of security equipment simulates how it would be deployed and functional in an operational setting. Thus, assessments can be made in a controlled environment to determine if the equipment functions well enough to be deployed in an operational setting with minimal impact on airport operations. Operational testing involves deploying equipment in an airport environment and using stream-of-commerce people to evaluate the effectiveness of the equipment. Both pre-operational and operational testing assures that the equipment meet specified detection criteria and characterize its impacts on airport operations. Statistical methodology is used to assure that sufficient and appropriate data is collected to determine if operational parameters affect the equipment and/or process results.

National Safe Skies Alliance (Safe Skies), a nonprofit organization that was founded in October 1997, performs pre-operational and operational testing for TSA and other federal agencies. On one of its first assignments, conducted in 1998, Safe Skies was asked to operationally test four "operator assist" X-ray machines. The machines indicated areas in carry-on luggage that had the characteristics of an explosive compound. None of the four machines from different manufacturers was successful enough to be deployed and thus was never released.

In 2003, Safe Skies conducted pre-operational testing of the eXaminer 6000 machines manufactured by L-3 Communications. The machines were rushed out, and a number were purchased by the FAA and stored awaiting deployment. Safe Skies' tests provided data that allowed the manufacturer to identify operational issues; they then made the necessary modifications to allow the eXaminers to meet the FAA's deployment criteria.

In 2007, Safe Skies tested the CastScope backscatter X-ray device manufactured by Spectrum, Inc. The CastScope was developed to screen passengers' prosthetic devices, casts, and braces, as these devices present an opportunity for threat concealment. Safe Skies demonstrated and tested the X-rays at different airports and conventions throughout the United States to understand human factor issues and evaluate how the CastScope performs when screening sensitive areas underneath a prosthetic device, cast, or brace.

---

[13] http://www.jpdo.gov/library/In_Brief_2006.pdf.

The backscatter images provide new information for the screeners as they screen and evaluate areas that have been extremely difficult in the past.

In recent years, use of biometrics has come to the forefront in the area of access controls. Operational testing of facial recognition, fingerprint, and iris scanners has supported manufacturers in the evolution of these technologies. The enrollment quality and human factors are features that have been enhanced through operational testing results.

In addition to the evaluation of technologies, operational testing encompasses the evaluation and optimization of current equipment within screening areas. Continual improvement of processes, passenger throughput, and the optimization of staffing levels are critical areas for ensuring that screening is effective and efficient. Safe Skies provides updated information on process, throughput, nuisance–false alarms, true alarms, characterization of items, processing times, etc.

Conclusions

The rapid growth of aviation over just one century has changed the world in countless ways. Even Orville Wright could not have envisioned an aircraft like the Airbus 380 that can carry over 500 passengers or the Boeing 787 Dreamliner with its composite construction.

Today it is common for travelers to use scheduled airlines, charters, air taxi service, and their own corporate and personal aircraft to get to more destinations faster than anyone might have dreamed before the Wright brothers' flight at Kitty Hawk. However, with this explosion of growth and advancement, there has been a continuing series of malicious events by individuals, groups, and even some nations to impede or destroy this industry through terrorism. Therefore, the mitigation of these events has brought about the growth of security programs by government and corresponding growth of a specialized security industry. This growth has been punctuated by the responses to major tragic events like Lockerbie and 9/11.

With bigger aircraft, more traffic in the air and on the ground, and increased congestion in air terminals, security needs will be even more compelling in the future. For these reasons, all parties can expect the emergence of technologies, methods, and procedures that have yet to reach the drawing board.

Important Definitions and Terms

Secured area means a portion of an airport, specified in the airport security program, in which certain security measures specified in the Code of Federal Regulations (CFR 49, Part 1542) are carried out. This area is where aircraft operators and foreign air carriers that have a security program under Part 1544 or 1546 of this chapter enplane and deplane passengers and sort and load baggage and any adjacent areas that are not separated by adequate security measures.

Air operations area (AOA) means a portion of an airport, specified in the airport security program, in which security measures specified in this part are carried out. This area includes aircraft movement areas, aircraft parking areas, loading ramps, and safety areas, for use by aircraft regulated under Part 1544 or Part 1546 and any adjacent areas (such as general aviation areas) that are not separated by adequate security systems, measures, or procedures. This area does not include the secured area.

Security identification display area (SIDA) means a portion of an airport, specified in the airport security program, in which security measures specified in this part are carried out. This area includes the secured area and may include other areas of the airport.

Sterile area means a portion of an airport defined in the airport security program that provides passengers access to boarding aircraft and to which the access generally is controlled by TSA through the screening of persons and property.[14]

References

Department of Homeland Security. 2008. Accessed at: www.dhs.gov.

Federal Aviation Administration. 2008. Accessed at: http://www.faa.gov/about/mission/activities.

Federal Aviation Administration, Office of Civil Aviation Security. *Criminal acts against civil aviation*. 1995. Accessed January 9, 2008, at: http://handle.dtic.mil/100.2/ADA296813.

Institute for the Analysis of Global Security. 2003. Accessed atwww.iags.net.

Metropolitan Planning Council, Chicago. 2008. Accessed atwww.metroplanning.org.

National Commission on Terrorist Attacks upon the United States. *The 9/11 commission report*. 2004. Accessed February 12, 2008, at: http://govinfo.library.unt.edu/911/report/911Report.pdf.

Otto, Daniel. *Economic impact study*. Strategic Economics Group and Iowa State University, 2004.

Sparks Bureau of Business and Economic Resources–Center for Manpower Studies. University of Memphis, May 2005.

Transportation Security Administration. *Recommended security guidelines for airport planning, design, and construction*. 2006 Accessed March 7, 2008, at: http://www.tsa.gov/assets/pdf/airport_security_design_guidelines.pdf.

---

[14] http://www.jpdo.gov/library/In_Brief_2006.pdf.

7

Maritime Security

John C.W. Bennett, J.D., LL.M.

Objectives of This Chapter:

- Provide an overview of maritime security issues at sea and in port
- Discuss the pre-9/11 security regime
- Introduce both international and U.S. post-9/11 maritime security regulatory regimes
- Investigate specific security regulatory requirements
- Perform a preliminary evaluation of the new security regulatory regimes

> *"Freedom isn't free, but security shouldn't be dumb."*
> —Participant at USCG's Small Vessel Summit, June 2007

Introduction

Corsair, Filibuster, Marauder, Freebooter, Sea Wolf, Privateer, Buccaneer, Pirate! By whatever name, raiders of the sea have struck fear in the hearts of peaceable mariners for almost as long as there has been maritime trade. To the foregoing list of the mariner's enemies must now be added the terrorist for whom the maritime transportation system presents multifaceted opportunities for mayhem in the name of the cause. Until potential terrorist interest in port and port facilities—the shoreside portion of the maritime transportation system—became a concern, depredations in ports did not attract the same level of public attention, perhaps because crimes there were amalgamated into the general category of crimes on land and because theft could occur without as much overt threat to port personnel. Whatever the reasons, thefts and other crimes against ports have generated no list of exotic synonyms for thief and such crimes figure little in historical texts, while piracy has a storied history.

Historical Piracy

It is recorded that the Assyrian king Sennscherib tried, in 694 B.C., to suppress piracy in the Strait of Hormuz, a choke point for ships carrying valuable cargoes such as gold, silver, copper, spices, and silks between the Middle East and India. Emperor Trajan of Rome also tried in the first century A.D., as did Persian king Shapur in the fourth century (Cawthorne, 2003) but to no avail....

In the Mediterranean, pirates regularly fell upon Phoenician merchants during that era. During classical Greece, their successors hid among the Greek islands and had their activities recorded by Thucydides and Herodotus. Alexander the Great tried to eliminate Mediterranean piracy in 330 B.C., but it continued to flourish to the point that depredations on Mediterranean trade threatened Rome itself with starvation in 68 B.C. Pompey was authorized to stamp out piracy by a law passed in 67 B.C. With a fleet of 270 ships, he cleansed the Mediterranean after a campaign that culminated in a major battle off Anatolia, in which 10,000 pirates were killed, 400 ships were captured, and many more ships were destroyed. Thereafter, the relative calm, enforced by military and naval patrols, lasted until the rise of the Barbary pirates with safe havens in the North African possession of the Ottomans. This threat to maritime commerce, the subject of several wars by the young United States and suppression efforts by Britain's Royal Navy, was not ended until the French took over Algeria in 1830 (Ibid).

In Southeast Asia, piracy long preceded the arrival of the European colonial powers. Rivalries among mobile communities and the resulting political instability fostered piracy, while existing trade provided the opportunity for plunder. The Europeans initially arrived in the form of trading companies and were not successful in suppressing piracy. One factor impeding suppression was overlapping claims that effectively allowed pirate sanctuaries free from enforcement actions. It was only towards the end of the 19th century, after the European governments took over colonial efforts from the trading companies and agreed to demarcations of their respective possessions (removing concern of provoking conflict between themselves as an inhibitor of the application of naval forces) that a combination of aggressive anti-piracy patrols in high-traffic areas, extension of effective governance to formerly contested areas, and pardoning participants who agreed to settle down in sedentary occupations made significant inroads (Teitler, 2002).

But the most well-known piracy occurred primarily in the Caribbean Sea in the 17th and early 18th centuries, having been romanticized in novels and films such as *Treasure Island* and *Captain Blood*. Initially, Spanish shipments of gold and other treasures from the Americas had attracted the interest of other European monarchs, who, however, were averse to paying for their own expeditions to the New World and wary of risking war through use of their navies. The way around these problems was to charter private expeditions by issuing these privateers letters of marquee, in essence licenses to plunder, in exchange for a share of the loot. A period of peace in Europe resulted in general naval demobilizations that had the twin effects of reducing the risk of capture for those resorting to piracy without official sanction and increasing the available manpower pool of unemployed seamen. When the War of the Spanish Succession broke out in 1701, pirates began to sail as privateers under government license again. When the war ended in 1713, the new generation of privateers it had brought forth had the skills and inclination for piracy in peacetime. Britain's Piracy Act of 1721 extended the penalties for piracy to any-one trading with pirates, curtailing the market for pirates' ill-gotten gains. This, coupled with more active enforcement activities that resulted in the deaths of the most famous pirate leaders, sent Caribbean piracy into decline. However, the end of the Napoleonic Wars in 1812 once again saw an oversupply of unemployed privateers while the numbers of cargo ships increased. Efforts by the Royal Navy and anti-piracy patrols by the small U.S. Navy led to the effective end of Atlantic and Caribbean piracy by the early 1830s (Cawthorne, 2003). When most major maritime powers renounced use of privateers as instruments of state policy in the 1856 Declaration of Paris, the cycling of legal wartime privateers into peacetime piracy was broken.

Maritime Security Issues

Modern Piracy

Once thought laid to rest, maritime piracy is now back in the news. Additionally, in the post 9/11 world, the possibility that some pirates might be allied with terrorists or that terrorist organizations might engage in pirate-like activities to advance their causes has caused considerable concern in homeland security circles. By contrast, until the specter of terrorism arose, criminal activity in ports was little noted outside the maritime and supply-chain industries. But the possibility that terrorists could use ports to facilitate their activities elsewhere, or even as venues for incidents, has received considerable atten-tion in both professional circles and the popular press.

The International Maritime Bureau (IMB), an agency of the International Chamber of Commerce (ICC), produces weekly, quarterly, and annual reports of incidents of depre-dations against ships throughout the world. The IMB's reports are accessible through the ICC's Web site at http://www.icc-ccs.org. As will be seen, most of the reported depreda-tions committed against ships and their crews are not "piracy" *stricto sensu,* as that term has come to be defined in international law. This has led the International Chamber of Commerce's International Maritime Bureau to formulate, for its statistical reports, a defi-nition for "armed robbery against ships," which means "any unlawful act of violence or detention or any act of depredation, or threat thereof, other than an act of 'piracy,' directed against a ship or persons or property on board such ship, within a State's jurisdic-tion over such offenses." The Bureau's reports combine acts of armed robbery against ships with acts of piracy, with the intent of covering all depredations against ships, per-sons, and property afloat except for cases of petty theft not involving the use of weapons. The IMB reports track events defined as "an act of boarding any vessel with the intent to commit theft or any other crime and with the intent or capability to use force in the fur-therance of that act" (International Chamber of Commerce, International Maritime Bureau 2007, Burnett). For simplicity, this chapter will ordinarily use the term *piracy* as encompassing all crimes within the IMB definition and the term *pirate* as including per-sons attempting or perpetrating such crimes. Dr. Eklöf, however, warns that use of this definition blurs important distinctions between types of maritime depredations that have very different characteristics and require very different countermeasures (Eklöf).

Today, piracy occurs around the world wherever socio-economic conditions result in seaborne trade in the vicinity of nations that are unwilling or unable to engage in effective countermeasures. A number of factors have spurred the return of piracy in recent years:

- More targets presented by an ever-increasing global fleet carrying dramatically expanding world trade
- Reduced naval deployments that might have had a deterrent effect resulting from the ending of the cold war, as well as reduced support for third-world navies that could provide localized countermeasures
- Inadequacies in the legal regimes and limited jurisdiction affecting piracy and other armed robberies at sea, as well as intergovernmental jealousies and distrust
- Industry changes, such as shipboard automation, that result in smaller crews, with the twin effects of fewer off-watch crew members available to serve as sentinels in high-risk areas and fewer personnel for successful boarders to control during an attack or hijacking
- Rising fuel costs having driven speed reductions from, perhaps, 22 knots to a more economical 14 to 18 knots, making it easier for pirates to board ships under way
- Falling costs of arms and technology that allow pirates to obtain more sophisticated weaponry, better assault craft, and tracking equipment[1]

Raids on ships continue to this day in the seas adjacent to various parts of the continents of Africa, South America, and Asia, as well as in island groupings such as Indonesia and the Philippines. Over time the number of incidents in a region has risen or fallen in response to local political, economic, and security conditions. Thus, incidents have been declining in the Straits of Malacca since the littoral nations have increased their cooperation and their security patrols, while the absence of any effective government in Somalia has led to a dramatic increase in both incidents and their severity off the coast of that country.

Piracy in the modern era includes a range of activities from opportunistic muggings, to organized gang attacks, to vessel hijackings.

- **Low-level armed robbery:** Ordinarily aimed at cash in the master's safe, the crew's personal effects, and shipboard supplies and items not firmly fastened down. The ship is usually pierside or at anchor. The perpetrators may be armed with no more than knives or machetes, although firearms are increasingly common and the scale of violence employed is on the rise. Common in South America and Africa.
- **Medium-level armed robbery:** Although aimed at the same type of booty as low-level armed robbery, the take is generally higher as the perpetrators of an individual attack are more numerous. Attacks generally occur from small, high-speed craft at night while the target vessel is underway in or near territorial waters. In addition to the risk of personnel injury to the crew, this type of attack could result in a grounding or collision, with consequential environmental disaster and/or blockage

---

[1] As a result of the requirement for Automatic Identification Systems onboard vessels covered by Regulation 19 of the 1974 International Convention for the Safety of Life at Sea, accelerated for some ships in the 2002 Amendments, governments are able to track suspect vessels at greater range. The systems also have had the unintended consequence of providing a means for pirates and terrorists, if so inclined, also to track vessels of interest to them at longer range without revealing their intentions.

of a critical strait used for international navigation caused by the pirates detaining or locking up the crew including the bridge watch. Prevalent in the Straits of Malacca (Indonesia, Singapore, and Malaysia) and other Indonesian waters.

- **Major criminal hijack:** Organized crime syndicates, using heavily armed operatives, professional mariners, and, often, an existing crew member as an "inside man," take the ship while underway for a buyer, who has already been identified and who may have commissioned the hijacking. The crew is disposed of—if they are lucky, by being left in life rafts; or by murder, if they are not. The ship is diverted to a predetermined port, where the cargo is offloaded and sold. The ship becomes a so-called phantom ship, being reregistered under a new name, and its services are offered to shippers. Cargos thus obtained are similarly diverted, stolen, and sold. The ship is then reregistered to start the cycle again. This form of piracy has most often occurred in Southeast Asia. Another form of hijacking has recently become prominent in the Red Sea, where gangs operating out of the failed state of Somalia[2] have hijacked ships, including some engaged in United Nations' food relief efforts as far out as 200 miles from shore, and held them and their crews for ransom, with initial demands of $1 million or so.

An actual dollar figure for the cost of piracy is ephemeral. One common figure is $16 billion a year, including losses of ships and cargoes, as well as rising insurance rates (Burnett et al.). This figure has been challenged; however, applying that figure to the IMB's report of 445 pirate attacks in 2003 results in an average loss of $38 million per incident; this average figure is particularly improbable given the minor nature of most reported incidents (Dragonette). According to the Deputy Director of the IMB, Jayant Abhyankar,

> *I would say that in 95 percent of the cases it is to steal property from the crew or the ship—you know, their personal effects, such as cash—and then make a quick getaway. This is what we call 'maritime mugging.' In the remaining cases, it is actually to steal the cargo on board and sometimes the ship itself —'hijacking,' we call it. But that's by and far much less frequent than in the first case.*
> —quoted in Hill

To be sure, reported piracy incidents are outnumbered by unreported cases. IMB statistics address only commercial shipping. They do not include attacks on yachts, small indigenous coastal cargo boats, and fishing vessels. Most attacks on merchant ships go unreported because owners and captains are not willing to incur the opportunity costs of tying their vessels up for lengthy investigations. Additionally, owners do not wish to face increased insurance rates or the implication that their crews did not execute the appropriate anti-piracy procedures (Burnett). The head of the IMB's Piracy Reporting Center estimates that more than 50% of cases are not reported (Lundquist). Other estimates between 10 and 30% are reported (Burnett). But, even with a severe underreporting problem,

---

[2] This presents an excellent example of the effect that local government attitudes and effectiveness have on the rise and fall of piracy in an area. Piracy declined to virtually nothing while Somalia was ruled by an Islamic government that condemned and suppressed the practice. When those rulers were displaced by warlords, piracy rebounded in frequency and intensity.

it appears that the loss piracy imposes on maritime trade is minimal relative to the value of goods shipped annually. This low financial impact, which ignores the trauma sustained by victimized mariners, whether physical for the murdered and wounded or psychological for those who survived attacks, and the physical remoteness of attacks[3] account for the general lack, until recently, of attention devoted to the problem by governments and the shipping industry. The incidence of piracy is on the increase. In comparison to 2007, it is up 20% in 2008. Nigeria has the largest number of events, representing 21% of the total. Many of these involve killings and loss of cargo.

Maritime Terrorism

In contrast to the widespread incidence of piracy, terrorism successfully directed at the international maritime transportation system thus far been a relatively isolated occurrence.[4] Some authorities, however, perceive a convergence between terrorists and pirates (Luft and Korin). In this view, increasingly violent pirates terrorize ships' crews, while piracy provides a cover for terrorism, at least for training or practice missions, and terrorists could adopt pirates' methods and tactics to seize control of ships to be used for their nefarious purposes (Banlaoi). Attention is also drawn to the changing nature of pirate attacks, increasingly being carried out with "almost military precision using sophisticated weapons and techniques" (Acharya). The March 2003 attack on the chemical tanker *Dewi Madrim* in the Straits of Malacca is frequently cited as an example of the convergence (Aegis Defense Services Ltd.). Ten pirates armed with automatic weapons and machetes boarded the vessel and tied up the crew. They manuvered the ship for an hour before leaving with the ship's cash and documents, equipment, and crew's effects. The thought is that ordinary pirates would have no need or desire to prolong their exposure by spending an hour at the helm. On the other hand, an Al Qaeda operation involving the use of a large vessel would require personnel with some ship-handling experience. Thus, the suggestion arises that this incident was a terrorist training exercise. There are reports of other such incidents, with the "pirates" questioning crews on how to operate ships but showing little interest in how to dock them (Luft and Korin). This view is contested by an authority on piracy for the U.S. Office of Naval Intelligence who, writing unofficially, argues that the pirates onboard the *Dewi Madrim* did not act out of character for Southeast Asia pirates, that there is no actual evidence it was a terrorist attack, and that such a hijacking would be an inefficient way in which to acquire training that would, in any event, be valid only for identical ships in identical conditions (Dragonette). Such skepticism appears warranted as it is unlikely that economically motivated pirates would have any interest in the suicide missions favored by the adherents of Al Qaeda. While there is no conclusive evidence of the nexus of any group of pirates with any terrorist organization, dread that pirates may turn into accomplices or trainers of terrorists strongly impacts threat perceptions, particularly in Southeast Asia (Raman).

This does not mean, however, that the tactics and techniques of the pirate would not be of substantial use to a maritime terrorist for some types of operations.

---

[3] Compare the 24/7 television coverage of aircraft hijackings with the brief mentions, if any, on the inside pages of newspapers that constitute the normal coverage of maritime piracy incidents.

[4] The count would be higher if one included attacks on military targets, such as the *USS Cole* in 2000, or on fellow nationals, such as bombings of Philippine inter-island ferries by separatists. Additionally, a number of plots have been thwarted prior to an attack.

In considering the types of terror operations, Professor Nincic has proposed a fivefold classification of maritime terror incidents for purposes of teaching and analysis:

- Hijacking and hostage taking
- Attacks on ships
- The use of a ship as a "vector," transporting terrorist personnel and/or materials for use elsewhere
- The use of a ship as a weapon, including both weaponizing an inherently dangerous cargo and carrying a weapon of mass destruction
- Sinking or disabling a vessel to block a chokepoint/port

Nincic apparently includes in her "vector" category the use of the maritime transportation system to raise funds, such as shipping legitimate (or illegitimate) cargoes owned by a terrorist organization on unsuspecting vessels. It might, however, be more useful analytically to treat the use of the ocean transport to generate revenues as a separate category.

A report by the Maritime Transport Committee of the Organisation for Economic Co-operation and Development (OECD) does so by including it under one of its four "Terrorist Risk Factors from Shipping," namely "Risk Factor: Money," which is further elaborated as "Risk factors: financing/logistic support," and includes using shipping enterprises both to generate funds for terrorist activities and to launder illicit funds acquired from involvement in activities such as the drug trade. The OECD is less complete than Nincic in other areas, however. Its other risk factors are those arising from

- Cargo: Transporting–smuggling people and weapons (Nincic's category of ship as a "vector")
- Vessels: Using a ship as a weapon, using a ship to launch an attack, sinking a ship to disrupt infrastructure
- People: Attacking ships to provoke human casualties, using maritime identities to insert operatives

In fact, ships owned by terrorist organizations have been used to obtain revenues through carriage of third-party legitimate cargoes at market rates, although such ownership could surely also facilitate pure "vector" operations. In this regard, it is well documented that the Liberation Tigers of Tamil Eelam (LTTE) run a profitable network of freight forwarders and 10 to 12 bulk freighters flying flags of convenience that operate openly in the world shipping market. Approximately 95% of the cargoes carried are legitimate goods that generate revenues used to support the LTTE's ongoing conflict with the Sri Lankan government. The other 5% of the cargoes is thought to be arms, ammunition, and other supplies for the conflict. On occasion, the LTTE has carried weapons and ammunition on behalf of other terrorist groups, which have paid for the privilege (OECD). As an example of terrorist use of the maritime transportation system to ship cargoes, legitimate or not, Al Qaeda is thought to be involved in the sesame seed trade (Nincic), as well as in the opium trade. The Revolutionary Armed Forces of Colombia (FARC) and a plethora of Asian terror groups are known to benefit from their involvement in the drug trade (Raman: Burnett). Terrorists have also resorted to piracy to "earn" income for their operations. Such activities would provide opportunities for training and rehearsal for other types of maritime terror operations as well. The Free Aceh Movement repeatedly used the ransoms it obtained for hijacked ships to buy weapons for its struggle against the Indonesian government (Luft and Korin). Filipino and other Southeast Asian terrorist organizations have also realized the cash potential of piracy (OECD; Burnett).

Incidents of terrorist ship hijacking and hostage taking include the 1961 seizure of Portuguese ship *Santa Maria* by political dissidents protesting that country's dictatorial regime and the 1985 *Achille Lauro* incident. In the latter case, Palestinian gunmen seized the Italian cruise ship and held the passengers and crew hostage under threat of death, in support of their demands for the release of some 50 Palestinians held in Israeli jails. They killed one disabled U.S. citizen and dumped him and his wheelchair overboard.

Attacks on ships have included Al Qaeda's small boat suicide actions against the French tanker *Limburg* in 2002, which resulted in one death and substantial damage to the ship, and against the *USS Cole* in 2000, which caused multiple deaths and disabled the vessel, as well as the 2004 bombing of the Filipino *Superferry 14* by the Abu Sayyaf group that resulted in almost 100 deaths and sank the ship. Although the latter two attacks were not against elements of the international maritime transportation system, they clearly indicate the types of attacks that terrorist organizations can mount against any kind of ship, as well as the results that may be expected from them. Periodically, reports of Al Qaeda interest in scuba diving lessons raise the specter of underwater attacks on vessels and infrastructure by terrorist frogmen (Hosenball).

Actual "vector" operations have included stowaways with terrorist ties. On October 18, 2001, port workers in Gioia Tauro, Italy, discovered a stowaway in a cargo container equipped with a bed, heater, toilet facilities, and water. The man had a cell phone, a satellite phone, a laptop computer, as well as airport security passes and an airline mechanic's certificate valid for four major U.S. airports. Released on bond after arraignment, the stowaway disappeared. The case demonstrates how terrorists can use the complex structure of free-flowing global commerce to their advantage. The container lot was chartered by a Dutch company's Egyptian officer and the container was loaded in Port Said onto a German-owned, Antigua-and-Barbuda flagged container ship (capacity almost 3,000 TEUs)[*] chartered by a company of yet another nationality. The container was to be transshipped in Italy for Rotterdam, where it was to have been transshipped to its final destination in Canada. Had the stowaway not been trying to enlarge the container's ventilation holes while port workers were nearby, it is probable that the container would have arrived at its final destination without incident (OECD).

A similar but less fortunate case was reported in a speech by U.S. Customs and Border Protection Commissioner Robert Bonner, who revealed that, in 2004, "two suicide bombers entered the port of Asdod Israel hidden in a cargo container and killed innocent people" (quoted in Doane and DiRenzo) and that suspected terrorists using false papers posed as mariners. In addition, it is generally believed that the materials used in the 1998 bombings of the U.S. embassies in Tanzania and Kenya were imported by ship (Burnett; OECD). In January 2002, Israeli agents arrested the *Karine-A*, registered in Tonga, in the Red Sea. She was carrying 50 tons of weapons and explosives to Palestinians from Iran (Ibid.). "Gun running" by the LTTE was previously discussed in the context of the use of the maritime transportation system by terrorist organizations for revenue generation.

---

[*]TEU is an International Standards Organizational (ISO) measurement for T (twenty-four) E (equivalent) U (units) regarding shipping containers. All ports are compared on the basis of the number of TEUs they handle yearly, regarding freight shipping containers. The more TEUs the larger the port container capacity, and their relative importance.

Fortunately, thus far at least, the use of ships as weapons remains hypothetical, although the 9/11 attacks highlight the possibility. A terrorist attack on the United States using a weapon of mass destruction (WMD) is the government's greatest fear. Terrorists have tried to obtain WMDs (chemical, biological, radiological, and nuclear weapons). Use of the maritime transportation system is an attractive means of delivery. The Congressional Research Service (CRS) reports that fabrication of a Hiroshima-type and -sized (15 kilotons) nuclear bomb would be within the capabilities of some terror organizations. Detonation of such a weapon in a major seaport would have the following estimated results: 50,000 to 1 million people dead, buildings out to 1 or 2 miles destroyed, fires started elsewhere, fallout spread over many square miles. Direct property damage would be in the $50 to $500 billion range; losses dues to trade disruption from $100 to $200 billion; and indirect costs of $300 billion to $1.2 trillion (Medalia).

Not possessed of a nuclear bomb, terrorists might try a radiological dispersal device or so-called dirty bomb,[5] which could create large radioactive plumes, cause short- and long-term health and psychological effects, and produce significant economic impacts due to the necessity for shutting down port operations—including evacuations, business losses, property losses, and decontamination efforts before life as normal could resume. The economic impact to the region of such an attack on the ports of Long Beach and Los Angeles has been estimated, in a rigorous academic paper, at between $300 million and $252 billion for a 15-day to 1-year shutdown of the ports (Rosoff and von Winterfeldt). Even if such an attack did not bring international trade to a standstill, it could also inflict multibillion-dollar damage to the economies of the rest of the world. A 2002 war game based on a coordinated terrorist plot to detonate both radiological and conventional bombs produced an estimated $58 billion impact by the time things returned to normal on day 92.[6] By way of comparison, two studies placed the cost of the 10-day lockout in the October 2002 labor dispute at U.S. West Coast ports at $19.4 million and $466.9 million, respectively (OECD).

Apart from WMDs, terrorists could weaponize any number of hazardous materials that routinely are carried by ships in internal and international trade. Fertilizer grade ammonium nitrate, when adulterated with fuel oil, can become a powerful explosive and was used as such in Oklahoma City federal building and first World Trade Center attacks, among others. Several major shipboard explosions have been linked to calcium hypochlorite (a bleaching agent used in swimming pools).[7] Finally, in this category, great concern has been expressed over bulk shipments of volatile liquids, including gasoline, kerosene, liquefied petroleum gas (LPG), and liquefied natural gas (LNG), with the latter two generating significant publicity and extreme security precautions in U.S. ports. A number of incidents, including a direct hit by an Exocet missile on an LNG cargo tank,

---

[5] In January 2003, British officials found documents in the Afghan city of Herat indicating Al Qaeda had successfully built a small dirty bomb as well as possessed training manuals on using the device (Rosoff and von Winterfeldt).

[6] In the game's simulation, only a conventional bomb exploded, although two radiological bombs were found.

[7] In July 2006, during debate on a port security bill, U.S. Senator Patty Murray used the example of one such ship explosion to highlight the danger of a terrorist bomb arriving in a U.S. port in a cargo container. "About 90 containers were blown off the side of the ship, creating a debris field five miles long. ... I want you to imagine this same burning ship a few feet from our shores—in New York Harbor or Puget Sound, off the coast of Los Angeles or Charleston, Miami, Portland, Hampton Roads, the Delaware Bay or the Gulf of Mexico. Now imagine that we're not just dealing with a conventional explosion. We're dealing with a dirty bomb that has exploded on America's shores" (quoted in Kimery).

suggest that terrorists would face significant difficulty in rigging a successful explosion on one of these ships (OECD). Notwithstanding extensive fear mongering on the issue, the absence of hard factual evidence leaves the actual extent of the vulnerability in doubt (Hooper).

The final category in Nincic's analytical quintet, sinking or disabling a ship to block a chokepoint, or, in the OECD's formulation, to disrupt infrastructure, also remains hypothetical for the moment. The desire of Al Qaeda to undermine the economies of the West is clear.[8] Communiqués soon after 9/11 revealed that one of the major motivations for attack was to inflict major economic losses on the United States. Coordinated sinkings in the Houston Ship Canal and the Lower Mississippi Waterway would block the ability to receive about 50% of U.S. oil imports, causing significant short- and medium-term impacts on the U.S. economy and, therefore, the world economy. With the concentration of world trade through a relatively small number of large container ports, attacks on major ports such as Hong Kong, Singapore, or Rotterdam could have devastating effects on the global economy as well (OECD).

Ignored in their analytical structures by both Nincic and the OECD, except in so far as it rates a mention as the means of attack on one or two ships, is terrorist use of smaller vessels to go on the attack. Yet, a number of experts, including the current commandant of the U.S. Coast Guard, consider it a likely, if not the most probable, maritime threat scenario (Allen; Bennett). The "same WMD [that fits into a shipping container] will surely fit into a 60-foot boat. Such boats can easily enter our home waters at leisure" (Bennett). Small vessels of less than 300 tons have been the vessels of choice for numerous terrorist attacks; they have also been the backbone of the smuggling industry for people, drugs, and other items. Nonetheless they have received little attention in security regulations in the years since 9/11 (DiRenzo III).

The Pre-9/11 International Legal Regime Relevant to the Security of the Maritime Transportation System

Up until the latter half of the 20th century, the principal threat to the security of shipping, excluding the actions of opposing nations during periods of armed conflict, came from pirates by whatever name, who in the pursuit of personal wealth, used force to take control of a ship or steal its cargo and who normally operated from another vessel or vessels. Threats to the security of port facilities were handled by the ordinary criminal law of the country concerned dealing with offenses against persons and property. As a consequence, the international law that was developed to protect the security of the maritime transportation system was, for most if its development, extremely limited in its scope.

Although most use of privateers as instruments of state policy was renounced by most of the civilized world in the 1856 Declaration of Paris, it was not until 1958 that the international community got around to agreeing on a definition and legal regime for piracy. The traditional international law for suppressing piracy codified in the 1958 Geneva Convention on the High Seas (GCHS) was quite limited, covering only certain maritime

---

[8] After the attack on the *Limburg*, Osama bin Laden warned: "By God, the youths of God are preparing for you things that would fill your hearts with terror and target your economic lifeline until you stop your oppression and aggression" (quoted in Luft and Korin).

depredations that occurred on the high seas. This narrow view of piracy and the limited legal regime were carried forward intact in the 1982 U.N. Convention on the Law of the Sea (UNCLOS).[9] In these treaties, piracy is defined as

- Illegal acts of violence, detention, or depredation occurring on the high seas or other place outside the jurisdiction of any State[10]
- By persons attacking from another vessel or an aircraft
- For private ends of the attackers

Under both conventions, on the high seas, warships and other authorized government ships of any nation may seize a pirate ship or a ship taken by pirates, and under their control the state effecting the seizure may exercise criminal jurisdiction over the pirates (GCHS Arts. 19 and 21; UNCLOS Arts. 105 and 107).

Thus, although this legal regime grants universal jurisdiction over piracy, maritime piracy is limited to depredations (1) committed on the high seas (including, for this purpose, exclusive economic zones) (UNCLOS Art. 58[2]), (2) for private ends, (3) by persons from another ship or aircraft. Excluded from the definition are acts (1) occurring in a nation's territorial sea, or its internal or archipelagic waters; (2) motivated by a political purpose, rather than economic gain, although occurring on the high seas; or (3) undertaken by corrupt military units[11] (unless the crew has actually mutinied). In these cases, jurisdiction normally resides in, and enforcement is up to, (1) the coastal state in whose water the depredation occurs; (2) the flag state of the vessel on which the "political statement" has taken place; or (3) the country to whom the military belong.

---

[9] Article 15 of the 1958 Geneva Convention provides the following:

Piracy consists of any of the following acts:

(1) Any illegal acts of violence, detention or any act of depredation, committed for private ends by the crew or the passengers of a private ship or a private aircraft, and directed:
   a. On the high seas, against another ship or aircraft, or against persons or property on board such ship or aircraft;
   b. Against a ship, aircraft, persons or property in a place outside the jurisdiction of any State;
(2) Any act of voluntary participation in the operation of a ship or of an aircraft with knowledge of facts making it a pirate ship or aircraft;
(3) Any act of inciting or of intentionally facilitating an act described in sub-paragraph 1 or sub-paragraph 2 of this article

The definition of piracy in Article 101 of UNCLOS differs only in formatting and capitalization:

Piracy consists of any of the following acts:

(a) any illegal acts of violence or detention, or any act of depredation, committed for private ends by the crew or the passengers of a private ship or a private aircraft, and directed:
   i. on the high seas, against another ship or aircraft, or against persons or property on board such ship or aircraft;
   ii. against a ship, aircraft, persons or property in a place outside the jurisdiction of any State;
(b) any act of voluntary participation in the operation of a ship or of an aircraft with knowledge of facts making it a pirate ship or aircraft;
(c) any act of inciting or of intentionally facilitating an act described in subparagraph (a) or (b)

[10] Although the piracy articles of UNCLOS maintain the 1958 Convention's references to the "high seas" and "a place outside the jurisdiction of any State" while other provisions of UNCLOS codify newer maritime jurisdictions for states, the balancing of coastal state economic rights and the international community's navigational rights contained in Articles 58 and 86 of UNCLOS results in the continued application of the traditional piracy regime in the Exclusive Economic Zone extending out to as much as 200 nautical miles from a coastal state.

[11] There have been reports of piracy committed by units of the Indonesian military (Burnett).

The shortcomings of this legal regime are readily apparent in everyday practice. Most depredations against merchant shipping reported (under the rubric "Acts of Piracy and Armed Robbery Against Ships") by the International Maritime Bureau occur in sovereign waters, usually in parts of the world where there is little capability or often little desire to enforce national laws.[12] Other weaknesses in the traditional piracy regime were highlighted by the *Achille Lauro* incident in 1985.[13] Although the hijackers did commit illegal acts of violence, detention, and depredation on the high seas, they were not crew or passengers of another vessel from which they launched their attack. Instead, they had embarked on the *Achille Lauro* as passengers. More importantly, their motive was not personal gain ("for private ends"), but rather to call attention to their grievances against Israel, achieve freedom for some 50 Palestinians jailed in Israel, and otherwise advance their political goals. With the rise of terrorism as a global phenomenon, the international legal regime of piracy, which does not address violence practiced for political purposes and which only applies on the high seas, provides insufficient protection for the maritime transportation system from the security threats of the 21st century.

The *Achille Lauro* hijacking, however, led to calls for development of new, and more general, international law to prevent unlawful acts that threatened the safety and security of ships and their passengers and crews. One response was the negotiation of the 1988 Convention for the Suppression of Unlawful Acts against the Safety of Maritime Navigation (the 1988 SUA Convention).[14]

The 1988 SUA Convention established the following as crimes:

- Seizure of, or exercise of control over, a ship by force or intimidation
- Acts of violence against persons onboard
- Damaging a ship or its cargo
- Placing onboard a ship a device or substance that can endanger safe navigation of the ship
- Destroying or seriously damaging or interfering with the operation of maritime navigational facilities
- Endangering navigation of a ship by knowingly communicating false information
- Threatening a natural or juridical person to commit an offence or to do or refrain from doing an act, if the threat is "likely to endanger the safe navigation of a ship"

Pointedly, the 1988 SUA Convention did not limit these crimes to any particular locations—the ship may be at anchor or under way; she may be in a nation's internal waters, archipelagic waters, or territorial sea, or any other part of the oceans.

The 1988 SUA Convention obligated its states parties to criminalize these activities under their laws and establish penalties appropriate to the seriousness of the offenses. Each state party may prosecute any alleged offender, regardless of nationality or location

---

[12] According to the IMB's director, Captain Pottengal Mukundan, "Piracy takes place in countries where you have economic problems and a weak maritime law-enforcement infrastructure ... countries like Somalia, where you have no national government and no law enforcement on a national basis" (quoted in Sekulich).

[13] The 1961 seizure of the *Santa Maria* had raised the same issues but far less public outcry.

[14] This negotiation also produced a protocol relating to unlawful acts against fixed platforms on the continental shelf.

of the event, who is found in its territory. If a state party is unwilling or unable to prosecute an alleged offender, it is obliged to transfer the person to another state party with jurisdiction that is willing to prosecute.

Thus, the 1988 SUA Convention attempted to broaden traditional international law: (1) by including any unlawful act against a ship or her crew or passengers, without regard to the juridical nature of the waters in which the ship was located or from whence the attack was launched; and (2) by providing universal jurisdiction for prosecution of alleged offenders wherever they might be found.

The Impact of 9/11 on the International Legal Regime

The use of hijacked aircraft as weapons themselves in the execution of the 9/11 plot prompted the realization that ships could similarly be used as "instruments or theatres of terrorism" (Mensah, p. 22). A resolution of the International Maritime Organization Assembly called for a review of existing international legal and technical measures with the intent to produce revisions to prevent terrorism against ships and improve security onboard and ashore. This led to revisions to the 1988 SUA Convention negotiated through the IMO's Legal Committee, while the organization's Maritime Safety Committee produced amendments to the 1974 International Convention for the Safety of Life at Sea (SOLAS), including the International Ship and Port Facility Security Code (ISPS Code).

The 2005 Protocol to the SUA Convention

Reconsideration of the 1988 SUA Convention produced the Protocol of 2005 to the Convention that amends it, as to states party to the Protocol, in a number of significant ways:

- Establishes a variety of new offences related to the intentional and unlawful carriage, use, or operation of biological, chemical, and nuclear (BCN) weapons aboard vessels or in the territorial waters of signatory states, with the intention to intimidate a population, or compel a government or international organization to carry out or abstain from any act against its wishes. The new offences extend to the use or carriage of radioactive or fissile material knowingly intended to be used in nuclear explosive activity, and to any equipment, materials, or software or related technology that significantly contributes to the design, manufacture, or delivery of a BCN weapon. It is also an offense to carry aboard persons known to have committed an offense by the SUA Convention, or any of nine treaties listed in the Annex.[15]
- Makes it an offense to unlawfully and intentionally injure or kill any person in connection with the commission of any of the offenses in the Convention; to attempt to commit an offense; to participate as an accomplice; to organize or direct others to commit an offense; or to contribute to the commissioning of an offense and requires states parties to take necessary measures to enable a legal entity to be

---

[15] The protocol exempts the transportation of any of these nuclear materials if material is transported to or from the territory of, or is otherwise transported under the control of, a state party to the Treaty on the Non-Proliferation of Nuclear Weapons.

made liable and to face sanctions when a person responsible for management of control of that legal entity has, in that capacity, committed an offense under the Convention.

- Covers cooperation and procedures to be followed if a state party desires to board a ship flying the flag of a state party when the requesting party has reasonable grounds to suspect that the ship or a person onboard the ship is, has been, or is about to be involved in, the commission of an offense under the Convention. The authorization and cooperation of the flag state, which may be declared in advance, is required before such a boarding. When a state party takes measures against a ship, including boarding, it must (1) not endanger the safety of life at sea; (2) ensure that all persons onboard are treated in a manner that preserves human dignity and in keeping with human rights law; (3) take due account of safety and security of the ship and its cargo; (4) ensure that measures taken are environmentally sound; and (5) take reasonable efforts to avoid a ship being unduly detained or delayed.
- Provides that none of the offences should be considered a political offence for the purposes of extradition. The Protocol requires states parties to afford one another assistance in connection with criminal proceedings brought in respect of the offences and establishes the conditions under which a person who is being detained or is serving a sentence in the territory of one state party may be transferred to another state party for purposes of identification, testimony, or otherwise providing assistance in obtaining evidence for the investigation or prosecution of offenses.

Even as amended to deal specifically with terrorists, the SUA Convention has severe limitations in combating terrorism. First, the amendments to the SUA Convention adopted by the 2005 Protocol are not yet in force. Even if the Protocol enters into force following ratification by a mere 12 states parties, many of its provisions would be effective only between the states parties to the Protocol. This limits their application on a global scale. Second, even if it were widely adopted, it only creates a structure within which its states parties can deal, *after the fact,* with those who commit or attempt to commit covered crimes of terrorism. Other than any possible deterrent effect[16] it does nothing to prevent an act of terrorism. This underlines the significance of the 2002 Amendments to SOLAS, and the ISPS Code in particular. They impose affirmative obligations not just upon governments, but also on many of the actors in the maritime transportation system. These include obligations on non-state actors that include taking measures that harden targets in the maritime transportation system and that could potentially prevent some terrorist acts.

Amendments to the 1974 Safety of Life at Sea Convention

In December 2002, the Diplomatic Conference on Maritime Security was held in London to formally adopt the new provisions for the International Convention for the Safety of Life at Sea, 1974 (SOLAS) that had been previously negotiated with incredible speed for an international treaty of such complexity. Sitting as the Conference of Contracting

---

[16] It is unlikely that the threat of legal prosecution, even if credible, would deter a motivated terrorist willing, or hoping, to die for his or her cause.

Governments to the International Convention for the Safety of Life at Sea, 1974, the representatives adopted Conference resolution 1, which contained several safety- and security-related amendments to existing provisions of the Annex to that Convention for vessels to which the various SOLAS regulations apply:

- Acceleration of the required installation and continuous operation of automatic identification systems (Regulation V/19)
- Permanent marking of the ship's IMO identification number on the exterior and interior of the ship by methods ensuring that the number is not easily expunged (Regulation XI-1/3)
- Issuance by the flag state of a vessel of a Continuous Synopsis record detailing defined information pertinent to the vessel's nationality, ownership, etc., which document is to be updated, including through changes of the vessel's flag, and kept onboard, available for inspection at all times (Regulation XI-1/5)

The principal intent of these amendments was to make it more difficult to convert hijacked ships to so-called phantom ships.

In addition, Conference resolution 1 created an entirely new regulatory regime in the form of SOLAS Chapter XI-2, entitled "Special measures to enhance maritime security," which incorporates 13 regulations.

SOLAS Chapter XI-2

The 13 regulations of SOLAS Chapter XI-2 have as their subjects as follows:

Regulation	Subject
XI-2/1	Definitions.
XI-2/2	Ships and port facilities to which the regulatory regime applies.
XI-2/3	Obligations of governments to set security levels for, and to provide security-level information to, ships flying their flags, to port facilities in their territories, and to foreign-flag shipping entering or in their ports.
XI-2/4	Requirements for companies and ships to comply with Chapter XI-2 and the ISPS Code, with ships:

- Having their compliance verified and certified.
- Complying with the requirements of the security level set by the port state, if higher than their flag state's.
- Responding promptly to a change to a higher security level.
- Notifying the appropriate authority if not in compliance with Chapter and Code or if unable to comply with the requirements for an applicable security level.

XI-2/5	The specific responsibility of companies to provide onboard information for port states concerning who is responsible for hiring crew and other persons working onboard and who is responsible for directing the employment of the ship, including any parties to any charter parties.
XI-2/6	Ship security alert systems.

XI-2/7	Actions by governments concerning threats to ships:

- Provide security-level information to ships in their territorial seas and those having communicated their intent to enter territorial seas.
- Provide a point of contact for such ships to obtain advice or assistance or to report security concerns.
- Provide information to ships and their governments concerning security measures, if a risk of attack has been identified.

XI-2/8	Shipmasters' overriding discretion to act as necessary to maintain the safety and security of the ship, including opting for safety when security and safety requirements conflict.
XI-2/9	Control and compliance measures allowing states to verify ISPS Code compliance, and correct or eliminate noncompliance, by ships in, or intending to enter, their ports.
XI-2/10	Compliance by port facilities, including port facility security assessments, port facility security plans.
XI-2/11	Alternative security agreements between Governments covering short international voyages on fixed routes between port facilities in their territories.
XI-2/12	Equivalent security arrangements under which governments may allow particular ships or port facilities or groups of ships or port facilities to implement others' security measures equivalent to, but no less effective than, those in Chapter XI-2 or the ISPS Code.
XI-2/13	Communication by governments to the International Maritime Organization (IMO), for dissemination to other governments, of information about

- Their maritime security authorities and points of contact.
- Locations of port facilities with approved security plans.
- Particulars on any recognized security organization to whom any authority has been delegated.
- Any alternative security agreements concluded.
- Any equivalent security arrangements authorized.

Most of these 13 regulations are fleshed out in the International Code for the Security of Ships and of Port Facilities, more commonly referred to as the International Ship and Port Facility Security Code or "ISPS Code," which was adopted by Conference resolution 2.

The ISPS (International Ship and Port Facility Security) Code The ISPS Code contains two sections, Part A and Part B. Part B was originally intended to supply advisory guidance, in contrast to the mandatory provisions contained in Part A. Thus, the phraseology in Regulation XI-2/4 "shall comply with the relevant requirements of . . . Part A, taking into account the guidance given in Part B." The "taking into account the guidance given in Part B" language also appears in other Chapter XI-2 regulations and throughout Part A. In general, Part A uses *shall* whereas Part B uses *should*, although both occasionally use *may*. Early on, however, the mandatory–advisory scheme was circumvented. The U.S. Coast Guard announced that it would require compliance with the provisions of Part B by all U.S. ships and by foreign flag ships calling at U.S. ports. The major international classification societies, which often audit compliance, then adopted the same position. The result is that for almost every ship to which the Code applies, Part B is mandatory and *should* must be read as *shall*.

The ISPS Code was designed (Section A/1.2)

- To establish a framework for international cooperation between governments, their agencies, and the shipping and port industries in order to detect security threats and take preventive measures against security incidents affecting ships and port facilities used in international trade
- To establish roles and responsibilities for the various players in the international maritime transportation system
- To ensure early and efficient collection and exchange of security-related information
- To provide a methodology for security assessments, in order to have ship security plans and port facility security plans, including procedures to react to changing security levels
- To ensure confidence that adequate and proportionate security measures are in place

To achieve those objectives, the Code establishes a number of functional requirements, including (1) collecting, evaluating, and exchanging information concerning security threats; (2) maintaining communications protocols for ships and port facilities; (3) preventing (a) unauthorized access to ships, facilities, and the restricted areas thereof and (b) the introduction of unauthorized weapons, incendiary devices, and explosives into ships or facilities; (4) raising the alarm for security threats and incidents; (5) developing, based on security assessments, ship and facility security plans that provide for the application of security measures designed to protect the ship or facility, persons, and property from the risks of a security incident; and (6) training, drilling, and exercising[17] for familiarity with those plans and their procedures (Section A/1.3).

CASE STUDY

SOLAS Regulation and ISPS Code Citation
The SOLAS regulations are referenced by their chapter, number, and subnumber, as for example, "Regulation XI-2/4," or "Regulation XI-2/6.1.1." The divisions of ISPS Code Part A are referred to as *sections* and written, for example, as "Section A/8.2" or "Section A/8.4.1." The provisions of Part B of the Code are indicated as, for example "Paragraph 4.5.1."

---

[17] The Code details the competencies required of the various categories of personnel, discussed *infra*. Sections A/13.4 and A/18.3 require periodic drills to ensure the effective implementation of ship and port facility (respectively) security plans. Paragraphs 13.6 and 18.5 require that drills test individual elements of the plan and be conducted at least every 3 months. Paragraph 13.6 further mandates a drill within 1 week after a personnel change results in more than 25% of a ship's crew having no prior participation in a drill on that vessel. Sections A/13.5 and 18.4 require the Company Security Officer and the Port Facility Security Officer, respectively, to participate in periodic exercises. Under Paragraphs 13.7 and 18.6, such exercises, which may be full-scale or live, tabletop simulation or seminar, or combined with other exercises, are to test communications, coordination, resource availability, and response. Exercises are to be conducted at least once a calendar year, but not more than 18 months apart. Curiously, the Code does not formally assign the responsibility for scheduling drills or exercises to any particular player.

Regulation XI-2/2 and Section A/3 set forth the applicability of Chapter XI-2 and the Code. They apply to the following:

- **Ships** "engaged in international voyages" that are passenger ships, including high-speed passenger craft; cargo ships, including high-speed craft, of 500 gross tons or more; and mobile offshore drilling units[18]
- **Port facilities** serving such ships in international voyages[19]

SOLAS Chapter XI-2 and the ISPS Code then (1) establish, as to such ships and port facilities, the obligations and responsibilities of the various stakeholders in the international maritime transportation system, from the halls of government to the deck plates, and (2) provide the means of organizing the security effort and the mechanisms to ensure compliance with requirements.

Obligations and Responsibilities

Contracting Governments

Under Chapter XI-2 and the Code, governments have a wide variety of rights and responsibilities.[20] In addition to those previously mentioned in connection with SOLAS Chapter XI-2 regulations, governments are responsible for issuing and renewing International Ship Security Certificates (ISSCs)[21] to ships flying their flags, after verification of compliance with the Code (Sections A/4.4, A/9.1, A/15.1, A/16.2, A/19.21). Governments

---

[18] Warships, naval auxiliaries, and other vessels owned or operated by a government and used only for government noncommercial service are specifically excluded (Regulation XI-2.3).

[19] It is up to the government to decide how much of SOLAS Chapter XI-2 and of the ISPS Code to apply to port facilities that are used primarily by ships *not* engaged in international voyages, but that are required occasionally (Regulation XI-2.2). Chapter XI-2 and the Code are not intended to apply to port facilities used primarily for military purposes (Paragraph 3.4).

[20] A Government may delegate the execution of many of its responsibilities a "recognized security organization," or RSO, including approval of *ship* security plans and amendments thereto, verification and certification, on behalf of the government, of ships' compliance with SOLAS Chapter XI-2 and the ISPS Code, and performance of port facility assessments (Section A/4.3, Paragraph 4.3). Governments may not, however, delegate their responsibilities to set security levels, to approve port facility security assessments and security plans, to exercise Port State control and compliance measures, or to establish the requirements for use of Declarations of Security (Section A/4.3).

To be designated as an RSO, an organization should be able to demonstrate, *inter alia*: relevant security expertise; knowledge of ship design, construction, and operations and/or port design, construction, and operations; capability to assess likely security risks and their minimization; knowledge the international and national maritime security regimes; and knowledge of current security threats and patterns; and knowledge of a variety of technical security issues (Paragraph 4.5).

[21] Section A/19.1 requires the flag state to carry out initial and renewal verifications of their ships' "security system and any associated security equipment" to ensure that they fully comply with SOLAS Chapter XI-2, the ISPS Code, and the ships' approved security plans, although the conduct of the verification survey may be delegated to an RSO. The flag state may specify the intervals for renewal (not greater than 5 years) verifications leading to a new ISSC. At least one intermediate verification must be conducted, but if only one is carried out, it must be between the second and third anniversaries of the ISSC. Intermediate verifications are endorsed on the existing ISSC. Issuance on an ISSC and intermediate endorsement thereof may be delegated to an RSO acting on behalf of the flag state. An ISSC ceases to be valid if a verification, including an intermediate verification, has not been completed within the required period, if a new company takes over operation of the ship, or if the ship changes its flag (the new flag state may issue an Interim ISSC valid for up to 6 months, provided a new SSA has been completed, a new SSP has been submitted for approval, verification has been arranged, and onboard security matters are in keeping with the Code).

determine the circumstances of ship–port interface or ship-to-ship interface that would require their ships and port facilities to enter into Declarations of Security.[22] Governments set the minimum period for which security-related records must be kept (Section A/10.1).

Companies

In addition to the obligations of shipping companies previously mentioned in connection with SOLAS Regulations XI-2/4 and XI-2/5, they must designate at least one Company Security Officer (CSO)[23] at the company level (section A/11) and a Ship Security Officer (SSO)[24] for each ship (section A/12). The company must ensure that the CSO and the SSO are given the necessary support to carry out their duties. And the company must ensure that the ship security plan clearly emphasize that the ship's master has the overriding authority to make decisions for the safety and security of the ship and to request assistance from the company or any government (Section A/6).

Ships

In addition to the obligations imposed on ships by SOLAS Regulation XI-2/4, they must carry a ship's security assessment (SSA) and a ship's security plan (SSP), with the latter being approved by, or on behalf of, its government (Sections A/8, A/9). A ship must also carry an International Ship Security Certificate issued by, or on behalf of, its government after verification that the ship is in compliance with the Code (Section A/19). Ships must maintain onboard, for the period designated by their governments, specified security-related records (Section A/10).

Port Facilities

Port facilities to which the Code applies must comply with it. In particular, each facility must hold a port facility security plan (PFSP), approved by its government, that is based on a port facility security assessment (PFSA) (Sections A/15, A/16). A facility must act on the security levels set by its government by implementing the protective measures

---

[22] A Declaration of Security (DoS) is an agreement "between a ship and either a port facility or another ship with which it interfaces." A DoS addresses all shared security concerns and specifies security measures each party will use during the period of interaction (Regulation XI-1.15) (Section A/5).

[23] The CSO should have knowledge of, and receive training in, appropriate areas related to security: administration; international and national regulatory regimes; ship and port operations and security; security techniques, methods, and equipment; recognition of threats; security training; assessment; and the like (see Paragraph 13.1).

[24] The SSO should have knowledge of, and receive training in, appropriate areas related to the same subjects as apply to the CSO. In addition, the SSO should be familiar with, or trained in, the ship's layout; the ship security plan; crowd management and control; operation of security equipment and systems; and testing, calibration, and at-sea maintenance of security equipment and systems (see Paragraphs 13.1, 13.2). Amendments to the 1978 Convention on Standards of Training, Certification and Watchkeeping that came into force January 1, 2008, have incorporated these requirements, necessitating that SSOs carry proof of their qualifications. A transitional implementation period (through July 1, 2009) will allow amendment of the regulations to require SSO training endorsements on mariners' licenses. Until U.S. regulations are amended, proof for U.S. mariners may either be by a course provider's course completion certificate (for training) or company certification (in the case of equivalent job experience) (MarEx Newsletter). The United Kingdom had, from the start, required all SSOs of UK-registered vessels to take training approved by the Maritime and Coastguard Agency. The UK has issued a new SSO training certificate for training received after January 1, 2008, while allowing the use of the previous certificate for approved training received before that date (Maritime and Coastguard Agency).

specified in the PFSP that are appropriate to the level, as well as any additional measures required by the government. A port facility security officer (PFSO)[25] must be designated (Section A/17.1).

Company Security Officers

The Company Security Officer (CSO) is responsible for ensuring that an SSA is carried out and an SSP is properly completed, maintained, and implemented for each company ship to which the ISPS Code is applicable (Paragraphs 8.1; 9.1; 11.2.1). Together with Ship Security Officers (SSOs), the CSO is responsible for procedures to assess the continuing effectiveness of the SSP and to prepare amendments to it after its approval (Paragraphs 9.5, 11.2.4). The CSO is to ensure the effective coordination and implementation of SSPs by participating in required exercises of the SSP (Section A/13.5). Other duties include (Paragraph 11.2)

- Advising ships on the level of likely threats
- Arranging for internal audits and reviews of security activities and for verifications of ships' compliance for ISSC purposes
- Ensuring that security problems identified through audits, review, inspections, and verifications are promptly dealt with
- Enhancing security awareness and vigilance
- Ensuring effective communications and cooperation between SSOs and relevant PFSOs
- Ensuring consistency between security and safety requirements
- Ensuring that any fleet security plans and sister-ship plans accurately reflect ship-specific information and that any alternative or equivalent arrangements approved for any of the company's vessels are implemented and maintained

Masters

The master retains the overriding responsibility for making decisions necessary, in his or her professional judgment, to maintain the safety and the security of the ship, including refusal to take on board persons, their effects, or cargo (Regulation XI-8.1). In conflicts between safety and security requirements, the master shall adhere to the requirements necessary to maintain the safety of the ship and may implement temporary security measures (which should be commensurate with the security level to the maximum extent possible). In such cases, he or she must inform the government whose flag the ship flies and, if applicable, the government in whose port the ship is or intends to be (Regulation XI-2/8.2).

Ship Security Officers and Port Facility Security Officers

Ship Security Officers (SSOs) and Port Facility Security Officers (PFSOs) have generally similar responsibilities applicable *mutatis mutandis* to their respective units, including the following (Sections A/12.2; A/17.2):

- Regular security inspections
- Implementation of the security plan, including amendments

[25] The PFSO should have knowledge of, and receive training in, appropriate areas related to security: administration; international and national regulatory regimes; ship and port operations and security; security techniques, methods, and equipment; recognition of threats; security training; assessment; and the like (see Paragraph 18.1).

- Coordination of security aspects of cargo handling with his or her counterpart
- Proposing modifications to the plan
- Enhancing security awareness and vigilance of personnel
- Ensuring adequate security training for personnel
- Reporting all security incidents
- Coordinating implementation of the security plan with CSOs and PFSOs
- Proper operation, testing, calibration, and maintenance of any security equipment

When a Declaration of Security has been requested, the PFSO and SSO should discuss appropriate security arrangements (Paragraph 5.2.1). Both types of security officers are expected to participate in required periodic drills and exercises (Sections A/13.4–.5; A/18.3–.4; Paragraphs 13.6–.7; 18.4–.6). A PFSO has the additional responsibilities of assisting an SSO, when requested, with the identification of persons seeking to board the ship and with other matters, as well as those of reporting to a competent authority when advised that a ship is at a higher security level than the port facility or that the ship is encountering difficulties in complying with the regime or in implementing measures or procedures detailed in the SSP (Sections A/14.5; A/14.6).

Ship and Port Facility Personnel with Specific Security Duties

Personnel with security duties are required to understand their security duties, as described in the security plan for their ship or facility, and have sufficient knowledge and receive training on an appropriate selection of security functions and methods (Sections A/13.3; A/18.2). They are expected to participate in required periodic drills (Sections A/13.4–.5; A/18.3–.4; Paragraphs 13.5–.7; 18.4–.6).

Other Ship and Port Facility Personnel

In order to ensure effective implementation of a security plan, all personnel must understand their roles in the plan at all security levels and certain baseline provisions of the plan (Paragraphs 13.4; 18.3). Periodic drills, which should each focus on a specific aspect of the plan (such as bomb threat procedures), are an important means of achieving the required comprehension.

Means and Mechanisms

The principal means through which the security efforts of companies, ships, port facilities, and their personnel are organized for delivery are the Ship Security Plan (SSP) and the Port Facility Security Plan (PFSP). Although governments are responsible for ensuring the compliance of their ports and flag shipping, the extension of the Port State Control system into the maritime security arena provides an important means of ensuring compliance on the part of visiting ships.

Security Plans

Ship Security Plans (SSPs) and Port Facility Security Plans (PFSPs) are fundamentally similar, with allowance for differences between ships and port facilities. Security Plans are to be based on underlying Security Assessments[26] and are to provide for actions at each of the three security levels. They require governmental approval (which may be delegated in the case of SSPs). They are to be in the working language of the ship or port facility, with the proviso that, if the working language of a ship is not English, French, or Spanish, the ship carry a translation into one of those languages. Plans must address the following:

- Prevention of introduction of unauthorized weapons and dangerous substances or devices
- Unauthorized access and restricted areas
- Response to security threats or breaches of security, including evacuation
- Response to governmental security instructions
- Organization and performance of security duties, including identification of and contact information for relevant security officers
- Procedures for auditing security activities
- Training, drills, and exercises associated with the plan
- Interfacing with port facility or ship security activities
- Procedures for periodic review of the plan and for updating it
- Procedures for reporting security incidents
- Inspection, testing, calibration, and maintenance of security equipment

In addition, SSPs must address procedures and guidance regarding the Ship's Security Alert System (SSAS). PFSPs must provide for the effective security of cargo and cargo-handling equipment at the facility; response to SSAS activation by a ship at the facility; ensuring security of information in the plan; and facilitating shore leave for ships' crews, ship personnel changes, and ship visitor access (Sections A/9, A/16). In Part B, the Code provides additional guidance addressing security measures that could be incorporated in security plans for action at each security level with respect to access, restricted areas, cargo handling, delivery of ship's stores, handling unaccompanied baggage, and monitoring security (Paragraphs 9.9 through 9.49, 16.10 through 16.54).

Port State Control in the Security Arena

Although the flag state and the state in which a port facility is located have the primary responsibility through their inspection and approval powers for ensuring the compliance of their ships and ports, Regulation XI-2/9, as amplified in Paragraphs 4.29 through 4.46, provides an additional means for ensuring ISPS Code compliance by ships. Port state officials may examine mandatory security-related records maintained onboard to verify that that the provisions of the ship's security plan are being implemented (Paragraph 10.1). This is true even though the plan itself is normally not subject to their inspection[27] (Section A/9.8).

---

[26] Ship Security Assessments (SSAs) and Port Facility Security Assessments (PFSAs) are written, risk-based evaluations of existing security measures and procedures, key operations that require protection, all possible threats to those operations and their likelihood, and weaknesses, including human factors, in infrastructure, policies, and procedures. They include an on-scene security survey (Sections A/8.4, A/15.5; Paragraphs 8, 15).

[27] If there are "clear grounds" to believe the ship is noncompliant and "the only means" to verify or rectify the noncompliance is to review the relevant portions of the plan, limited access to the specific sections of the plan related to the noncompliance is allowed, but only with the consent of the ship's master or its government. Even here, a number of plan sections are considered confidential and not subject to Port State inspection absent agreement of the flag state (Section A/9.8.1).

As to foreign ships already in port, the port state's control is initially limited to verifying that the ship has a valid ISSC, "which if valid shall be accepted, unless there are clear grounds for believing the ship is not in compliance" (Regulation XI-2/9.1.1). Under Regulation XI-2/9.1.2, if there are such "clear grounds" or if no valid ISSC is produced, the port state "shall" impose one or more of the following control measures, set forth in Regulation XI-2/9.1.3:

- Inspection of the ship
- Delaying the ship
- Detention of the ship
- Restriction of operations, including movement within the port
- Expulsion of the ship from the port

Lesser administrative or corrective measures may be applied in lieu of or in addition to the foregoing control measures.

In the case of a foreign ship intending to enter one of its ports, the port state government is permitted by Regulation XI-2/9.2 to require certain security-related information, including, *inter alia*, the validity of the ship's ISSC, the ship's security level, the levels it operated in its past 10 port calls, and other "practical security-related information," such as any use of Declarations of Security and security measures taken while in ship-to-ship activity with a vessel not subject to the Code, as well as crew and passenger lists and cargo descriptions. If, notwithstanding the information provided and follow-up communication, the authorities of the port state have "clear grounds" for believing the ship is noncompliant, the port state may take steps in relation to the ship, including the following:

- Requiring rectification of the noncompliance
- Requiring the ship to proceed to a specified location in the port state's sovereign waters
- Inspection of the ship, if in the territorial sea of the port state
- Denial of entry into port

When informed of the steps the port state intends to take, the master of the ship may withdraw the intention to enter port, in which case, the port state may not apply its intended steps.

In either circumstance, the control and compliance measures or steps taken must be "proportionate" and shall be imposed only until the noncompliance has been corrected (Regulation XI-2/3.4). Denial of entry or expulsion may be imposed only where there are "clear grounds" to believe that the ship "poses an immediate threat to the security or safety of persons, or of ships, or other property and there are no other appropriate means for removing that threat" (Regulation XI-2/3.3). Measures or steps are considered "proportionate" if they are "reasonable and of the minimum severity and duration necessary to rectify or mitigate the non-compliance" (Paragraph 4.43). The issue of "clear grounds" is far more complex and of considerable concern to ship operators, who stand to lose considerable sums from delays and diversions if the concept is abused by port state officials.

Part B of the ISPS Code devotes two paragraphs, one a page long, to the explication of "clear grounds." The phrase means "evidence or reliable information that the ship does not correspond with the requirements of chapter XI-2 or part A of this Code." It may arise from the boarding officer's "professional judgment or observations gained while verifying" the ship's ISSC. And "professional judgment" may provide "clear grounds" even though the ship has a valid ISSC (Paragraph 4.32). Examples, usually citing the application of port state officials' "professional judgment," include the following (Paragraph 4.33):

- An invalid or expired ISSC
- Evidence or reliable information of "serious deficiencies" in required security equipment, or arrangements
- Reports, deemed reliable in professional judgment, clearly indicating noncompliance
- Evidence or observation that the crew is not familiar with essential security procedures or cannot carry out security drills, or that such procedures or drills have not been carried out
- Evidence or observation that key crew members are unable to establish communications with key personnel with security responsibilities
- Evidence or reliable information that the ship picked up persons or materials from a ship or port facility that is either in violation of the Code or not required to comply with the Code, without having implemented appropriate security measures or procedures
- A consecutive Interim ISSC, if professional judgment indicates that one purpose for obtaining such a certificate was to avoid full compliance with the Code

The issue of port state control and the application of control and compliance measures are of great concern to ports as well as shipping lines and their customers, because of the potential costs involved. The following examples are instructive.

On August 8, 2002, random X-ray examination, along with subsequent nonintrusive tests, of a container arriving at the Port of Miami from Israel, revealed apparent munitions. A portion of the port was shut down while the bomb squad went to work. They found household goods of an Israeli citizen, including two flower pots: one made from a spent 155-mm artillery shell, the other made from part of an exploded test missile. The costs resulting from the partial shutdown of a major U.S. port are unknown (Bryant, 4/2005).

On September 10, 2002, radiation detectors alerted during a routine examination of the container ship *Palermo Senator,* recently arrived in Port Elizabeth, New Jersey, from Spain. A security zone was established around the vessel, which was escorted to an anchorage, where it was fully inspected by personnel from several federal agencies, who ultimately found that the radiation stemmed from natural emissions of clay tiles from Italy. The vessel incurred demurrage costs well above $50,000 during the delay for inspection. Costs for the inspection effort or to other port users are unknown (Ibid.).

On July 29, 2004, the U.S. government received an anonymous e-mail stating that a harmful biological substance (unspecified) was in one of five containers of lemons onboard the *CSA Rio Puelo,* scheduled to arrive the next day in Port Elizabeth, New Jersey, from Argentina. The ship was detained offshore. Although the containers were quickly located, representatives of the 40 or so federal, state, and local agencies that had become involved disagreed on what to do, with state and local officials insistent that the vessel and its suspect cargo be kept offshore until there was no risk of danger. The containers were fumigated before the ship was permitted to dock and unload on August 6. No harmful biological substance was ever found, and it seems likely that the e-mail probably originated with a competitor of the exporter or importer of the lemons. The lemons, valued at $70,000, were spoiled. Demurrage costs for the ship have been estimated at in excess of $150,000. Costs of the agencies involved or of parties associated with the rest of the containers onboard are unknown, but the total costs for the incident may have been about $1 million (Ibid.).

On January 26, 2005, personal dosimeters of an inspection team on board the container ship *Toledo* at the Port of Los Angeles indicated the presence of radiation. A specialized team was called in from Nevada. The ship and cargo were delayed almost a full day, until it was determined that the source of the radioactivity was a device used to test fire detection and suppression equipment. Costs of the incident are unknown (Ibid.).

United States Legislation and Regulations

Impelled by the same visions of a 9/11 in the maritime sector that led to the 2002 SOLAS Amendments and the ISPS Code, the U.S. Congress passed the Maritime Transportation Security Act of 2002 (MTSA) not long before the adoption of the international regime. Some key provisions of MSA included the following:

- Vulnerability assessments of facilities and vessels
- National, area, vessel, and facility security plans
- Biometric transportation security cards
- Maritime Safety and Security Teams
- A maritime security grant program, with funding to be disbursed on a "fair and equitable basis"
- A foreign port assessment program
- Regional Maritime Security Advisory Committees
- Automatic Identification Systems for many vessels operating in U.S. waters

More recently, in October 2006, the Security and Accountability for Every Port Act (commonly known as the SAFE Port Act) was enacted to add to this port security framework, by creating and codifying new programs and initiatives, and amending some of the original provisions of MTSA. Much of the Act's focus was on broader issues beyond the scope of this chapter notwithstanding their consequential impact of port and vessel security, including (1) codification of the Container Security Initiative (CSI) and the Customs-Trade Partnership Against Terrorism (C-TPAT), two programs administered by U.S. Customs and Border Protection (CBP) to help reduce threats associated with cargo shipped in containers throughout the supply chain; (2) establishment of the Domestic Nuclear Detection Office (DNDO) to conduct research, development, testing, and evaluation of radiation-detection equipment; (3) implementation schedule and fee restrictions for the Transportation Worker Identification Credential (TWIC, formerly the biometric transportation security card); (4) requiring phased-in radiation scanning of all containers entering U.S. ports; and (5) requiring additional data be made available to CBP for targeting cargo containers for inspection. Of more focused impact on port security, the

Act requires (1) inclusion of a salvage response plan in Area Maritime Security Plans to lend emphasis to recovery after a security incident; (2) interagency operational centers where agencies organize to fit the security needs of the port area at selected ports; (3) inspection of port facilities for plan compliance twice a year (at least one unannounced) instead of once; (4) implementation of a long-range vessel tracking system, a Port Security Training Program, and a Port Security Exercise Program; and (5) allocation of port security grants based on risk.

To implement the MTSA, the U.S. Coast Guard published Title 33 Code of Federal Regulations Subchapter H—Maritime Security, its original implementing regulations as final rules on October 22, 2003, which have since been amended on several occasions. The Subchapter, frequently known as the "MTSA regulations," is further divided into Parts 101 through 106 by topic.

Part	Topics
101	Maritime security—General: General (definitions applicability, etc.), MARSEC Levels, Communications & Reporting, Control & Compliance Measures, Miscellaneous
102	(Reserved)
103	Area Maritime Security: General, Designation & Authorities; Area Maritime Security (AMS) Committees; AMS Assessment; AMS Plan
104	Vessels: General; Vessel Security (VS) Requirements; VS Assessment; VS Plan
105	Facilities: General; Facility Security (FS) Requirements; FS Assessment; FS Plan
106	Outer Continental Shelf (OCS) Facilities:[28] General; OCS Facility Security (OCSFS) Requirements; OCSFS Assessment; OCSFS Plan

Designed in part to be the U.S. implementation of the 2002 SOLAS Amendments and the ISPS Code, Title 33 Code of Federal Regulations Subchapter H has one purpose, to "align, where appropriate the requirements of domestic maritime security regulations with" those international standards (Section 101.100[2]). The federal regulations applicable to vessels and port facilities therefore track the structure and content of many of the international provisions quite closely, although there is often greater detail.

One important difference lies in the applicability of the regulations. Based on the MTSA, the regulations of Part 101 apply, unless otherwise specified, to "vessels, structures, and facilities of any kind, located under, in, on, or adjacent to waters subject to the jurisdiction of the United States" (Section 101.110). The reach of the regulations encompasses far more vessels than are covered by the ISPS Code, and includes smaller foreign cargo ships trading with the United States. The United States has also chosen to apply requirements similar to the ISPS Code to many port facilities in addition to those serving ships covered by the ISPS Code on international voyages, even on an occasional basis.

Part 104 of the MTSA regulations, which contains the detailed provisions for vessels, applies not only to the vessels governed by the ISPS Code, but also to foreign cargo ships

[28] The ISPS Code does not apply to OCS facilities. Paragraph 4 of Conference resolution 7, however, encouraged Contracting Governments to ensure that security regimes applicable to mobile offshore drilling units and fixed and floating platforms on their OCS or in their Exclusive Economic Zone "allow interaction with those applying to ships covered by chapter XI-2 of the Convention that serve, or operate in conjunction with such units or platforms."

greater than 100 gross tons[29] and a whole host of other vessels, most of which are included by reference to some category regulated by some other regulation, such as "Barge subject to 46 CFR chapter I, subchapters D or O." In all, some 10,000 U.S. vessels, over 90% of which are non-SOLAS vessels, are covered (Gilmour).

Foreign vessels subject to the ISPS Code that have a valid ISSC are deemed to be in compliance with these U.S. regulations, except for those dealing with (1) MARSEC level coordination and implementation (Section 104.240); (2) Declarations of Security (Section 104.255); (3) additional requirements for passenger ships and ferries (Section 104.292); and additional requirements of cruise ships (Section 104.295). Although this expansive definition of applicability reaches non-SOLAS foreign vessels, or ships of a country that has not adhered to SOLAS, jurisdiction is only asserted if they are destined for or departing from a U.S. port or other U.S. jurisdictional area (Section 104.105). In such cases, however, the vessel is required to have an approved Security Plan.

Part 105 of the regulations, which contains the detailed provisions applicable to facilities (other than outer continental shelf facilities), applies by its terms to facilities that (1) are subject to regulation for dangerous cargoes, liquefied hazardous gases, and bulk transfer of oil and other hazardous materials; (2) receive vessels certified to carry more than 150 passengers; (3) receive vessels subject to SOLAS Chapter XI; (4) receive foreign cargo vessels greater than 100 gross tons; (5) receive U.S. cargo vessels greater than 100 gross tons (except facilities receiving only commercial fishing vessels); and (6) are barge fleeting facilities receiving barges carrying bulk cargoes regulated under tank vessel or dangerous cargo regulations. All other facilities, with the exception of government facilities used primarily for military purposes and certain oil and natural gas production, exploration, or development facilities, are still subject to the provisions of Part 101 that deal with general matters, MARSEC levels, communications, and reporting.

Another difference is that the MTSA regulations provide a list of control and compliance measures for facilities that are not in compliance, a subject on which the international regime is silent. Measures available to the authorities include the following:

- Restrictions on facility access
- Conditions on operations
- Suspension of operations
- Lesser administrative or corrective measures
- Suspension or revocation of the facility security plan approval, with consequence of making the facility ineligible to operate in, on, under, or adjacent to U.S. jurisdictional waters

Part 103, dealing with area maritime security, has no counterpart in Chapter XI-2 or the ISPS Code. The regulations contained therein are designed to bring together the numerous federal, state, local, and private stakeholders with interests in port security in meaningful and useful geographic subdivisions of the nation's port system. Area Maritime Security Committees provide the forum, the Area Maritime Security Plans they develop provide the methods, and the Interagency Operational Centers required by the

---

[29] The ISPS Code does address ships below Convention size, albeit briefly. Paragraph 4.46 provides that they "are subject to measures by which States maintain security." Such measures should be taken "with due regard" to Chapter XI-2 and Part B. It is not clear that requiring security plans and security officers for smaller foreign-flag ships falls within this scope.

SAFE Port Act provide the means for achieving unity of effort from all the stakeholders to properly secure their ports (Doane and DiRenzo, 2007).

Terminology provides another point distinguishing the United States and international regimes. Parts 104 and 105, which provide regulations specific to vessels and facilities, follow the general form and substance of the ISPS Code, but frequently use "owner or operator" instead of "the Company." While both regimes have a "Company Security Officer," in U.S. parlance *vessel* replaces *ship*, and *facility* is used in place of *port facility*. This means there are Vessel and Facility Security Assessments and Security Plans instead of Ship and Port Facility Security Assessments and Security Plans, and there are VSOs and FSOs instead of SSOs and PFSOs.

The qualifications and duties of the American CSO, VSO, and FSO are, for the most part, the equivalents of those of their international counterparts, albeit ordered and worded differently. The principal difference is that American security officers are required to be familiar with the Transportation Worker Identification Credential (TWIC) Program and have duties regarding its implementation.[30] Personnel with security duties and other personnel have essentially the same qualifications under both regimes.

The U.S. requirements for drills and exercises generally track those of the ISPS Code, but are laid out in greater detail. They both must test the proficiency of assigned personnel at all MARSEC levels. Responsibility for scheduling drills is laid specifically on the VSO or FSO. Exercises are to be "a full test of the security program" involving "the substantial and active participation" of "relevant company and vessel security personnel," in the case of vessels and of FSOs for facilities (Sections 104.230, 105.220).

The American regulations lay out the categories of records required to be maintained and go into great specificity as to their content. Unless otherwise provided for particular cases, records must be kept for at least 2 years (Sections 104.235, 105.225).

With respect to procedures and security measures that the ISPS Code lists as required contents of security plans, the U.S. regulations take a somewhat different approach. Most of the same procedures and measures are covered, but the U.S. specifies the format for a security plan with similarly worded headings (Sections 104.405, 105.405) and treats the procedures and measures directly as matters the "owner or operator" must ensure are actually implemented (see Sections 104.240 through 104.290, 105.230 through 285). Notable modifications to the structure of these obligations are incorporating unaccompanied baggage measures under access control (Sections 104.265, 105.255); explicitly adding bunkers to the delivery of ship's stores (Sections 104.280, 105.270); setting forth extensive security measures specifically for newly hired personnel; and the articulation of special requirements applicable to passenger vessels and ferries and their associated facilities (Sections 104.292, 105.285); cruise ships and their terminals (Sections 104.295, 105.290); and facilities handling "Certain Dangerous Cargo" (Section 105.295).

---

[30] After many delays, the TWIC Program has started issuing biometric identification cards for maritime workers in an expanding number of ports. With the exception of certain government workers, a TWIC will be required for unescorted access to designated restricted areas in ports. Obtaining a TWIC requires submittal of an application, with fingerprints, and undergoing a background check. There is an extensive list of disqualifying felonies, some of which may be subject to waiver. The extensive regulations governing the program are at 49 CFR Subchapter D, starting at Part 1570.

Broad-Brush Evaluation: Is the Maritime Transportation System Now More Secure?

The question then arises whether these two elaborately structured regimes have had an impact on the actual state of security of the maritime transportation system since they went into effect on July 1, 2004. Two early commentators suggested that neither governments nor the private sector had done enough to prevent attacks on the maritime transportation system. One suggested that the international community was "simply going through the motions" (Kulisch quoting Stephen Flynn). The other noted that over 50% of ports certified as ISPS Code compliant were "only compliant on paper, whilst remaining poorly equipped to deal with crime and terrorism" (van Unnik).

Official statistics, on the other hand, show increasing Code compliance by shipping. Between July and December 2004, the first 6 months the ISPS Code was in force, the U.S. Coast Guard found 1,155 security deficiencies in foreign flag shipping subject to the Code, resulting in 51 major control actions. Just under half of the deficiencies related to access control and restricted areas. In 2005, the Coast Guard conducted 9,117 ISPS inspections, with only 115 security deficiencies associated with 51 major control actions. Over half of the violations related to access control and restricted area. In 2006, the last full year for statistics currently available, major control actions had dropped to 35, based on 73 security deficiencies, of which over half again related to access control and restricted areas (U.S. Coast Guard). The secretariat of the Paris Memorandum of Understanding (MOU), an organization that seeks to harmonize the Port State Control actions of Canada and various European countries, also reports a drop in security deficiencies found in visiting ships between 2005 (817) and 2006 (735) (Paris MOU).

Over the same period, deficiencies at U.S. port facilities declined somewhat in 2005 and then rose almost as much in 2006. About one third of approximately 3,200 facilities were identified as having deficiencies (usually multiple) each year. Access control matters predominated in all 3 years, while record-keeping violations overhauled restricted area issues in 2006 (GAO 2008). The data are maintained differently for facilities and vessels, with the result that many of the far more numerous violations reported for facilities are not comparable to the significant violations associated with major control actions against vessels. On the international front, the compliance of most ports subject to the ISPS Code appears adequate, based on assessments by the U.S. Coast Guard under the International Port Security Program. Additionally, only 10% of reports made SSOs to the Baltic and International Maritime Council reveal problems relating to port facility ISPS Code compliance (Timlin).

But has even excellent compliance with the ISPS Code actually contributed to increased security? Implementing the tighter security procedures of a ship's security plan ought to make the ship less vulnerable to, at least, low-level armed robbery, and such attacks appear to have decreased, although the underreporting of piracy make the statistics somewhat suspect. More reliable, however, is the U.S. Coast Guard report that there was a greater than 50% drop in stowaways in the first 6 months that the ISPS Code and MTSA regulations were in force (Gilmour).

Similarly, one would expect that the security measures taken to keep terrorists out of a port facility would also keep thieves out, lowering losses from theft experienced by port facilities. Three months after the MTSA came into force, a Department of Homeland Security official, while speaking at a cargo security forum, referred to a study that showed that implementation of MTSA requirements had lowered a New Jersey port facility's theft losses by 30% (Doan). A more recent article, however, suggests that "the paper process" of implementing the ISPS Code has not made ports and terminals much more secure than they already were, citing discussions with cargo insurance underwriters, who had not experienced any reduction in claims for cargo theft subsequent to July 1, 2004. "[I]f port cargo thefts have not decreased, how secure have ports and facilities become against infiltration of terrorists?" (Austen).

Whether or not a decrease in cargo theft can be documented, it seems likely that implementation of the ISPS Code and the MTSA regulations has, to some degree, "hardened the target" through standardized procedures for coordinating security measures and international cooperation on maritime security. It would be unrealistic to expect more than modest gains. The open nature of the maritime transportation system ensures its vulnerability. Some of the most fearsome threats originate elsewhere in the supply chain[31] and must be addressed by measures that secure the entire supply chain, not just vessels and port facilities. But even within the purely maritime context, the ISPS Code has fundamental weaknesses as a counter-terrorism regime. It does not deal with terrorist use of the shipping assets to raise or launder revenues. Nor do its provisions counter terrorist use of the maritime transportation system to insert operatives disguised as mariners, although the TWIC in the United States and TWIC's eventual international equivalent will help a great deal in this regard.

Additionally, the Code does not require measures to protect ships from seaborne attack. Further, it does not apply to the likely means of such attacks—small vessels, which could also stage attacks on targets in ports. The MTSA regulations do apply to many of these non-SOLAS vessels, both foreign and domestic, but not with the small craft that have been the terrorists' vessel of choice in two attacks on ships (*USS Cole* and *M/V Limburg*). The United States proposed expanding its notice of arrival regulation to include all foreign commercial and recreational vessels and those U.S. vessels returning to the United States after visiting a foreign port, as well as requiring those vessels to be equipped with an Automatic Identification System (AIS) (Gilmour). Singapore has already mandated AIS for ocean-going non-SOLAS vessels. In addition, harbor craft are required to comply with the Harbor Craft Security Code. Routes for small craft entering and leaving the port have been designated in order to keep the craft away from key sensitive installations, and a low-cost version of an AIS was being tested. An integrated surveillance and information network permits the authorities to have a real-time picture of the port (Tong). Singapore has an estimated 4280 non-SOLAS vessels operating in the harbor on a daily basis (Japan International Transport Institute). In contrast, Japan has an estimated 849,250 domestic non-SOLAS vessels, and 16,500 foreign, non-SOLAS ships that call at ports annually (Ibid.). In the United States there are over 13 million recreational craft with virtually unrestricted access to the nation's military and civil harbors (Goward). These figures suggest that both Japan and the United States would have considerable difficulty in applying the Singapore model. In any event, AIS only works as a tracking technology when it is turned on, which is to say that it only works with compliant vessels (Bennett). Finally, the ISPS Code is preventive in nature; it does not address mitigating the damages of an attack after

---

[31] The radiological weapon in a container is the prime example. Policy makers' obsession with this scenario is understandable, given the consequences of a successful attack in this vein. The monomania, however, has led to neglect of noncontainerized cargoes, which many in the industry consider more vulnerable as a means of delivery (Parfomak and Fritelli).

it has occurred (McNaught). In the United States, the SAFE Port Act mandates that a salvage recovery plan be part of the Area Security Plan, as a means of focusing on recovery and continuity of operations. Other countries could legislate similarly.

Conclusion

Perfect security is unattainable (Giermanski and Lodge). The ISPS Code and the MTSA regulations are steps in the right direction to help protect one aspect of the global supply chain—the maritime transportation system. They are not stand-alone documents. They form part of a larger security effort that includes various supply-chain security programs, as well as efforts to develop secure identification means for mariners and others in the maritime transportation system. They do serve to raise security consciousness about maritime terrorism. Security awareness must be part of every organization's culture, both in the form of noticing something out of place that provides an indication of a potential incident and in the form of recognizing that an appropriate security program is not simply a hole into which money disappears, but an opportunity to generate cost savings and operational efficiencies.

References

Acharya, Arabinda. Securing the seas in Southeast Asia. *INTERSEC*, 15(6), 187–188, June 2005.

Aegis Defense Services Ltd. Piracy terror connection. *Cargo Security International*, 1(2), 14–16, December 2003.

Allen, Thad. USCG commandant opening remarks at 25th IMO Assembly. MarEx Newsletter. Last accessed November 23, 2007, at: http://www.newsletterscience.com/marex/readmore.cgi?issue_id+270&article_d=2734&1=1&s=29159.

Austen, Chris. Increased port security: Burden or benefit to port operations? *Proceedings of the Marine Safety & Security Council,* Vol. 63, No. 1, 69–72, Spring 2007.

Balaoi, Rommel C. Maritime terrorism in Southeast Asia: The Abu Sayyef threat. *Naval War College Review*, 58(4), 62–80, Autumn 2005.

Bennett, Capt. Robert F., USCG. Close the gaps. *U.S. Naval Institute Proceedings*, Vol. 138, No. 8, pp. 42–46, August 2007.

Burnett, John S. *Dangerous Waters: Modern Piracy and Terror on the High Seas.* New York: Dutton, 2002.

Bryant, Dennis L. Lemons, tiles & false assumptions. *Marine Reporter & Engineering News*, pp. 17–20, April 2005.

Cawthorne, N. *A History of Pirates: Blood and Thunder on the High Seas.* Edison, NJ: Chartwell Books, Inc., 2003.

Code of Federal Regulations. Title 33, Chapter I, Subchapter H, Maritime Security (as amended).

Convention on the High Seas. Done at Geneva April 29, 1958; entered into force September 30, 1962, 450 U.N.T.S. 82. Last accessed November 30, 2007, at: http://sedac.ciesin.org/entri/texts/high.seas.1958.html.

Convention for the Suppression of Unlawful Acts against the Safety of Maritime Navigation. Done at Rome March 10, 1988; entered into force March 1, 1992.

DiRenzo, Joseph, III. Important gains registered at small-vessel security summit. *Dom Prep Journal*, 3(7), p. 12, July 2007.

Doan, Douglas (DHS business liaison for Border & Transportation). "Remarks," presented at the Special Government Q & A Panel, 3rd Eye For Transport North American Cargo Security Forum, Washington, D.C., October 18–19, 2004.

Doane, Christopher, and Joseph DiRenzo. Area maritime security committees: A unified effort for securing U.S. ports. *DomPrep Journal*, 3(8), 23–24, August 2007.

Doane, Chris, and Joseph DiRenzo, III. Disrupting maritime terrorism: What has been accomplished since July 2004? Maritime Security Sourcebook, Supplement to Maritime Reporter, pp. 8–11, March–April 2005.

Dragonette, Charles N. Lost at sea. Letters to the Editor, Foreign Affairs, Vol. 84, No. 2, March–April 2005. Available on line at: http://www.foreign affairs.org/20050301faletter84267/charles-n-dragonette/lost-at-sea.html.

Eklöf, Stefan. Piracy in Southeast Asia: Real menace or red herring?" Last accessed February 24, 2008, at: http://www.zmag.org/content/showarticle.cfm?ItemID=8445.

Giermanski, Jim, and Peter Lodge. Tolerating reasonable risk. *HS Today*, 4(7), p. 8, July 2007.

Gilmour, Rear Adm. Thomas (chief, Marine Safety, Security, and Environmental Protection, U.S. Coast Guard). "Remarks," delivered to the Seminar on Maritime Security Measures of Non-SOLAS Vessels, London, May 10, 2005. Available online at: http://www.japantransport.com/conferences/2005/05/st_gilmour.pdf.

Government Accountability Office (GAO). *Maritime security: Coast Guard inspections identify and correct facility deficiencies, but more analysis needed of program's staffing, practices, and data.* Report GAO-08-12, February 14, 2008.

Goward, Dana A. Maritime domain awareness: The key to security. *U.S. Naval Institute Proceedings*, pp. 20–24, April 1990.

Hill, Don. Maritime piracy grows as threat to world commerce. Radio Free Europe–Radio Liberty. Last accessed February 24, 2008, at: http://www.rferl.org/featuresarticle/2004/07/442835f0-ef2a-4938-b0bb-ed9978c17f24.html.

Hooper, Craig. The peril of power: Navigating the natural gas infrastructure. *U.S. Naval Institute Proceedings*, pp. 40–44, June 2006.

Hosenball, Mark. Look out below: The terror threat from the sea. *Newsweek*, p. 5, March 28, 2005.

International Chamber of Commerce, International Maritime Bureau. *Piracy and armed robbery against ships: Annual report 1 January—31 December 2006.* Barking, Essex, UK. Available through: http://www.icc-ccs.org.

International Maritime Organization. International Ship & Port Facility Security Code and SOLAS Amendments 2002, 2003 Edition, London.

Japan International Transport Institute. *A study on maritime security measures for non-SOLAS vessels.* Prepared for the Seminar on Maritime Security Measures of Non-SOLAS Vessels, London, May 10, 2005. Available at: http://www.japantransport.com/conferences/2005/05/seminar_report.pdf.

Kimery, Anthony. Dangerous cargo. *HS Today*, pp. 14–19, January 2007.

Kulisch, Eric. Baby Steps' U.S. maritime security is less than advertised, security experts warn. *American Shipper*, pp. 88–89, October 2004.

Luft, Gal, and Anne Korin. Terrorism goes to sea. *Foreign Affairs*, 83(6), 61–71, Nov.–Dec. 2004.

Lundquist, Edward Piracy: Criminal at best; terrorism at worst. *Maritime Reporter and Engineering News*, pp. 17–20, August 2006.

MarEx Newsletter. U.S. Coast Guard issues notice regarding mandatory STCW/ISPS Requirements. Accessed November 9, 2007, at: http://www.newsletterscience.com/marex/readmore.cgi?issue_id=268&article_id=2698&l=1&s=29159.

Maritime and Coastguard Agency (UK). Marine information note, MIN 309 (M). November 30, 2007.

Maritime Transportation Safety Act of 2002, Pub. Law 107-295, 116 STAT. 2064. Signed by the president November 25, 2002, codified at 46 U.S. Code section 70101 et seq.

McNaught, Fiona. Effectiveness of the international ship and port facility security (ISPS) code in addressing the maritime security threat. *Geddes Papers*, pp. 89–100, 2005.

Medalia, Jonathan. Terrorist nuclear attacks on seaports: Threat and response. Congressional Research Service Order Code RS21193, updated January 24, 2005.

Mensah, Thomas A. The place of the ISPS Code in the legal international regime for the security of international shipping. *WMU Journal of Maritime Affairs*, 3(1), 17–30, April 2004.

Nincic, Donna J. *Maritime terrorism: Developing case studies for teaching and analysis: An interim report.* Paper given at 6th Annual General Assembly of Int'l Assn of Maritime Univs., Oct. 24–26, 2006, at Malmo, Sweden. Preliminary version available (last accessed February 24, 2008) at: http://www.iamu-edu.org/generalassembly/aga6/s1-nincic.php.

OECD (Organisation for Economic Co-operation and Development), Maritime Transport Committee. Security in maritime transport: Risk factors and economic impact. July 2003.

Parfomak, Robert W., and John Fritelli. *Maritime security: Potential terrorist attacks and protection priorities.* Congressional Research Service Report to Congress, RL 33787, January 9, 2007.

Paris MOU on Port State Control. *Port state control: Steady as she goes.* Annual Report 2006. Available at: http://www.parismou.org/ParisMOU/Organisation/Annual+reports/default.aspx.

Protocol of 2005 to the Convention for the Suppression of Unlawful Acts against the Safety of Maritime Navigation. Done at London October 17, 2005 (not yet in force).

Raman, B. Future threat: Terror from the sea. REDIFF News, September 7, 2007. Last accessed November 24, 2007, at: http://in.rediff.com/news/2007/sep/06raman.htm.

Rosoff, H., and D. von Winterfeldt. A risk and economic analysis of dirty bomb attacks on the ports of Los Angeles and Long Beach. *Risk Analysis*, 27(3), 533–546, 2007.

Sekulich, Daniel. Pirate tales, the conclusion. Posted Oct. 16, 2006. Last accessed February 24, 2008, at: http://oceantitans.blogspot.com/2006_10_01_archive.html.

Teitler, G. Piracy in Southeast Asia: A historical comparison. *Maritime Studies (MAST)*, 1, 67–83, 2002. Last accessed November 14, 2007, at: http://www.marecentre.nl/mast/documents/GerTeitler.pdf.

Timlen, Thomas. The ISPS Code: Where are we now? *Cargo Security International*, 5(3), 14–15, April–May 2007.

Tong, Yeo Chow. Beyond ISPS. *Cargo Security International*, 3(1), February 2005.

U.S. Coast Guard. *Port state control in the United States.* Annual Reports 2004, 2005, 2006.

United Nations Convention on the Law of the Sea. Done at Montego Bay, December 10, 1982; entered into force November 16, 1994, 1833 U.N.T.S. 3. Last accessed November 30, 2007, at: http://www.un.org/Depts/los/convention_agreements/texts/unclos/unclos_e.pdf.

Technology Applications to Transportation Security

8

Computer and Transportation Systems Security

Peter V. Radatti, Ph.D.

The Objectives of This Chapter Are to Introduce, Discuss, and Provide a Basic Understanding of the Following:

- That security is flawed
- Holistic security design
- That physical security and locks have limited value
- The trade-off between security and accomplishment
- That the attacker has the advantage
- Cameras, computers, and physical security
- Computers and physical security
- Securing computers
- That holistic security design is transparent and passive
- Radatti's rules of computer security
- Disaster recovery
- That time management is part of computer security
- The conclusion: that security is a promising career for those who like challenges

> *Bellum omnium contra omnes,*[1] *"War of all against all."*
> —Thomas Hobbes (1588-1679)

---

[1] War of all against all. Read Thomas Hobbes (1588–1679), as his theories have a bearing on security. The natural state of men, before they were joined in society, was a war of all against all. This idea places people in a pre-social condition and theorizes what would happen. Hobbes theorized that people enter a social contract, giving up some of their liberties for security. This is considered a test for the legitimating of a state in fulfilling its role as "sovereign" to guarantee social order.

Security Is Flawed

It is interesting to note that very few people seem to see the significant failures in security that surround us. These failures are not limited to any specific type, such as physical security or computer security. They are systemic, potent, and enduring. Most of the efforts that have been made in security just do not work well enough to stop a real threat in a comprehensive manner. In fact, most efforts do not even address the first mover advantage that most attacks have.

Here are some examples of systemic failures:

1. Physical security that relies upon mechanical keys and locks.
2. Physical security that relies upon human guards at gateposts. These are important, but guards at remote monitoring sites are more important.
3. Computer security that relies upon antivirus scanners. Viruses and software attacks are only one aspect, and the first mover advantage holds here.
4. Computer security that relies upon system–network configuration or firewalls. This includes full isolation of a network.
5. Internet transaction security that relies on PKI–PKE encryption. Specifically the PKI (Public Key Infrastructure) portion.
6. Employee trust that relies upon initial background checks upon hiring. An initial background check has no value 1 year later.[2]

Notice that the keyword in this discussion is *relies*. For the purpose of security, anything that uses the word *relies* is flawed. Using this term indicates a pivot point in implementation. As everyone knows, a pivot can be worked in two directions, not just the desired one. Consequently, what is needed is a holistic, multilayered system that does not rely upon any specific concept.

Holistic Security Design

Holistic, multilayered designs are interesting because they tend to check themselves. Each part of the design "pulls and pushes" against the other parts, producing a tension that demonstrates any problems prior to a failure. These are not systems that can fail dramatically like a steel cable under tension but rather have the ability to react instantly with power, like a marshal artist's power stance. Like a marshal artist, systems that do their work well appear to be effortless but take years of devotion to master. This is in direct conflict with the "sell them something" attitude that often prevails in the security industries. Very few want to take the necessary time to learn how to build security systems correctly. In the past, even less were willing to pay for a correctly implemented system.

The September 11, 2001, attack against the World Trade Center in New York City was a wake-up call to the security industry. No one group or organization took responsibility; however, there was more than enough blame to share. Even simple things were done improperly. In the second tower, for example, workers were told to return to their offices. This caused additional deaths for no justifiable reason. This was a procedural failure in a system that was not holistic.

---

[2] Where legal, continued monitoring of an employee's credit score is extremely valuable.

At the moment, the security industries are evolving due to increased spending caused by the September 11 attacks. However, this will not last in perpetuity, since many people have short-term memories and there are a million other things to competitively spend money on. Optimistically, what might last is the increasing emphasis on security that never prevailed in the past. This gives us an important opportunity and ability to change security from a fundamental industry into an advanced profession. As professionals we can bring the next paradigm change to the security industry, incorporating holistic systemic design.

Physical security as practiced today has changed little in hundreds of years. Electrical alarm systems have existed almost as long as commercial electrical service, and the Roman and Chinese empires effectively used nonelectrical alarm systems, not to mention deliberately designed, squeaky floorboards.[3] Passage key locks have existed for thousands of years. Fences, earth works, walls, and barriers of all forms are prehistoric. The basic concepts have not changed. What has changed is the world and the relevance of these methods.

Most alarm systems assume a thief. Not only that, but they generally assume a dumb thief. Luckily, most thieves are dumb.[4] Notwithstanding the television dramas that show technologically advanced thieves, it is generally not true in real life. This provides us with assurance that most alarm systems will stop most thieves. What they will not stop are people with large, well-funded, technologically advanced organizations. This profile reflects almost all nations, including rogue nations and well-known international terrorist organizations. An alarm system will not stop thieves of high intelligence.[5] An example of a slightly more intelligent than average thief is the Pennsylvania Pocono summer home thieves. They cut a hole in the wall of homes to bypass the alarms on the doors and windows. Three-dimensional spatial entry (surface, aerial, and subsurface) offer opportunities.

Transportation hubs offer significant targets of opportunity because of their concentration of high-value goods and people. This includes airports involving both commercial and freight, seaports, bus and train terminals, and truck terminals. Existing systems handle traditional thieves somewhat well. However, there is room for significant improvement as witnessed by the fact that stealing still takes place.[6] These protection systems are geared for thieves. They are not geared toward terrorist acts or stealth military operations.

A technologically advanced organization will have no problem dealing with an alarm system. Its members also will have no problem in dealing with visible cameras, since there are many ways to bypass cameras.[7] In this case, the concepts and plans that deal with this type of operation were created during the cold war between the United States–Western Europe and the Soviet Union. The basic concept is to cause delays in the operation of the perpetrator that would increase the probability of detecting and reacting to them. This is accomplished using every method known: alarms, passage locks, cameras, lighting, guards, and even the design of hallways and road networks. The problem is that these concepts assume that stealth and time are issues. In our modern

---

[3] Look up the role of geese as an alarm system for the Roman army.

[4] If you are a thief, please stop reading this now and turn yourself in to authorities!

[5] Thieves of high intelligence are involved in white-collar crime and are not the subject of this chapter.

[6] Many operators assume a standard percentage of theft based upon volume and incorporate it into their operating costs. I always found this attitude silly, but it does exist.

[7] At the very least, a bullet or a can of spray paint will disable a visible camera. The real issue is how an organization responds to a camera outage.

world, these are not as important as they once were. The September 11 attacks depended upon surprise and confusion. We could have knocked the aircraft out of the air but did not. Consider that one of them impacted the Pentagon, and consider what the normal defense of the Pentagon must include. We did not act because we were surprised and unprepared.[8] Our level of preparedness for this terrorist scenario was underoptimized.

Physical Security and Locks Have Limited Value

To accomplish our goals of securing transportation terminals we still depend significantly upon key locks. We set up inspection points and believe that a bad guy will go through the inspection point because the way around it is secured with high-security locks. In many cases these locks are governmentally approved. The entire concept of mechanical key locks is a failure and has been so for at least the last 50 years. For as long as there have been locks, there have been lock picks. Modern small locks such as those for securing filing cabinets, as well as desks and equipment lockers, can be picked in less than 15 seconds without trouble by someone with 1 day of training. This is not news; it has been well known for a long time.

What appears to be real news, at least in the United States, is that most high-security mechanical locks have marginal utility because they are vulnerable. In fact these locks cannot be picked using a standard lock pick. About 50 years ago, a new type of lock pick was invented. News of this pick moved slowly throughout Europe and interested groups.[9] This new method works best with high security pick-proof locks. These locks can be opened by someone with only a few minutes of training in less than 2 seconds, and he or she can do it in such a way that it may not be noticed by anyone.[10] This method has two names: One name is bump key; the other is 999 key. The method is simplicity itself and works about 80% of the time.[11] Thanks to the Internet, anyone can learn about and obtain bump keys. Perform a Google search on the term bump key to learn more. You can also visit www.youtube.com to watch a video on this topic.

What this means is that most mechanical key locks have no value in the face of an informed, but otherwise unintelligent, perpetrator. All security plans that depend on the existence of a key lock are now in jeopardy.[12]

As people consider these problems, they will also see that converting from analog to digital will solve many of these problems. Digital locks,[13] cameras, and alarm systems are much harder to trick or bypass because they can perform evaluation on their inputs, not

---

[8] Assuming that the Pentagon has a rooftop missile launch system (a guess based on public documents), the fact that it did not fire indicates a serious error in procedure, training, or command. If it did not have such a system, then I hope it does now.

[9] There are lock-picking clubs. There is even an annual international competition. They are a great source of information on what works and what only looks like it will work when it comes to locks.

[10] This method does not leave the small scratches that may indicate a lock pick was used.

[11] There is a set of five keys that open most of the locks in the United States. There is a different set for Europe and another different set for Asia. A medium-sized key ring can open 80% of all the locks in the world.

[12] There are locks that are bump-key proof, but the basic concept that mechanical key locks are no longer a good form of protection does not change. Mechanical locks are obsolete.

[13] Not all electronic locks are created equally. Some are fully worthless. Generally, the electric lock end of the system is separate from the electronic portion that evaluates if the lock should open or not. The electric lock part can sometimes be easily opened, so care must be taken when selecting both ends of an electronic lock. In addition, I suggest that electric locks that rely upon a battery to keep a door secure in the event of a power failure not be considered. I also consider magnetic locks to have no value, since there is a little-known but effective way of opening these without tools.

just simply report them. A computer can easily tell the difference between a live camera view and a prerecorded camera view. A computer can easily detect if someone is attempting to bypass a digital lock by picking. Digital alarm systems can report much more information than status. This brings the computer into focus as a critical point for the security of facilities, including transportation hubs. That opens facility security to all the problems of computer security.

The Trade-Off Between Security and Accomplishment

The most important thing that must be understood is that everything is a trade-off. More security means less ability to do things, since security is restrictive. Less security means more ability to do things including stealing, sabotaging, and killing. Consider it as the playground seesaw: On one end is security and on the other is the ability to get things accomplished. You need to balance the two ends, but it will never be possible to balance them equally, nor should you desire to. To balance them equally, you will accomplish little security while creating frustration among those who need to get things done. Generally, the board is balanced a little in the direction of getting things done. That means that what you select to stack on the "security" side must be the most important of the items available for security. You need to allow people to get their job done while maximizing security. Of course, even here there are trade-offs.

In a shipyard that is importing consumer goods, security may be more lax than at a naval yard servicing war ships. This introduces the concept that what you are protecting helps to determine the acceptable costs of security. If you are protecting only televisions, you might not care if a percentage are stolen; however, there are two other factors that come into play when determining the level of security that is appropriate. The first is the level of loss or pain that the organization is willing to accept. The second, and perhaps the more dangerous, is the secondary effects of a security breach. For example, if you want to make sure no unregistered weapons are entering a port, the secondary effect of a breach is that those weapons could be used to kill someone. Another example is that your port may be damaged by outsiders and you may therefore deny use of the port not only to the companies that use it but also to the consumers who rely on importation of goods from that port. This represents a multilayered economic breach, since over 90% of all valued goods utilize the maritime system that fans out to involve other transport modes. Think of this as the ripple effect. In security terms this is called a *denial of service attack*. This holds true for all transportation nexuses including airports, seaports, bus terminals, highways, etc.

The Attacker Has the Advantage

There is another critical concept that must be understood: The attacker always has the advantage, that is, the first mover advantage. An attacker possesses secrecy and can pick and choose where and when to attack. You, on the other hand, must be more inclusive, protecting all of your area, all of the time. Attackers know when, where, and how they are going to attack. Potential victims do not know this. Attackers can use something new that has never been seen before. You have to always be prepared for the unknown. The problem is not that people can attack you with known attacks or attacks that are not known to have been used but have been thought of. The problem is that they can attack you in areas and ways that you don't know and never considered. It is here that a potential catastrophe exists. It is here that holistic security concepts can provide you a safety net to allow you to react to an attack of this type within time.

Let us now consider physical security, that is, guards, guns, and gates. First consider guards. When I was attending school I worked as a guard in a factory and in railroad yards during weekends. There are very few jobs as boring as being a guard. Guards spend 99% of their time being bored and 1% of their time in terrifying situations. At the same time they are usually underpaid and receive little respect from the people they work with and for. Guards are not police. They generally do not have the power to accomplish much, except react to conditions.

When I was a guard in the railroad yards there was a problem with a well-known outlaw motorcycle gang stealing car parts from the trains. Here I was, a young man in a badly fitting uniform in a bad-smelling plywood guard shack down in a railroad yard in the middle of the night trying to keep a motorcycle gang that was well known for killing people from stealing cars. No one was going to be afraid of me. What I did was find a railroad tie-down chain with a couple of hooks on the end to carry. Now you might think I was carrying this as a weapon. No. A chain was not going to stop a dozen guys with guns and knives. What I did was rattle the chain as I walked around the yard. The criminals could hear me coming and they would leave. What this did was change the rules of the game. They didn't want to see me as much as I didn't want to see them. By advertising the fact that I was coming, I limited the amount of time they had to steal. The railroads were happy. Stealing and the costs associated with it were kept in check. The perpetrators were happy; they didn't take a chance of going to jail and still got something. I was happy; I got to live.

Having guards alone is not enough. You have to use them in smart, critical ways. Sitting in a box or walking around a railroad yard is not good enough. There have been thousands of years of people doing it this way and thousands of years of the bad guys getting around it. Consider what you need to accomplish with the guards and then set up their actions according to the goals. Remember that not everyone is going to be afraid of your guards.

These are the primary purposes of guards:

- To advertise that there is someone with the responsibility for security present, thereby providing a deterrent
- To restrict the actions of an attacker
- To think—guards have the ability to react to unknown circumstances

Of course if the guard looks unkempt or is sleeping or so bored that he or she doesn't appear as any kind of a threat, this will encourage an attacker and have no value at all. We won't even talk about the fact that guards are people and possess all the shortcomings of people everywhere.

With all the problems mentioned with guards, there are easy ways to resolve these problems, and there are significant benefits to having guards. The first and most important benefit to having guards is that they can think. Do not underestimate this, but do back it up with appropriate training and written procedures. It is also important to hire guards wisely. Do background checks. Lots of criminals will apply for guard jobs, which would be putting the fox in charge of the henhouse. Pay them well and make sure they have smart, well-fitting uniforms. Fire anyone who doesn't fit in. Respect them and make sure they know it. Don't allow others to disrespect them. They are your first and last line of defense. Boredom and lack of training are serious problems that can destroy even the best of guard teams. You need to have constant training and solicit them for constant improvement in the security process. You also need a *red team*. A red team is an outside group of people who pretend to be attackers and are paid to break in.[14] The guards know about them and are always looking for them. It will keep them on their toes and help them break the boredom. It will also find most of your physical security weaknesses. There are a lot of companies that provide red team services.

People operate guns. That would be your guards and the local authorities, normally police. Provide good-quality guns and training. Get the local police involved in the training of your guards so that they feel comfortable when they approach your facility during an emergency. Make sure the local police think of your guards as an asset, not a danger. Have frequent interaction between the local police and your security staff. This should not be steak and beer parties but actual working meetings and exercises where they all work together.[15] The goal is to get them used to working together.

These days it has become popular to have unarmed guards. Sometimes they are issued sidepieces but without ammo. I strongly discourage this, as the only power an unarmed guard has is to call the police. In addition, your guards should be armed with the latest semiautomatic pistols and rifles. Shotguns should also be kept on-site but locked.[16] These need to be kept in good working order and replaced with newer models when necessary. You need to remember that the attackers will have taken a good look at your guards and know what weapon systems they possess, including the type of ammunition they are carrying. If these armament systems your guards have are substandard and ineffective, the deterrent value is reduced.

Gates and fences are for the control of honest people. As you have already learned, mechanical locks are easy to bypass, and fences or walls are easy to cut. Alarm systems are basically part of your fences–walls and doors–gates. They are supplemental but do not replace a holistic design. A well-designed alarm system will slow down and detect intruders, which is critical. At the same time, special attention must be paid to how alarm wires are run and if the alarm system components are active or passive. You should use both. You also need to consider if the components are normally silent or if they produce an "all clear" heartbeat. There are advantages and disadvantages to both types.[17]

Walls and fences can be designed for easy and clear access or for difficult and obstructed access. There are places for both in a holistic security design. As a general rule, you want easy and clear access for wherever large groups of people will congregate. You especially need an easy-to-see-and-access exit so that a stampede does not happen during an emergency. On the other hand, in areas that are not open to the public, hallways designed to restrict how far someone can see or how fast someone can run will aid the local team, who already know the terrain and have the advantage of camera placement eliminating blind spots. Additionally, a reinforced hallway that allows only one person to pass will restrict the firepower an attacker can bring to bear, while providing maximum protection for defenders. Remember that an attacker who has to turn many corners is worrying about who may be waiting for him or her.

---

[14] Make sure the local police know about a simulated attack in advance, and make sure no one is hurt.

[15] You should also have the steak and beer parties, but after the work is done. Don't forget to invite me.

[16] Shotguns have the ability to really stop an emergency in mid-track. At the same time, they have the ability to cause a lot of unplanned damage. Use of a shotgun should be restricted to individuals who receive special training. Think of them as your ultimate response once an emergency has started.

[17] Generally I dislike wireless systems, but there are some places where they are appropriate.

Cameras, Computers, and Physical Security

Cameras are a very powerful home team advantage.[18] Not only can they act as a deterrent, but they also allow one person to see and coordinate efforts at many areas at the same time. When used in conjunction with wireless communications to the guards, this is very powerful. Cameras provide a documented record of "what happened." As mentioned, older cameras can be disabled or "tricked," but modern cameras are no longer passive devices displaying images on a screen watched by an operator. They provide digital images, which can be processed. Some of this processing can detect these "tricks." There are many advantages to digital computer image processing, the least of which is a significant cost savings in operator expenses.[19] Another advantage of digital processing of video is recognition. These systems not only can do face recognition but also vehicle license plate recognition, as well as alert an operator to potential problems. The recorded record of what happened will be useful not only as a deterrent but also in legal actions and analysis of what went wrong or well.

When designing a camera system, you need to consider the value of both visible and hidden cameras. You also need to consider if the hidden cameras should be known or unknown. Generally I am in favor of having both very visible cameras and hidden cameras, because you get the advantage of each. One advantage of hidden cameras occurs only if it is generally known that there are hidden cameras but their location is a well-guarded secret. That advantage is that no one can calculate blind spots that allow him or her to operate unseen. I am in favor of making sure that everyone in a facility knows there are hidden cameras, but not the number or location of them. This is easy to do by requiring all employees to sign an acknowledgement that both types of cameras exist at your facility.[20]

Another consideration on camera system design is the software used to process images. The best systems will detect "tricks" and provide additional features such as alerting the operator using face recognition of individuals of special interest. This could be people you don't want in your facility, but it could also be highly important people that you want to pay special attention to.[21] There are many other features to these types of systems, and you will need to select those features you want from a large array of options.

Our discussion of cameras demonstrated a powerful concept: visible and hidden security. This concept can be applied to everything, including alarm systems, building design, and security staff.

Hidden security staff is an interesting topic because there are a lot of ways of implementing this functionality. Generally, hidden staff is an adjunct to visible staff, not a replacement. There are two ways of implementing this concept. The first is hidden staff on-site; the second is staff who are hidden by being off-site. Las Vegas casinos are reported to use the first method but with an interesting twist. Their camera systems are monitored by hidden staff on-site but with their own entrance; they never mingle with the other staff

---

[18] They are also a source of unending amusement for the camera operators, which tend to help them keep a sharp eye.

[19] Modern systems display only images that need to be viewed. An empty room is generally not displayed. This allows fewer operators to monitor any given number of cameras.

[20] Don't try to do this on the cheap by telling people there are hidden cameras if they do not exist. Eventually people will find out and you will lose credibility.

[21] Such as your boss when he is coming down the hall toward your office.

members of the casino. The advantage of doing this is that you do not have to worry about communications between the casino and the operators. One disadvantage is that, no matter how much you try to separate these two groups, you are not going to.[22] There is also the large expense of basically keeping a separate building within your building. Finally, these groups are located in the same geographic region. They are going to meet even if only at the supermarket. The primary disadvantage of remote monitoring is one of communications. I think this problem is mostly moot due to the large expansion of fiber-optic long-distance cable, which was installed thanks to the growth of the Internet. At the same time, satellite communications have dropped in cost and are unaffected by anything that might happen to cables on the ground. Finally, a holistic design of using the Internet, leased lines, and satellite will ensure that your data are delivered. Remote monitoring has none of the disadvantages of local monitoring, and if you cannot find a company that you like to provide these services, you can always create them yourself at a remote location.

Another way of providing hidden staff on the ground is to have staff who have assigned stations or areas of responsibility as well as staff who are assigned to wander in unpredicted areas and at unpredicted times. If you think carefully about this subject you will see that it has many aspects and there are many ways to implement this concept.

While we have discussed having visible and invisible staff, we have spoken about them in a preventative manner. The same holds true when responding. You can have both visible and invisible staff during a response to a problem. Of course, the role of off-site staff is easy to understand. They not only can coordinate staff on the ground but also can bring in more staff such as local, state, and federal authorities or private staff, as appropriate. Depending upon how your facility is designed, they can initiate a multilevel response such as starting generators, locking specific doors, opening or closing gates, closing valves or pumps, stopping traffic from entering your facility, and many other activities that would otherwise occupy your on-site staff. It is also possible to have contracts with security companies that provide additional staff on demand. One of the advantages of off-site hidden staff during a response is that they are not in the middle of the fight. They can remain calm and respond to the situation as required. Depending upon their training, these additional people may be a great help.

Communications and power are similar from a security point of view because they are normally carried on wires that have to enter your facility. The chance that someone will block cell phone or satellite communications is low, but the chance that someone will cut a wire is high. Of course, a remote monitoring site will detect a cut wire and elevate the security status of the facility while calling for repair. Power is the same with the exception that you can always generate power on-site, provided you are prepared for such a contingency.

The concept of redundancy and hidden security can play a role in communications and power. When planning on power for your facility, you should have it enter the property from two different areas and if possible from two different power company circuits. If one power company circuit fails, your facility continues to operate on the secondary circuit. If the secondary power cables coming into your facility can be hidden, it makes it much harder for someone to sabotage them. If both circuits fail, you continue to operate on locally generated power. In no case should you ever lose power totally. The same

---

[22] Especially if an operator is attracted to someone he or she sees on camera.

thing can be done with communications cables. The backups to communications cables are the previously mentioned cell phones and satellite communications that need to be integrated and reliable under all possible conditions.

Computers and Physical Security

At this point you should clearly understand that infrastructure and physical security rely heavily on computers. Some of these are computers that you control, and some you do not. Even management of your staff and of material transiting your facility is most likely done on a computer. Modern video monitoring systems are computer driven.

Computers allow us to do things that were impossible or not cost effective in the past. This caused expansive growth in use of computers in everything from door locks to ovens to air traffic and port control systems. They have roles in places unexpected to the average person. Computer systems are layered in extremely complex structures that are hard to comprehend. In fact, your average person may not even realize that computers have been embedded in many traditionally analog devices. An example is cell phones. Computers are important in almost all disaster response programs. Computer security breaches can cause disasters. Consider the power failure in North America that occurred August 14, 2003, that was attributed to a transmission circuit protection device that functioned correctly. The remainder of the power grid did not function properly.

Secure the Computers

This places computers at the center of priority for attacks. They are both the lynchpin to success and a central point of failure. They are a high target objective. There are lots of smart people in the world, and it is much easier to break into a computer system than to build a good one. Here are some facts about computers:

- They will do what they are told.
- They can be told to do things that you don't want.
- They can be subverted in ways that make detection difficult.

Computer security today is generally a failure. It gets in the way of getting the job done; therefore, it is normally bypassed or weakened. Even if security systems are implemented correctly, they are often bypassed for mission or operational reasons or are inadequate.

The largest single hole in computer security exists in all systems. It exists even in well-designed systems. There is no way to eliminate this problem, but you can reduce its scope. That problem is the "unforeseen."

For the purpose of this presentation, I will use the word *virus* loosely to mean all forms of attack software including viruses, worms, bots, Trojans, bombs, and all other forms, both known and unknown.

Most people feel that the installation of an antivirus–antispyware program will protect them from virus attacks. This assumption is false.

Virus, spyware, and all other scanners can detect only attacks that they already know about. They do not directly protect against zero day attacks, nor do they protect against stealthy, low-profile attacks in which a sample was never collected for analysis. A virus can be designed that makes collection a low probability.

There are methods in place that attempt to address the zero day problem in the form of behavior blockers and code emulators. These methods are only partly successful and cannot address viruses that were designed specifically to avoid this type of detection.

Heuristics also plays a part in zero day defenses, but the word *heuristics* has become a marketing buzzword and no longer has any real meaning. When it has meaning, it is because a manufacturer has defined it with a specific meaning. A good example is virus family detection. Computer viruses normally come in families of related but distinct programs. Each virus has a detection algorithm assigned to it. If there are enough viruses, an analyst will review all of the algorithms and design one new algorithm to replace them. This new algorithm will normally detect all of the known viruses, plus any new ones created in that same family.

Computers normally are the guards that protect themselves. That is, they run their own protection software such as antivirus. This is the same as putting the chickens in charge of guarding the chicken coop. There have been viruses that specifically target and corrupt antivirus programs to help spread the virus.

There are solutions to this problem!

If none of these methods is 100% effective, is there an answer? Yes. In addition to using a combination of a scanner and any of the other methods already mentioned, there is one tool that can detect almost all forms of attack, including software and human attacks: baseline integrity.

I will discuss the baseline tools that I am most familiar with, which are the products I designed.[23] Do understand that this is not a hidden sales pitch; I expect the user to source whatever product is best for his or her specific design.[24] I am going to discuss these specific tools because, when I designed them, I implemented what I thought was the optimal solution. One implementation is the CyberSoft CIT[25] tool that provides a report listing all files that have been added, deleted, modified, or duplicated. It also reports any files whose cryptographic signature has been flagged as dangerous. If you think about it, these are all the things that can happen to a file. There really is nothing else, so by reporting on these functions for all files in a computer system, it reports on everything that has happened in that computer.

A tool like CIT does not rely upon signatures, emulation, behavior, or heuristics. The only way to hide from this tool is not to touch the file system. Any attack that is solely memory resident can be removed by rebooting the system. In addition, this report provides the most important of all computer security reports, aggregate data. Using aggregate data, you can detect insider attacks from spies, as well as outsider attacks from hackers or workers who are not doing what they should, and all forms of software attack that touch the file system.[26] The only drawback is that an administrator must read the report. This drawback is not much, since very few files are actually touched during any 24-hour period; therefore, a trained administrator should be able to review hundreds of computer reports in only a couple of hours.

---

[23] www.cybersoft.com, www.cyber.com, www.radatti.com.

[24] Of course I will not object if you buy a program designed by me!

[25] I have always believed that program names should be short and descriptive. CIT is the Cryptographic Integrity Tool.

[26] For more information on aggregate data, see the VSTK Training Manual, Volume 1, Chapter on CIT, located at www.cyber.com.

There is a special version of baseline integrity called *self-healing*. The CyberSoft self-healing system is called Avatar.[27] It was designed for a "lights out" battlefield operation, with the assumption of proactive hostile intent. Avatar and all self-healing systems will put the baseline back where it belongs once an unauthorized modification is detected. This preserves the operations capacity of the system and disables the attack. In an emergency, preserving the operational capacity is critical. During an emergency your foe wants you to lose systems because that is exactly when you need to rely upon them most.

Another method of protection that is often used against both hackers and viruses is physical separation in an isolated environment. This is also known as compartmentalization. It helps against outsider attacks. It will not stop a virus attack or an insider attack. I have seen fully compartmentalized computers in which access is restricted to a few people. In these instances there are no external network connections, and movement of media is strictly controlled, however, they have had full-blown virus infections. This is a common enough problem that it is not a surprise to people in the industry. Bricks and fences do not provide protection against virus attacks or insider attacks. Compartmentalization does not solve the problem and may make solving problems, once an attack is noticed, even harder. The target is of too great a value to ignore.

There is an argument that a high-security computer system such as Trusted Unix or SELinux will provide protection against virus attacks. This is false. Dr. Fred Cohen proved that the fastest way to gain privilege in any system, including trusted systems, is by use of viruses. Tom Duff also proved this, in addition to being the first person to observe a virus infection originating from backup media. Only common computer operations are needed for attack software to be successful. A correctly configured, hardened, isolated, trusted computer system is not immune.

In Britain, there was a concerted effort to produce educational systems that could not be subverted by attack software. They invented and fielded the Acorn computer. This computer held the entire operating system and all programs in ROM memory. They were still infected. Today very few devices use ROM memory; they instead use Flash RAM. This would make it even easier to attack. Properly designed, secure hardware is not protection against software attacks.

There is a classification of software attack that is not often discussed but occasionally seen: a targeted attack. A targeted attack is a program that is targeted against a specific entity or role. For example, a virus could be released that looks for all systems whose name ends in .gov or .mil, or for all systems in a specific IP block for which public information identifies the owner. They can also be programmed to keep a very low and slow profile. This in effect is a sleeper virus. An attack can be targeted to a specific system but exist in the wild! This was projected by Sung Moo Yang in 1997.[28]

Sleeper viruses can be programmed to be opportunistic. They can exist for long periods of time before becoming resident in their target objectives. They also can become more active upon receipt of a trigger. Consider a virus that attacks cell phones. It has existed for years. Such a virus could be programmed to wait for a specific date before triggering a payload. What would happen if all cell phones in an area were knocked

---

[27] Reference the same training manual as for CIT for more information.

[28] This paper can be found in the white papers section of www.cyber.com.

out during a terrorist attack? At minimum, it would increase panic. At worst, it could also disable the cell phone network.

In addition to software attacks, there are basic computer security paradigms that other computer security professionals and I have been working on. We feel that significant changes need to be made in computer security implementation. Specifically, the user ID and password paradigm need to be replaced. We are not alone in this feeling. What is needed is a new paradigm that replaces the user ID–password paradigm with a method of identifying the actual user and the role he or she is performing. Passwords only identify that someone or something[29] knows the password. This new way will provide more protection not only against software attacks but also against hackers, spies, and unauthorized actions by authorized users.

The number-one problem for security is the human. Most security is dependent upon the user. How do we get users not to share passwords or be vulnerable to human engineering?[30] A survey by SonicWALL indicates that 44% of users do not memorize their passwords but record them. One of the most common places to record passwords is under the keyboard or in a nearby place. Humans can give up the system unknowingly and bypass security.

User education is not the solution. We cannot make users experts.

The solution to the password problem that most people think of first is biometrics. Biometrics does not work unless layered as part of a holistic solution. Biometrics is easy to capture and replay. Once input, a biometric signature is only a binary stream.

PKI is a failure, which means the way most people do online transactions such as banking and stock trading is not safe. It also means that secure networks that reply on PKI/PKE[31] encryption are not secure. A user ID–password or biometric login that relies on PKI to secure the transmission between access point and service provider should not be considered secure, and all information transmitted via PKI should be considered at risk of having been captured.

Another method of access is the use of a device, such as a handheld access device. Once the make and model of any device is known, specific targeted attacks can be devised. For example, an access control panel can be drilled and signals injected. It is the same with all electronics, including wall-mounted panels. Wires are normally just below the drywall and are easy to detect and perhaps inject signals. While this may sound wild and unreal, it is a real problem. Unattended gasoline pumps and bank ATM systems are attacked all the time using false faceplates known as skimmers. There are small devices that look like part of a keyboard cable that connect between a computer and the keyboard and will capture everything typed, including passwords.[32]

The following are factors of identification:

- Something you know (user ID and password)
- Something you know and something you have (security token device)

---

[29] Software attacks have been known to emulate actual users and use user ID–passwords to access restricted information.

[30] Also known as spoofs.

[31] PKI = Public Key Infrastructure, PKE = Public Key Encryption. Many people think they are the same thing, but they are not. You can use PKE without PKI.

[32] Skimmers are commonly available in black markets throughout Asia. Key capture devices are available at many computer stores and on the Internet for around $30. Key capture devices have legitimate uses and are legal.

- Something you know and something you have and who you are (one of your biometric identification factors—fingerprint, iris scan, facial characteristics)

The priority of factors is important. You should use what you have (security token) and who you are (biometric) first. The primary purpose of something known is to allow the users to signal between normal and duress situations. From a practical viewpoint, that means they have two passwords. For example, if their password is bunnyluv, their duress password could be bunnyluv9. It is also a backup in case their security token device is captured and their biometric identification is spoofed.[33]

Holistic Security Design Is Transparent and Passive

A complete solution must be designed from the top down for security! All signals on cables must be encrypted, or playback can happen anywhere.

Security systems should be transparent and passive, not requiring overt actions by users for security to accomplish its job.

As a user approaches a closed area, the security system should sense the user's security token and verify the user's biometric identity, unlock the door, and possibly even open the door. After the user passes, the security system should lock the door behind the user. Computer systems should sense the user's security token, verify the user's biometric identity, and unlock the system when approached by the user, and the system should lock as the user departs. All security interactions should be passive—without overt user action associated with security as the user accomplishes his or her job.

The paradigm of unique user IDs for login does not match real-life situations. For example, multiple users will use the same terminal without logon–logoff sequences between users; an operator may work a position and pass it to the next shift without a logoff sequence to preserve the desktop's state. The NSA[34] has separated identity from role in SELinux. With this construct, a person in the role can start the role's session and someone else in the same role can continue it. Actions are accounted to both the role and the person in the role. Moving from the user ID and password paradigm to a two- or three-factor identity security system causes this mismatch to stop. In many current systems with the paradigm of unique user IDs for login–logoff, security is circumvented by the sharing of user IDs and passwords to accomplish the mission. To implement real security with two- or three-factor identification, separation of role from identity becomes critical.

In the past, mostly antisocial teenagers performed virus and hacker attacks, but today they are in a minority. Organized crime is a more significant factor. All major nations, including hostile nations, have cyberattack and defense units. We already know that some terrorist groups have the technology for cyberwarfare.

All this can be implemented today. Most already have been implemented as separate parts and need only system integration.

---

[33] For more information, do an Internet search on "spoof biometric lock." You can also watch an interesting episode of the television show *Mythbusters* which demonstrates spoofing a biometric lock and computer login using a plastic thumbprint over a real thumb. The thumbprint was captured from something that was touched.

[34] The U.S. National Security Agency.

Radatti's Rules of Computer Security

Here are my rules for computer security:

1. The minimum cost of computer security failure is directly related to the value of the information lost, the loss of the system's operational capability, operational corruption, and the cost of cleanup.
2. The outcome and financial costs associated with a computer security breach are always higher than predicted and unknowable by all parties involved.
3. If a computer security solution is common enough that an attacker can study it, then its value may be greatly diminished.
4. The more interconnected a system or network, the greater the opportunity for a security breach.
5. Nothing is completely secure, but you can make it difficult to penetrate.
6. The job of computer security is to allow only the good guys to do their jobs effectively and efficiently.
7. Computer security is a never-ending task that requires constant vigilance.
8. There is no relationship between the cost of security and its effectiveness.
9. When designing security, it is critical not to forget things like wires, simple physics, and most important of all, the users.
10. A $1 million computer security system protected by a $20 door lock is worth only $20.
11. Intelligent people are always available, on both sides.
12. Formally prepare for a security breach before it happens.
13. Authorized users can and will perform unauthorized actions.
14. An exhaustively tested system is already obsolete and insecure. Constant vigilance requires a method of constant improvement without months of testing.

Disaster Recovery

One of the concepts in computer security that can also be applied to facilities' security is the concept of a disaster recovery site. Management of a disaster recovery site in the computer world involves either using offline media such as backup tapes or continuous synchronization to a duplicate set of computers at a remote location. While these concepts do not at first seem like they can be applied to transportation hubs, they can. Let's discuss a large imaginary commercial airport called ALPHA. The other airports used by ALPHA during weather closures are also large commercial airports but are not convenient to customers who would have to drive 2, 3, or more hours to get to them. These other airports are not a good choice for a disaster recovery site because they already are busy with their own traffic and because they are too far away for customers who would then have to drive home from that location. Extended use of these other airports not only would overtax their ability to operate but also would increase the road traffic significantly. What is interesting is that ALPHA operates in an area where there is a large number of military air bases that are closing. At least one of these bases, BETA, has more than one runway and has equipment that may be superior to that of ALPHA. These bases are closer to ALPHA than the other commercial airports are, have no traffic, are modern, and most likely could be had by the ALPHA airport authority for free.

Part of what operating a disaster recovery site means is the ability to operate for an extended amount of time. Plan for a minimum of several months to several years while the primary site is rebuilt. When planning for a disaster recovery site, consider the fact that all your rolling stock that survives the initial disaster can be relocated as long as a plan is in place to do so. The New Orleans flood is a perfect example of how not to do anything correctly. While there were plans in place, no one knew about them, and those who knew about them didn't follow them. Entire fleets of buses were flooded and destroyed while people in nursing homes drowned due to lack of transport. Local politicians did not take responsibility for controlling the situation. Commercial bus lines that sent aid were turned away. This is what will happen to you, if you do not implement plans and then distribute and practice those plans. The companies that supply jet fuel, sanitation services, food services, cleaning services, mechanical services, and all the other things that keep ALPHA in operation need to be aware of BETA and need their own plans for relocating within a specified period of time to BETA. For example, you must have jet fuel available at BETA on its first day of operation. If you feel that you can operate at other facilities for 5 days, then the jet fuel company must have its stock moved and operational at BETA within 5 days. Jet fuel for an airport is a Critical Path Management (CPM) issue. Most vendors will be happy to work with you on these plans because it means that they remain in business during the emergency.

Another issue that needs to be considered is that your facilities' security needs to be trained in advance for both ALPHA and BETA. Your normal BETA security team will be supplemented by team members from ALPHA while a small group of ALPHA team members remain at the now deactivated ALPHA site. You want your ALPHA and BETA teams to be comfortable with both sites. The same is true of your off-site monitoring. While the workload at BETA increases to what would be normal for ALPHA, the same is true for ALPHA. The workload at ALPHA will decrease to what is normal for BETA.

One of the things that people post–September 11 have to think about is the fact that some critical infrastructures are easy targets for terrorist attacks. Of course, all the planes taking off and landing are easy targets for a shoulder-launched missile, but that would affect only one, two, or three planes. On the other hand, the control tower has a great line of sight to not only the entire field but also outside roadways and properties not controlled by ALPHA. Where there is a proper line of sight, a control tower provides an unlimited view of potential terrorists. What would happen to your facility if the tower were knocked down? If you did your homework in advance, the tower would not contain people; the people would be relocated to a hardened bunker somewhere else on the property. Cameras can be located in the tower, on the tower, and on all the other buildings on the property. In fact, some airports have large office–hotel towers nearby, which could also have cameras and other electronics located therein.

Let's discuss train and bus terminals. With advanced planning, a bus terminal can be relocated within an hour or two, most likely nearby. This is because busses are independent and can move at will. This is not true with a train. Trains run on tracks and have schedules. People know that a specific train will be in a specific location at a specific time. This makes trains soft targets. Fixed routes, such as bridges, tunnels, and other hard-to-repair transit areas, become high-priority targets. To a certain extent, remote cameras can monitor these high-priority targets, but there is very little than anyone will be able to do besides reroute the trains. Sometimes that is enough, especially in places like the eastern United States, where rail lines are well built and rerouting is possible. At the same time, train terminals may appear to be hard to move, but that is not actually true. Rail lines often travel past many abandoned industrial properties. These properties, due to their nature, often have rails going into and out of them, in addition to good access to roads, parking, industrial utilities, etc. In fact, railroads are unusual in that they can generate their own electricity and move large volumes of material quickly. Anywhere a large open space is available, an inflatable building or tent could be erected quickly. Train engines are power plants, and wireless communications means that phone, Internet, and private networks can be in place quickly. Other than rerouting trains to bypass the out-of-service area, a train station could be operational in one day. If you think about this, you will see that this type of plan can be put in place for everything.

In addition to having a standby facility, there is an alternative solution for when this is not possible. Hardened buildings are something that most people think were abandoned with the cold war. They bring to mind missile bunkers and other military facilities. Not true. Using modern materials, science, and engineering, any building can be built as a hardened building without its appearing as such. Modern blast-proof ceramics and glass are available in any color, shape, and size. They can look like bricks, stone, marble, stucco, or any other finish desired. This type of material can be added to the outside of a building and make it not just bullet proof but bomb proof. In fact, whenever making plans for physically hardening a building or structure, always consider making it blast proof instead of bullet proof. The Oklahoma City bombing of the Murrah Federal Building in 1995 is a good indicator of the real threats. Being bullet proof is a given with bomb proofing, but the reverse is not true.

Using standby mesh nets means that communications in an area do not have to be disrupted if the primary network dies, even assuming that someone managed to destroy hardened cables. In the event of a total loss of local communications capability, these facilities can be provided locally. It is even possible to have a "pop-up" cell phone site that becomes active only if the normal cell phone service is disrupted. Telephone, Internet, and television service can be provided by satellite in such a way as to be transparent to the facility occupants. There is no reason why high-quality, high-speed communications should ever be disrupted by anything that happens outside a hardened building. With off-the-shelf technology, the work of integrating it into one seamless system is significant, so plan on using people with experience doing this type of work.

Air filtration can make a building immune to attack vectors that are based in this medium. Even if an attack is not air-based, there may be significant air pollution caused by the attack. I do not think anyone can forget the television scenes of clouds of white dust hanging in the air after the World Trade Center towers fell on September 11, 2001. That pollution was long lasting and entered buildings that were closed. The particle size of dust along with its chemical makeup can cause damage not just to facilities but also to people who inhale the material. Sometimes the health risks come years later. For that reason, a hardened building must maintain positive pressure at all times with filtered air. The air pressure inside the building is greater than the air pressure outside the building, disallowing outside air and its contents to enter the building except via the filtration system.

Humans must have water to survive. Water filtration and storage can make a building independent of the need for outside water. The fact is that water filtration as a preventative measure is a good idea even when no attack has taken place. In 1993, a waterborne cryptosporidiosis made 403,000 people ill and hospitalized 4,400 people in Milwaukee, Wisconsin. In 2001, North Battleford, Saskatchewan, Canada, between 5,800 and 7,100 people, became ill from the same pathogen. In 2005, there were outbreaks in Seneca Lake State Park in New York and Gwynedd and Anglesey in North Wales. This happened again in 2007 in Galway, Ireland; North Walsham, Anglia, England; Montgomery County, Pennsylvania; and 20 counties in Utah. This was just one pathogen. There are thousands of possible pathogens and millions of possible chemical contaminates that can make water unfit for human use. Even if the building has a deep groundwater well, the water still must be filtered before use. In some parts of the world, it is common for spring water to contain heavy metal contaminates including arsenic. Even though the water is from deep underground, that does not mean it is uncontaminated.

All buildings have to have electricity to be fully functional. A hardened building is even more sensitive to electrical vulnerability. The air and water filtration systems require power. The communications systems and internal control systems of the building all require power. Air-conditioning and heat are required for the comfort and health of the inhabitants. I was in New York City during the great power blackout of 2003 and saw what happened to large buildings when their generators ran out of fuel. You do not want this to happen to your facility. Modern turbine electrical power plants are small and efficient and can provide air-conditioning and heat as a by-product of making electricity. Fuel for 30 days of normal operation of the building can be stored on-site. Sometimes that means an underground storage tank, and sometimes it means tanks in a sub-basement. Either way, this is a critical system that must not fail or run out of fuel. Everything depends on the ability to have power available, which is why this part of the plan must be well designed and executed.

In areas where flooding is possible, the basements and first floor should be designed in such a way as to not allow the admission of outside water. This is especially important if the sub-basements are used to house support equipment.

Modern food management means that the building occupants might actually eat healthier and better-tasting foods during a crisis than they normally would eat daily. Food service does not have to limit itself to dried beef and crackers but can include many luxury foods such as premium ice cream, cakes, fresh fruit, and assorted other foods. Remember that the people in this building will be well aware of what is happening outside and may in fact feel trapped. It is necessary to remember people's emotional needs when designing security, and luxury foods are a critical item.

Of course, waste management planning is also an issue that must be considered. Outside sewage disposal may not be available during the emergency. Many cities rely upon electric pumps to move sewage. If the electricity outside the facility is out, on-site storage must be planned.

Buildings of this type can be designed to be fully self-sufficient for 30 days or longer with no outside utilities required. Such buildings are currently expensive, but consider what is inside them. Would a disaster control center for a state best be housed in a building such as this? In at least one state, the disaster control center was located on a fault line because the land was cheap. A govermental entity would have to hope their disaster is not an earthquake. Consider what else should be housed in building of this type, including: Air traffic control centers, emergency standby computer centers, financial clearinghouses such as federal reserve banks, armory depots, emergency supply depots, electrical grid control centers, and, in fact, all the utility control centers and hospitals. The list can go on for a long time, and that does not even consider such entities as embassies, housing for visiting dignitaries, high-risk buildings like the White House, federal agency headquarters, corporate headquarters, or houses for famous people who might be attractive targets. Even small buildings such as pump houses can be hardened in this way.

While it is not always possible to harden an existing building to the extent that a building constructed for the function is, it still is possible to harden existing structures to a point that is very useful. Consider not only hardening buildings such as airport control towers but also blast-proofing tunnels, bridges, critical pipelines such as fiber-optic arteries, and any other high-value objects where the cost of hardening justifies the expense. Buildings of this type are not just something that might be available sometime in the future. They exist today and are something you can consider when putting together your security plans. Smart, sustainable, secured building technology and design are rapidly evolving to fill a major need.

Time Management Is Part of Security

The single most important thing you must have when dealing with a crisis is planning. Planning is something done in advance and includes training. NASA was able to put a man on the moon by using extensive planning. The effort that went into that historic venture was significantly greater than anything you will be called upon to do in crisis planning; however, the methods used by NASA are available to you. The most important sources for these methods are handbooks, Gantt charts, and critical path management (CPM). Handbooks are self-evident. They tell people what they need to know during a crisis. If you decide to make your handbooks available via computer, you also need to make them available in a bound printed-paper form. The old fashioned three-ring binders with a printed-paper handbook never need batteries to be read.

Gantt charts are a way of planning events in a time frame. While this may not seem important, it is one of the most important things you will ever do when planning for crisis management. I am sure you remember the old saying about closing the barn door after the horse has left. In this case, a Gantt chart of closing the barn door before the horse leaves would have prevented that event.

Your crisis has a timeline. If you have advance notice, there is a pre-crisis timeline that needs to be planned. Then there is the actual crisis time, which can last from minutes to months, followed by post-crisis. Each event in this timeline must be planned, in order and on time. For example, one of the first things that should happen post-crisis is the refitting of the building so it is ready for the next crisis. Fuel needs to be brought in and waste needs to be removed. In addition, there are critical path management issues that need to be noted as such on your Gantt charts. These are events that must happen in order for other events later in the timeline to take place on time. An example of a critical path management issue is inspection of all stored materials to ensure they are ready for use. If any material is not ready, it becomes a down-the-line crisis in itself. This inspection becomes a critical path management issue because it prevents a crisis later in the plan.

Conclusion

In conclusion, physical and computer security have been merging for many years, and we can expect this to accelerate. In addition, old concepts in physical security such as locks and gates are obsolete and need to be replaced by holistic security concepts. Out-of-the-box thinking, such as having critical structures hardened and/or having replacement structures at the ready, helps reduce the effects of security failures. Security, both physical and computer related, is one of the greatest games in the world where the stakes are human lives, futures, and fortunes. This "game" is deadly serious and will be won by the best "players" on either side. Security engineering is a rewarding field for those who want a challenging job with the payoff of making a positive difference in the world. Integrated systems security involving multiple disciplines will be required to have the most effective infrastructure and preventative system. The challenge or threat will always remain and must be constantly addressed.

9

Intermodal Transport Security Technology

Robert Sewak, Ph.D.

Objectives of the Chapter:

- Provide an overview of the rationale and basis for pursuing cargo container security
- Offer a historical and economic foundation upon which to justify intensifying maritime security
- Provide a review of selected initiatives, designed to help achieve greater supply-chain security
- Discuss the various methods and technologies being investigated and developed to that end

Hurricane Katrina, the Northeast power-grid failure, and the collapse of the I-35W Bridge should have served as wake-up calls to the neglect and weaknesses in the critical infrastructure of the United States. It seems that, at almost every possible level, the tendency has been to react only after the fact to the need to "do something"—to guard against threats, both natural and man-made.

Introduction

Obviously, the devastation of 9/11 was an insufficient warning of the fact that "... our shores are not immune to the horror we know today as global terrorism." Because, even to this day, one of the most precarious and highly volatile pieces of the national infrastructure puzzle exists in the intermodal transportation system—where the unassuming cargo container has a principal role yet remains unconscionably exposed.

Global commerce is utterly dependent on the movement of shipping containers. Containers, in terms of value, carry about 95% of the world's international cargo. More than 48 million full cargo containers move between major seaports each year, and containers move through the countries of the world daily on trains, trucks, and barges. And, while it is

FIGURE 9.1 Standard Cargo Containers

possible for containers to transport nuclear materials, drugs, biological matter, arms, and chemicals, as well as criminals and/or terrorists, less than 2% are subjected to in-depth inspection. So, with almost all the world's commerce flowing through the infrastructure of the maritime systems within the global supply chain, it's not just the United States that needs to be vigilant and prepared; it is every country that trades goods and engages in commerce internationally.

The first container vessels, built in 1968, had a capacity of 2000 20-foot containers. Both containers and vessels have grown substantially since then. Containers have doubled in size (with the term *twenty-foot equivalent,* or *TEU* established), and newer vessels have a capacity of 8,000 teu's. Vessels of 14,000 teu's are planned, while even larger containers (40- and 45-foot equivalents) are being put into service. Modern river barges have a capacity of several hundred containers. And container traffic is by no means confined to water: Practically all goods that arrive or depart by ship from a port facility have been or will be transported by truck or rail. Trucks dominate land transport around the globe.

Significantly, 50% of maritime container traffic is handled by only 10 international operators, and 75% is handled by 20 operators. Maersk Sealand, a Danish company, is the largest container shipping company in the world (no major U.S. companies have a dominant presence in the business). Competition among shipping lines is high, and customers

FIGURE 9.2 The Largest Container Ship Afloat

may move easily from one company to another to find the quickest and least costly way to ship their goods. The cost of container transport is low: The shipping cost of a full container from Lyon, France, to the East Coast of the United States is on the order of $2,833, while the cost from Los Angeles, California, across the Pacific to Korea is about $4,250.

The container transport system is complex and involves a number of different players, while the basic documentation for each container, a simple bill of lading that is created by the shipper and specifies the content of that container, is both the beginning and the end of the document trail. After a container is packed, it is sealed by the shipper, normally with a simple mechanical tamper-indicating seal used primarily for reasons of liability for the transport company. The World Customs Organization made this note, "High security manual or mechanical seals can play a significant role in a comprehensive container security program. But it is important to recognize that container security starts with the stuffing of the container and that seals do not evidence or guarantee the legitimacy of the container's load." After a container is sealed, it is transported to a container terminal such as San Diego, California, which serves as a hub for transshipment. For example, estimated inland container traffic arrives at San Diego by road (86%), train (12%), and barge (2%).

Because an overwhelming majority of finished goods, raw materials, and component parts are transported in cargo containers (along with bulk cargo in tank ships) that move by sea, just as this containerized cargo system became more efficient and inexpensive, its weaknesses became exposed—economic and strategic weaknesses, should the system fall prey to disorder, disruption, or disaster. If a weapon of mass destruction (WMD) were detonated in a container, the steady operations of container terminals and seaports around the globe—as well as fleets of countless ocean carrier vessels—would suffer dramatically. The risk, at the world trade level, would increase experientially should such an occurrence take place in a U.S. seaport or somewhere inside the U.S. intermodal transportation system.

A single container could be offloaded from a container ship directly to a truck chassis and be out of the port facility gate in a matter of minutes. Containers loaded with cargo at a manufacturing site, say in China, can be delivered directly to a wholesaler, retailer, or distributor, say in Pittsburgh, without the container ever being opened. A container ship at any state-of-the-art port facility can now be loaded or unloaded in a matter of hours—a job that, just 20 years ago, would have taken days to accomplish. Containers are stacked eight high, one on top of another, in the cargo area of these huge ships until the maximum load limit has been reached. Then they are chained down to prevent movement while the ship is at sea. To allow access to the stacks, approximately 8 inches of free space is made available between them. One container ship will most likely carry a variety of cargo: electronics, hard goods, soft goods, finished goods, and raw materials; however, all anyone can "see" are these stacks upon stacks of huge and, except for color and markings, identical metal boxes.

Consider an event, such a "bomb in a box," which causes the U.S. government to shut down all the country's ports and borders, even temporarily—as was done with air transportation on 9/11—as a demonstration of the greatest harm not coming from an external force and what they are capable of doing to us, but what we are capable of doing to ourselves in reaction or response to that event. It's almost impossible to imagine what the result of such a shutdown of the whole system would be like in the aftermath, especially with the anxiety of not knowing "what's next" still present. The horrific impact of death, damage, and injuries would no doubt be paled by the disastrous economic toll that would ensue.

Certainly as a nation we should be committed to hardening every aspect and element of our critical infrastructure, making them capable of withstanding the impact of such a catastrophic event, whether a natural disaster, an industrial accident, or a terrorist attack, such that we can rapidly return to a relative state of normalcy.

In the area of container security, Congress has passed and the president signed a homeland security bill that calls for 100% screening of all containers, while in overseas ports and bound for the United States, to be in place within 5 years. It goes without saying that a requirement like this is heavily burdened with enormous operational, economic, and technological hurdles, not mention the nightmares of international politics, regulations, and governance; however, successful implementation would provide one of the most meaningful positive initiatives ever undertaken to secure the complete supply chain around the globe.

History

For the United States, the oceans have served as battlefields, defensive barriers, food providers, highways of commerce, sources of energy, and science laboratories. However, in more recent times, the realm of maritime traffic has become a vulnerable harbinger of fear and threat, virtually an exposed accessway for potential enemies. The point in fact is that an overwhelming majority of the vessels that visit these shores are engaged in proper and lawful activity and present no threat to the United States, in addition to which their passage is vital to the U.S. economy. What has come to light is the inordinate and complex task of attempting to identify an extremely small number of potential threats that might exist somewhere inside this large, complex, and economically critical system of global commerce. And, as if the job of identifying and neutralizing these possible threats were not daunting enough—as well as doing so as far away from these shores as possible—there is the equal dilemma of ensuring that whatever is done not disrupt or interfere with the speedy and ceaseless flow of international commerce.

In 2007, more than 200 million container transits will take place between the world's ports, making the maritime industry one of the most time and money sensitive in the world. Huge investments have been made in ships, crews, and technology. And, with so much at stake, little is spared to obtain maximum speed, using computer-generated routes to gain the advantage of wind, current, and weather so as to aid both speed and fuel efficiency; and to coordinate time of both arrivals and departures with the availability of dock space, equipment, and people to handle loading and unloading around the clock. No stakeholder in the entire supply chain can afford to have these assets sitting at a pier, riding anchor, or waiting to get cleared.

What this amounts to is that containerized shipping has become and will continue to be the principal way of transporting goods and material throughout the world. So, with all that technology and sophistication, why do we have no reliable, efficient, effective, and acceptable way to find out where these boxes are and what's inside them—from anywhere on the globe—for any of the host of stakeholders involved?

The Threat

Anyone who wants to move contraband has figured out that the safest way to hide it is do so in plain sight and move it through the normal stream of commerce—possibly from country to country, and often hidden in cargo containers. So if and when the next attack

FIGURE 9.3 The Port of Elizabeth, New Jersey

takes place, it very well might, or most likely will, come from inside some ordinary, routine, unexceptional, and commonplace aspect of everyday life.

There are as many theories for countering this threat as there are possibilities; for instance, air travel security has improved and increased since 9/11; certain hardening has taken place around select power facilities; large public events have taken on new and tighter security measures; and security in and around government buildings, banks, and certain office structures has been strengthened. Missing from this ever-expanding list, however, are the estimated 16 million shipping containers that are freely moved about or nestled ubiquitously throughout the country. Containers are literally everywhere: in freight yards, warehouses, container terminals, and neighborhood parking lots as well as on truck chassis, railcars, and river barges. To grasp the gravity of this issue, consider this: Approximately 2 miles outside New York City, just a bit south on the New Jersey Turnpike, are Newark's airport and Port Elizabeth, where one can see stacks of hundreds of containers, eight high, just a few hundred yards from the highway and the active airport runways. Once a container clears the port of entry, it can go almost anywhere under a cloak of invisibility and with little or no scrutiny.

Almost 5 years ago, the Department of Homeland Security (DHS) issued its Comprehensive Approach to Port Security, citing the following nine key factors for implementing the strategy for securing a vessel—its cargo, crew, and ship:

1. International Ship and Port Security Code (ISPS)
2. Container Security Initiative (CSI)
3. 24-Hour Advance Manifest Rule
4. Customs–Trade Partnership Against Terrorism (C-TPAT)
5. 96-Hour Advance Notification of Arrival
6. Mariner documents
7. Offshore strategic boardings

8. High-interest vessels

9. Integrated Deepwater System (IDS)

The DHS cited the following five factors for protecting the port:

1. 2002 Maritime Transportation Security Act (MTSA)

2. Port security committees

3. Sea marshals

4. Maritime safety and security teams

5. Armed helicopters

The DHS also cited these five factors for port security at the facility and infrastructure:

1. Operation Safe Commerce (OSC)

2. U.S. Automated Targeting System (ATS)

3. Radiation, chemical, and biological screening

4. Port vulnerability assessments

5. Dangerous cargo handling

Within the preceding manifesto was a "story" called "Watching Container #3091778," in which a fictional cargo container of auto parts shipped from a supplier in China through the ports of Hong Kong and Los Angeles was followed. The container was shipped from a Chinese manufacturer to a large auto parts supplier in Riverside, California. The interesting part of this story is that the container was subjected to several of the layered strategy elements of the port security plan—from before it left, through the voyage, and when it docked at the U.S. port. The most logical, effective, appropriate, strategic, and important measures, however, are those that take place *offshore*!

Container Security Measures

Over the course of the past 5 years, a number of the initiatives from the preceding lists were launched, several of which are discussed in the following sections.

Customs–Trade Partnership Against Terrorism (C-TPAT)

C-TPAT is a public–private partnership designed to strengthen the global supply chain by the voluntary agreement of private sector partners to adopt a wide range of security measures. For example, at the point of stuffing, containers and trailers are to be sealed using an approved high-security seal, and an inspection is to be conducted to assure the physical integrity of the box and the protection of containers and truck trailers against introduction of unauthorized material and/or persons. Additional requirements include personnel security, procedural security, information technology security, physical security, and security training–threat awareness.

C-TPAT is open to a wide variety of industries and partners within the international commerce community, including importers and exporters; freight consolidators; air, land, and sea carriers; air freight consolidators; port–terminal operators; warehousing and distribution operators; and foreign manufacturers. In return for their participation, C-TPAT members are to receive certain benefits intended to reduce the level of scrutiny that participants' shipments will be subjected to upon entering the United States. Members with C-TPAT certification are to have the risk profile on their shipments lowered, therein reducing the likelihood of extensive documentation and/or physical inspection. In addition, they are to receive access to Fast lanes at the Canadian and Mexican borders for expedited cargo processing through. C-TPAT is built upon a protocol of self-reporting and self-policing policies. C-TPAT members get certified by Customs and Border Protection (CBP) based solely on self-reported compliance with mandated security measures and are vetted, in part, based on their prior history regarding violations and compliance with customs regulations. The Government Accounting Office (GAO), along with other critics of C-TPAT, question whether CBP has in place sufficient procedures and personnel to verify whether or not C-TPAT members are indeed compliant with mandated security measures.

Container Security Initiative (CSI)

At this time and around the globe in between 50 and 60 seaports, teams of both CBP and Immigration and Customs Enforcement (ICE) personnel are assigned to collectively account for about 90% of the containerized freight destined for the United States, per today's CSI initiative. Under the program, the host nation's customs officials work with CBP personnel in examination of high-risk containers located in foreign seaports before they are loaded on vessels bound for the United States. Statistically, CSI ports scan less than 1% of containers, and fewer than that are ever opened and inspected. The three key CSI components are as follows:

- Identifying high-risk containers. This is done using automated targeting tools, based on advance information and strategic intelligence, on containers considered to pose a potential risk for terrorism.
- Using large-scale X-ray and gamma-ray machines and radiation-detection technology to prescreen high-risk containers without slowing down movement of trade.
- Prescreening and evaluating containers by doing so as early in the supply chain as possible, preferably before containers get loaded onboard.

The 24-Hour Advance Manifest Rule

All cargo shipping carriers, except bulk carriers and approved break bulk cargo, must provide detailed cargo descriptions and complete consignee address information 24 hours before the cargo is loaded (at the foreign port) for shipment to the United States. Prior to this rule, a great percentage of cargo manifests would simply declare the container's content as "freight—all kind" or "general merchandise." This type of information is useless when trying to determine if a container poses an element of risk. Failure to comply with the 24-hour rule can result in a "do not load" notice, as well as other penalties.

Automated Targeting System (ATS)

ATS is a system that uses enforcement and commercial databases and that cross-references manifest information provided through the 24-hour rule. Designed to detect irregularities and identify cargo that could be high risk, the system analyzes data and rates the order of risk by selective rules and algorithms and then determines the need for additional scrutiny. When certain thresholds are reached, cargo may be targeted for further action by CBP, which can include physical inspection.

The 100% Screening Requirement

The concept of 100% radiation screening being required for all U.S.-bound containers was argued over for years; however, in August 2007 the Homeland Security Bill ended the discussion and made the requirement a reality. The law embraces the recommendations of the 9/11 Commission as signed by the president calling for radiation screening of 100% of U.S.-bound maritime cargo before loading at foreign ports, to be implemented within 5 years. However, the secretary of the Department of Homeland Security can extend the deadline in 2-year increments in the event of insurmountable technical or other hurdles. The far-reaching action by Congress with its 100% screening requirement outpaced the Commission's desire for an intensified effort to track and screen high-risk cargo through more practical and sophisticated means, such as a layered method of integrated intelligence gathering and engagement of both foreign governments and the international private sector to assist in the targeting and scrutiny of potentially dangerous cargo.

The Bush administration, DHS leadership and component agency heads, and numerous business groups including the Chamber of Commerce, the National Retail Federation, and the International Cargo Security Council, as well as—not surprisingly—shippers, all strenuously opposed this component of the bill. They contended that the requirement is beset with problems, not the least of which is the fact that the technology to perform the screening may not exist, that it is not clear precisely what is to be scanned, that how the cost will be allocated has yet to be addressed, and that the process may delay the flow of inbound goods. Some of these arguments have merit; others do not.

Secure Freight Initiative (SFI)

This is a pilot program designed to test high-volume scanning at six ports in Pakistan, Honduras, Britain, Oman, Singapore, and South Korea. Containers arriving at participating ports are scanned with both nonintrusive radiographic imaging and passive radiation-detection equipment placed at terminal arrival gates to screen incoming containers. Relay containers—that is, containers being transferred from ship to ship—also are scanned. Sensor and image data concerning U.S.-bound containers are transmitted in near real-time to the National Targeting Center, where they are combined with other available risk data to improve risk scoring and targeting of high-risk containers, thus enhancing the opportunity to conduct further scrutiny of suspect cargo while still overseas. A side benefit of SFI is that it will probably serve as a good indicator of the practicality of the 100% inspection requirement.

Others initiatives include the following:

Operation Safe Commerce (OSC): a joint industry–government project implemented by the Transportation Security Agency (TSA) to demonstrate advanced techniques and technologies for tracking intermodal shipping containers through use of various seals and tracking technologies.

Smart and Secure Trade Lanes (SST): a supply-chain security initiative focused on deploying an end-to-end security solution across multiple global trade lanes. SST is one of the largest commercial cargo security programs currently in operation, with over 70 companies involved.

Megaports Initiative: intended to provide early detection of possible illicit trafficking of nuclear materials through foreign ports. Under the program, the National Nuclear Security Administration would install radiation-detection equipment at foreign ports in order to provide foreign governments with the ability to screen incoming, outbound, and transshipped cargo while posing only a minimal threat of delay to port operations. Some 70 ports of interest based in 35 countries have been identified; however, today this initiative is operational only in Greece, Bahamas, Sri Lanka, Spain, Singapore, and the Netherlands.

The Problem Persists

Based on the preceding information, one might assume that a great deal of progress has been made in the process of making the world's maritime commerce and the global trade and transportation system more secure; however, such is not the case. First, for the most part each of the programs has gone along independently, with little or no coordination between them. Add to that the continuing funding issues, questionable claims of efficiency and performance, and lack of governance and cooperation, and you see that the problem gets bigger.

On March 28, 2006, before the Senate Homeland Security and Governmental Affairs Committee, Dr. Stephen Flynn of the Council on Foreign Relations, and author of the books *America the Vulnerable* and *The Edge of Disaster*, proffered a scenario as to how a resolute terrorist could thwart these various initiatives. Flynn's scenario went something like this: A load of sneakers, manufactured for a name-brand company, gets stuffed into a container in Indonesia. Somewhere along the route to the port, the driver takes a detour and the container gets breached—without disturbing the mechanical seal. Some sneakers are removed, and a dirty bomb wrapped in lead shielding is put in their place. At the port, a coastal feeder ship picks up the container and carries it to Jakarta, where it is loaded onto an Inter-Asia ship for transport to Hong Kong... where it gets loaded onto a giant trans-Pacific container ship bound for Vancouver. Because the shipment came from a trusted, name-brand company and a member of C-TPAT, CSI inspectors in both Hong Kong and Vancouver do not have "cause" to identify (mark) the container for further inspection. The construction of the destructive device prevents it from being detected by any radiation screening the container passes through. From Vancouver, the container gets loaded directly onto a Canadian Pacific railcar, where it moves to a rail yard in Chicago. When the container reaches a pre-selected distribution center in the Chicago area and is opened, a triggering device, attached to the door, detonates the bomb.

What Flynn's scenario points out is that there is no way to determine where the security compromise took place, and therefore, a logical presumption would be that the entire supply chain had been compromised and everything and everyone would be presumed to present a risk for a follow-up attack. Further, all the existing container and port security initiatives would be compromised by the incident. Governors, mayors, and others charged with public safety, along with the American people, would lose faith in the risk management strategy that the government had put in place and doubtless demand an immediate 100% screening of all cargo entering the United States—and once again, the United States would have self-imposed a worldwide embargo.

Technological Solutions

Currently, there are two types of scanning systems being tested at selected seaports around the world to monitor cargo containers. The first and more widely used is radiation portal monitoring (X-ray), designed to detect the existence of radioactive material in a container. The second is the gamma ray, which scans the contents of a container in search of dense material that could indicate the presence of explosives or explosives shielded with lead. Though these technologies offer increased security, neither can detect the presence of biological or chemical agents. Nor can they establish whether a container has been breached at some time from its point of stuffing to its final destination.

"Smart" containers offer a way to increase the global supply chain's security for the "good guys" while making it more difficult, risky, and problematic for the "bad guys" (terrorists, criminals, etc.) who traffic in stolen goods, weapons, counterfeit goods, drugs, humans, and more. Today there are essentially two types of smart containers:

1. The first employs radio frequency identification (RFID) tags that get installed inside the container (or on its door) and are connected to a sensor that can detect when or if the door was opened.

   RFID systems have some inherent and important limitations. Foremost is that the system transmits only in response to a query initiated from a separate transceiver—allowing for a long period of time between a breach and the report of that breach. Next, being able to detect the opening or closing of a container's door indicates that the container has not been breached by some means other than the door. Finally, RFID systems by themselves typically do not provide continuous geo-location information.

2. The second is a much more sophisticated system, designed to use a group of sensors capable of detecting any number of different "events," including change in temperature, shock, breach of the container, speed, position, date and time, and so on. These systems typically use satellite communications technology or a combination of radio frequency technologies (RFID, GSM, GPRS, GPS, and satellite) to communicate their information.

FIGURE 9.4 Cargo X-Ray Scanning

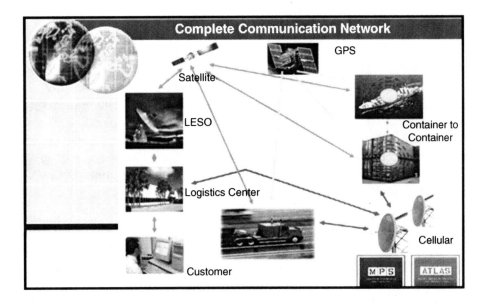

FIGURE 9.5 AEGIS Communication Network

A truly "intelligent" or "smart" container should have capabilities that exceed the preceding. It also should be able to

- Have its security system at its point of stuffing
- Provide for electronic capture of manifest data from stuffing, through departure, to delivery–destination
- Detect a breach anywhere within the body of the container
- Report the breach in real-time or near real-time and provide a time stamp and geographic position
- Provide geo-location information throughout its movement anywhere, when polled, and provide an automatic position report when it is off or stopped along a designated route of travel
- Report any "other" event, according to an established protocol, including but not limited to change in temperature, shock, speed, etc.
- Recognize and record opening and/or closing of the container at all points along the route, including final destination
- Be capable of interfacing with the different software programs used by the shippers, freight forwarders, and carriers

Intelligent containers should and could provide multiple benefits on a number of levels and a number of fronts. For example, security: A smart container can serve as a major deterrent to terrorists and/or acts of terror. Additionally, smart containers can serve as an ideal forensic tool by providing detailed information and records, as well as data logging, in the event that a successful attack is launched. Commercially, every stakeholder in the supply chain can and will benefit by having a much greater level of visibility and transparency in the processing, movement, and status of their cargo.

Container Tracking Technologies

Since that fateful 9/11 wake-up call, a host of different and diverse technologies, all especially designed for container security, have been tested, tried, and evaluated separately and in all manner of combinations, including GPS, satellite messaging, sensors of every possible nature from radioactivity to odor, cellular messaging, RFID, Bluetooth, Ultra Wideband, Wi-Fi, optical character recognition (OCR) technologies, and ZigBee. To some extent, these technologies either have been or are being employed, albeit in varying degrees and with varying levels of success or satisfaction in the emerging market of electronic container tracking. Following is a partial listing of some that appear to be gaining recognition and acceptance:

> **GPS:** The U.S. global positioning system (GPS) is operated by the U.S. government and consists of 24 Navistar satellites, which broadcast positioning data to earth-based receivers. GPS data are distributed nearly everywhere, so position–location can be known in all parts of the world; however, for real-time location data, a system that sends the data back from the container is also necessary. Because of the way containers are stored, GPS signals are frequently blocked, severely limiting use of GPS in container tracking. However, a handful of firms are working on devising more advanced methods for obtaining the signal.

> **Satellite and Cellular Communications:** Long-range communications between the container and central servers can be facilitated by satellite and cellular messaging systems. Satellite is quite expensive compared with cellular systems but has virtually universal coverage, so it can relay status messages and GPS data from nearly anywhere in the world. Satellite messaging systems will no doubt be utilized in developing regions that lack cellular infrastructure. Low-Earth Orbit (LEO) satellites are most likely to be used for sending short data messages because their cost is relatively inexpensive compared with that of traditional satellite services.

A number of commercial telematics service providers are using short-packet data services to relay information from North American trucking fleets to proprietary resource and enterprise management systems. These data transfers operate over analog cellular networks. Digital cellular networks, mainly GSM–GPRS technology, are being considered for container tracking, as nearly everywhere coverage is available in Europe.

> **RFID:** The most significant deployment of electronic container tracking has been put into practice with radio frequency identification (RFID). A number of traditional seal manufacturers have developed electronic cargo seals that use RFID technology in an effort to have security verification become more easily automated. Ultra-high-frequency tags and readers represent the majority of RFID applications for containers. In marine and port terminals, microwave frequencies are also present. RFID use for large numbers of containers will require significant outlay of capital because the functionality of the RFID network necessitates infrastructure at every chokepoint where cargo status is important. This can translate into a high initial cost but more reasonable costs thereafter.

> **Wi-Fi:** Wireless fidelity (Wi-Fi) can provide intermediate-range data transfers at ports, in marine vessels, and in container yards and terminals. Wi-Fi already plays a large role in some of these fixed locations as an RTLS (real-time locating system). While

this technology helps locate assets in a yard, such as heavy equipment and cranes, the role will expand for Wi-Fi to provide rapid data access. Wi-Fi–enabled RFID readers, handheld or in a fixed location, can transfer data from RFID-tagged containers to a central location for data storage and verification.

Ultra Wideband, Bluetooth, ZigBee, and OCR: Ultra Wideband (UWB) is a wireless communications technology that differs fundamentally from all other radio frequency (RF) communications in that it achieves wireless communication without using an RF carrier. A handful of UWB vendors are targeting the container market but are in the initial testing stages and appear to be focused on specialized and more high-end security solutions.

Bluetooth has gained momentum in other arenas but as yet not been practically applied to containers. Similarly, ZigBee has been discussed as an application; however, only limited plans and applications have proven successful employment, and insofar as determining when and how it gets to market, may well be dependent on the inventive thinking and novel employment of methodologies conceived by those who toil in the fields of innovation.

The International Standards Organization (ISO) has decreed that all containers have a number emblazoned on the outside of the container. This standardization has allowed OCR technology to thrive at ports and container yard facilities.

Maritime Piracy Technology

Due to the increase in the Annual International Piracy Index, maintained by the International Maritime Bureau (IMB), prior-to-port vessel and cargo screening technologies are developing. The government of Nigeria, which in 2008 has been the recipient of the most

FIGURE 9.6 Sea Sentinel System

pirate attacks (21%), has awarded a $750 million contract for installing a Sea Sentinel system. It will consist of five offshore platforms, each containing a helicopter pad and living quarters for a 15 person crew for identifying passing ships via sonar and scanning hulls. In addition, there will be 160 smart buoys distributed to support the Sea Sentinel system network. Radar tracking will be deployed 300–600 miles offshore and integrated with the offshore platform security surveillance networks. It will serve the Nigerian Port of Lagos and Port Harcourt and will be installed by Sea Away, a growing seaport security firm.

Summary

The day is long past when mere debates on the merits of political, economic, operational, or technological issues were meaningful. Today combined stakeholders around the world that constitute this critical industry must join together in implementing dependable, reliable, and viable solutions to the known weaknesses in the existing system. Technology exists that is capable of protecting global commerce at the most sensitive points along its path through the supply chain, as well as in adopting both intelligent container systems and placement of scanning and radiation screening in ports where the highest percentage of internationally transited containers come and go.

As previously stated, intelligent containers have both security and commercial benefits. They remove all questions as to whether or not the goods stuffed in the container at the point of origin are indeed the same goods in the same quantity as at the point of destination. The intelligent container generates the electronic equivalent of a receipt that documents content and evidence of shipping. These data are provided by an electronic key accessible only by authorized persons at origin and destination. The intelligent container provides the shipper and carrier unprecedented visibility into the location and status of assets and cargo. Even more compelling, intelligent containers, with their ability to detect a breach at any time or place and transmit an alert in real-time or near real-time, offer a substantial deterrent to any or all desiring to use the container, or the system, for transport of contraband or to use the container as a weapon.

The objective and facility to enhance the security of the world's trade and commerce are responsibilities that must be shared by the stakeholders and are not limited to governments, intermodal shippers, transporters, forwarders and distributors, seaports, and marine terminals. Recognized globally, cargo container security is one of the principal security vulnerabilities in the supply chain, and enforcement measures should be governed and controlled under a set of rules and regulations that are consistent, predictable, and transparent. To accomplish such a task, a clear definition of each party's role and responsibility must have universal support from the participants and include both administrative and regulatory measures that can be uniformly applied to make enforcement meaningful and effective.

This chapter references a selection of specific, actionable measures designed to enhance the security of cargo, cargo containers, and facilities. A number of these initiatives have been implemented as government requirements for some of the following reasons:

1. Security measures are expensive and could very well interrupt the flow of commerce. If one company forced inconveniences on its customers or caused them to incur additional costs, and a competitor wasn't doing the same thing, the scales

of business "justice" would be tipped in favor of the latter, in addition to which compliance with those security measures more likely than not would be haphazardly implemented.

2. Whatever the measure, rule, regulation, or standard, if it provides a valid enhancement to security, everyone involved should be required to do it.
3. The industry needs to be assured that all measures meet, conform to, and satisfy government requirements and objectives.

All parties must be required to play by the same rules on a level playing field.

This chapter commented on various forms of technology, including how these technologies may be applied and when such technologies could be most effectively applied in the chain of international commerce. The present infrastructure, through many of the initiatives noted herein, is being enhanced with improvements and expanded deployment of nonintrusive container inspection equipment. However, this arena requires a clear focus on making sure that these technologies are properly deployed and utilized, not just in the ports of the United States but in all ports of loading and discharge around the world.

The enormous flow of containers that move around the world acts as a lubricant for the world's economy. Current estimates for inspections say that less than 2% of the content of all containers is checked to verify that what is supposed to be inside is actually what is there. The global transport system is a critical part of the global economy—which also causes it to be vulnerable.

As mentioned earlier, containers can and are being used by lawbreakers to transport all manner and form of contraband, including human beings. This issue of the illegal trafficking of goods and people assumes much more grievous proportions when viewed in light of terrorist threats and activities. Terrorists could easily use a cargo container to transport weapons or hazardous materials or to literally turn that container into a weapon of mass destruction. Therefore, the threat of terrorists using containers poses an inestimable risk to the people of the world and their economies. To determine the size of this risk and the most effective ways to reduce it, a system-wide analysis should be made including the following points:

1. An integrated assessment of global threats, risks, and existing and potential security measures regarding costs and benefits for same
2. Weighing all technical and nontechnical factors, along with the potential of their interdependency serving as weak point in the system
3. Testing all possible implications of implementation for feasibility in relation to political, legal, economic, cultural, and other facets
4. Having withstood the analysis, development, and implementation rigors, major investments from the major stakeholders that are yet to be identified

An "intelligent" container should be able to perform three basic functions:

1. Monitor and report any access to the inside of the container by any and all means of access through any part of the container and indicate whether the penetration is authorized
2. Report its position throughout the supply chain, as technologically possible, to whoever needs the information and is willing to pay for it
3. Send manifest data and other data from the container itself

Conclusion

The market requires container tracking simply because there is no one player universally responsible for everything in the container's transportation process. Containerized transport involves a number of handoffs, with many complex interactions between manufacturers, shipping lines, ports, marine vessels, dray operators, and a host of other players along the transport chain.

Intermodal transport gets even more complex, with containers moving between sea, air, and land on ships, trains, planes, and trucks. Then even more complexity is added when containers travel over international boundaries with different laws regarding transport liability. The primary feature that electronic container tracking presents to the entire system is a reliable audit trail, so the end-users and shippers (end to end) can know the exact point where an event took place or where something went wrong anywhere along or within the entire supply chain.

Unquestionably, two of the most significant aspects are ensuring the security of the container and streamlining the supply chain. Along with those, the greatest factors in the container security arena are loss prevention and reduction in the threat and acts of terrorism. These two security issues need to be addressed simultaneously.

Electronic container tracking simply cannot experience widespread adoption without some form of government mandate. The fastest growing efforts have been in asset utilization at ports and container terminals, many of which do not apply technology to the containers themselves but rather to the assets that move them. Most of these technologies will no doubt play a role in tracking actual containers someday, but for large-scale adoption, implementation, and use, don't look in the near term.

Today, the fact that security and control within supply-chain logistics are vital to the security of every trading nation in the world, as well as to the maintenance of operational efficiencies and flow of international trade and commerce, is universally

FIGURE 9.7 Methods of Intermodal Transport

acknowledged. Intelligent containers can provide information automatically on movement of cargo from origin to destination as well as detect, record, and transmit surreptitious breaches into a sealed container destined for a U.S. port of entry. The claim that the capability to produce intelligent containers does not exist should no longer be made.

References

ABI Research. *RFID border security markets*. Oyster Bay, New York, 2005.

Arquilla, John. The forever war. *San Francisco Chronicle*, p. 6, January 9, 2005.

Associated Press. Authorities find 32 Chinese migrants in ship container at LA port. *San Jose Mercury News*, January 16, 2005.

Balfour, Frederik. Fakes! *Business Week*, pp. 54–64, February 7, 2005.

BearingPoint. A secure global supply chain: Evaluating the return on investment. McLean, Virginia, 2004.

Biological terrorism. *The Economist*, May 24, 2005.

Blake, Scott. "Local company's offshore platforms to combat piracy. Florida Today, May 15, 2008.

Bonner, Robert C. Statement of Robert C. Bonner, Commissioner, U.S. Customs and Border Protection, Hearing before the Permanent Subcommittee on Investigations, Senate Committee on Homeland Security and Governmental Affairs, Washington, D.C., May 26, 2005.

Bonner, Robert C. Statement before House Appropriations Committee, Subcommittee on Homeland Security. Washington, D.C., March 15, 2005.

Bowman, Michael. Congress told U.S. port security improving, but still deficient. *Voice of America*, May 18, 2005. Available at: www.VOANews.com.

Bowman, Steven R. *Biological weapons: A primer*. CRS Report for Congress, RL31059, Congressional Research Service, Library of Congress, Washington, D.C., July 24, 2001.

Cargo Security International, Is it Safe? February–March 2007, James M. Giermanski.

World Trade Organization. *2006 trade report*.

Clarkson Research Services Limited. *Container Intelligence Monthly*, 8(10), London, UK.

Federal Emergency Management Agency. *Are you ready? A guide to citizen preparedness*. Washington, D.C., September 2002.

Flynn, Stephen E. *America the Vulnerable: How Our Government Is Failing to Protect Us from Terrorism*. New York: HarperCollins, 2004.

Frittelli, John F. *Port and Maritime Security: Background and Issues for Congress*. CRS Report for Congress, RL31733, Congressional Research Service, Library of Congress, Washington, D.C., December 5, 2003.

Frontline Solutions. *"Smart" container success will depend on government mandates*. Peterborough, NH, February 1, 2005.

Garfinkel, Simson. RFID rights. *Technology Review*, November 3, 2004.

Gerin, Roseanne. Security opens new doors: Guarding nation's ports and commerce sparks growing IT market. *Washington Technology*, September 13, 2004.

Greenspan, Alan. Committee on Financial Services hearing. February 28, 2001.

Harrald, John R., Stephens, Hugh W., and Johann Rene van Dorp. A framework for sustainable port security. *Journal of Homeland Security and Emergency Management*, 1(2), 2004.

Harris, Shane. Detecting the threat. *Government Executive Magazine*, July 15, 2002.

Homeland security: Summary of challenges faced in targeting oceangoing cargo containers for inspection. General Accounting Office Report and Testimony, GAO-04-557T. March 31, 2004.

Inbar, Daniel, and Michael Wolfe. New smart container studies find: Deployment of maritime smart containers will improve U.S. economy and profits as well as security. Homeland Security.

Kates, Brian. Harbor fears high, terror funding low. *New York Daily News*, December 21, 2003.

Knobler, Stacey L., Mahmoud, Adel A. F., and Leslie A. Pray (Eds.) *Biological Threats and Terrorism: Assessing the Science and Response Capabilities*. Washington, D.C.: National Academies Press, 2002.

Koch, Christopher. Remarks before the Journal of Commerce's 4th annual Trans-Pacific Maritime Conference. Long Beach, CA, March 9, 2004.

Koch, Randy. *Secure commerce: Securing your global supply chain*. Unisys White Paper, Unisys Corporation, Blue Bell, PA, 2004.

Kurtz, Howard. Terrorism stunt angers U.S. officials. *Sydney Morning Herald*, September 13, 2003.

Lane, Earl. Port security: Concern lingers over new scanners; their use is growing, but whether they can detect materials used to make devastating weapons is a worry. *Newsday*, August 17, 2004.

Lawler, Andrew. The unthinkable becomes real for a horrified world; research and development to combat terrorism; statistical data included. *Science*, 293(5538), p. 2182, September 21, 2001.

Lee, Hau L., and Michael Wolfe. Supply chain security without tears. *Supply Chain Management Review*, January 1, 2003.

Lipton, Eric, with Matthew L. Wald. U.S. to spend billions more to alter security systems. *New York Times*, May 8, 2005.

Lipton, Eric. Loopholes seen in U.S. efforts to secure overseas ports. *New York Times*, May 25, 2005.

Lok, Corie. Cargo security. *Technology Review*, June 2004.

Looking for a big market? Try container security. Sunday Business via News Edge Corporation, June 3, 2005. Available as of April 2, 2006, at: securityinfowatch.com/article/article.jsp?id=4289&siteSection=384.

Martha, J., and S. Subbakrishna. Targeting a just-in-case supply chain for the inevitable next disaster. *Supply Chain Management Review*, September–October, pp. 18–23, 2002.

Powell, Alvin. New Harvard report: Chilling warnings on nuclear terror. *Harvard University Gazette*, March 20, 2003. Available as of April 2, 2006, at: www.news.harvard.edu/gazette/2003/03.20/11-warnings.html.

Powell, Peter H., Sr. Testimony before the Subcommittee on Trade of the House Committee on Ways and Means. Washington, D.C., June 15, 2004.

Rae Systems. Securing the supply chain: Container security and sea trial demonstration results. Sunnyvale, CA, January 2005.

Research Corporation and North River Consulting Group. San Jose, CA, and North Marshfield, MA, 126, April 2004.

Securing the global supply chain: Customs–trade partnership against terrorism (C-TPAT) strategic plan. U.S. Customs and Border Protection, April 2005.

Scalet, Sarah D. Sea change. *CSO Magazine*, September 2003.

Sheffy, Yossi. Supply chain management. *Defense Transportation Journal*, 58, September–October 2002.

Stowsky, Jay. Secrets to shield or share? New dilemma for military R&D policy in the digital age. *Research Policy*, 33(2), March 2004.

Shipping Statistics and Market Review, June 2006.

Slovic, Paul. *The Perception of Risk*. London: Earthscan Publications, 2000.

Testimony before the Senate Homeland Security and Governmental Affairs Committee, March 28, 2006, Stephen E. Flynn, Ph.D., Jeanne J. Kirkpatrick, Senior Fellow for National Security Studies, Council on Foreign Relations.

The Journal of Commerce Online. Smart containers and the chain of custody. *Traffic World*. February 12, 2007, James Giermanski, Chairman, Powers International.

The fragile state of container security. Written testimony before the Senate Governmental Affairs Committee, March 20, 2003, Stephen E. Flynn, Ph.D., Jeanne J. Kirkpatrick, Senior Fellow for National Security Studies, Council on Foreign Relations.

The National Academies, Transportation Research Board. The intermodal container era: History, security, and trends. *TR news*, 246, September–October 2006. Available at: www.trb.org.

U.S. customs emphasizes rapid compliance with advanced manifest rule. *American Shipper Online*, 6(235), December 3, 2002. Available at: www.americanshipper.com.

U.S. Department of Transportation, Research and Innovative Technology Administration, Bureau of Transportation Statistics. *Transportation indicators report*. August 2002. Available at: http://www.bts.gov/publications/transportation_indicators/august_2002/.

U.S. Department of Transportation, Maritime Administration. *Vessel calls at U.S. and world ports 2005*. Washington, D.C., April 2006. Available as of December 2006 at: www.marad.dot.gov/marad_statistics.

U.S. Department of Transportation, Federal Highway Administration, Office of Freight Operations. *Freight facts and figures* 2006. Washington, D.C. Available at: www.ops.fhwa.dot.gov/freight.

U.S. Department of Transportation, Research and Innovative Technology Administration, Bureau of Transportation Statistics. *North American freight transportation*. Washington, D.C., June 2006.

U.S. General Accounting Office. Container security: Expansion of key customs programs will require greater attention to critical success factors. GAO-03-770, Washington, D.C., 2003.

U.S. Government Accountability Office. Container security 2002. Current efforts to detect nuclear materials, new initiatives, and challenges. Statement of JayEtta Z. Hecker, Director, Physical Infrastructure Issues, before the Subcommittee on National Security, Veterans Affairs, and International Relations, Committee on Government Reform, U.S. House of Representatives, GAO-03-297, Washington, D.C., November 18, 2002.

U.S. Government Accountability Office. Homeland security: Preliminary observations on efforts to target security inspections of cargo containers. Statement of Richard M. Stana, Director, Homeland Security and Justice, before the Subcommittee on Oversight and Investigations, Committee on Energy and Commerce, U.S. House of Representatives, GAO-04-325T, Washington, D.C., December 16, 2003.

U.S. Government Accountability Office. Maritime security: Substantial work remains to translate new planning requirements into effective port security. GAO-04-838. Washington, D.C., June 2004.

Van de Voort, Maarten, O'Brien, Kevin A., Rahman, Adnan, and Lorenzo Valeri. "Seacurity": Improving the security of the global sea-container shipping system. The RAND Corporation, MR-1695-JRC, Santa Monica, CA, 2003.

Veiga, Alex. Radiation detectors to scan all incoming cargo at LA port complex. *Associated Press*, June 4, 2005.

Wein, Lawrence M., Wilkins, Alex H., Baveja, Manas, and Stephen E. Flynn. Preventing the importation of illicit nuclear materials in shipping containers. Stanford University and Council on Foreign Relations, Palo Alto, CA, and New York, 2004.

Willis, Henry H., and David S. Ortiz. Evaluating the security of the global containerized supply chain. The RAND Corporation, TR- 214-RC, Santa Monica, CA, 2004.

Wolfe, Michael. Technology to enhance freight transportation security and productivity. Appendix to: Freight transportation security and productivity. North River Consulting Group, North Marshfield, MA, 2002.

World Shipping Council. Liner shipping: Facts and figures. Washington, D.C., n.d.

Wrightson, Margaret T. Maritime security: Enhancements made, but implementation and sustainability remain key challenges. Testimony before the Committee on Commerce, Science and Transportation, U.S. Senate, GAO-05-448T, Washington, D.C., May 17, 2005.

10

Transportation Security: Applying Military Situational Awareness System Technology to Transportation Applications

William S. Pepper IV

Chapter Objectives:

- To investigate available military situational awareness systems
- To thoroughly discuss the key characteristics of situational awareness system technology
- To describe key elements of situational awareness system technology
- To illustrate practical application of this technology to a Transportation Security example

Introduction

Prior to September 11, 2001, no one suspected that any means of transportation could be used as a weapon of mass destruction. After September 11, however, the need to better secure our airports and all other means of transportation, especially seaports, became widely recognized. Nevertheless, action was slow to come. The Madrid and London train bombings of 2004 and 2005 further demonstrated the continuing vulnerability of transportation systems. The reasons for not immediately addressing these needs are myriad, including, but not limited to, funding, jurisdictional limitations, policy, intelligence sharing, and many others. These issues notwithstanding, there are also technology issues. Among these are Situational Awareness, intelligence gathering, information dissemination, network connectivity, data fusion, and others. Over the last few decades, the military has addressed these and many other technical issues as have commercial ventures that require the manipulation and dissemination of large volumes of data. By adapting military and commercial technology and practices, such as Situational Awareness,

Communications Network Management, and Data Fusion and Correlation tools, the Transportation Security problem may readily be addressed. For example, one can build statistical models to predict the likelihood of possible threats within the transportation environment. Using data from each of these models, data can be further correlated to support the process of real-time decision making. This chapter will discuss these three elements of technology and provide an overview of their application to airports and seaports in support of Transportation Security.

Situational Awareness

Situational Awareness (SA) is concerned with the movement of people, goods, and information by means of air, sea, and land. If a secure environment is to be maintained, the operational space must be closely observed to verify that something or someone is really a threat. "Operational space" in this sense may be a specific locality such as a city or town, a geographical area such as a city or town, an infrastructure entity such as a seaport, airport, or train station, or an expansive entity such as a nation's electrical or rail infrastructure.

Integrated Situational Awareness

The provision of Command and Control(C^2) capability is essential in today's battlefield environment. In fact, the arsenal of tools provided to the warfighter of today supports the concept of what has come to be known as Net-centric warfare, formerly known as the concept of Command, Control, Communications, Computers, Intelligence, Surveillance, and Reconnaissance(C^4ISR). Another common usage more applicable to Situational Awareness in a Transportation Security environment is C^3I (Command, Control, Communications, and Intelligence). This will be the primary focus of this chapter. In these applications, intelligence is gathered from a variety of data sources including various types of sensors, such as video cameras, access control systems, and data mining for various databases available to the Transportation Security systems. Civilian and military customers have been making tremendous investments in wireless connectivity, increased bandwidth, and surveillance technologies. These additional capabilities have not been accompanied by a corresponding improvement in situational awareness. The Situational Awareness/Common Operating Picture (SA/COP) capability is considered a key element in the development of effective Transportation Security.

The key to Situational Awareness in this environment is the implementation of Standards-based C^3I architecture and framework, built around a generic Observe-Orient-Decide-Act (OODA) decision-execution cycle. Sometimes referred to as an "OODA loop," this cycle is illustrated in Figure 10.1.

The OODA loop illustrated here is a simplification of the concept originated by military strategist Col. John Boyd of the United States Air Force. For the purposes of this chapter, the loop is simply stated, Observe, Orient, Decide, Act. Keep in mind that each step of the loop may be iterative. Many attempts may be required for the successful completion of each step. The corollary of the OODA loop in an emergency management scenario is illustrated in Figure 10.2 below. Where the OODA loop ends with the action step, in a

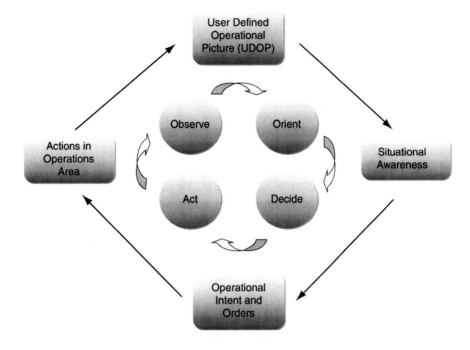

FIGURE 10.1 Observe-Orient-Decide-Act (OODA) decision-execution cycle

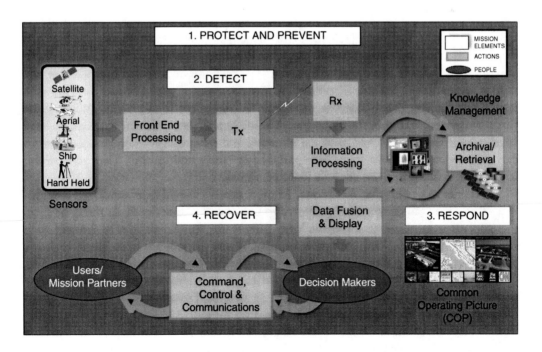

FIGURE 10.2 Transportation Security Situational Awareness Operational Model

Transportation Security scenario, the process ends with the Recovery Step. These are the mission elements of Transportation Security:

- Prepare and Prevent – Assess the vulnerability of infrastructure to terrorist attacks, and develop strategies designed to protect them. Train, model, simulate, and document our response to any emergency, attack, or incident.
- Detect—Apply the necessary sensors, comms (communications capabilities), and information processing capabilities to identify the existence of Chemical, Biological, Radiological, Nuclear, and Explosive (CBRNE) agents or other threats and convey the occurrence to government agencies.
- Respond—Quickly react to an incident and manage the response with the necessary resources and intervention to appropriately contain the impact of any emergency.
- Recover—Restore normal operations and commerce after the occurrence of an incident. Plan to promote rapid recovery from terrorist attacks to ensure prompt restoration of transportation, utilities, critical communications, and information systems as well as contain and remove hazardous materials.

The primary focus of this chapter is concerned with the technology elements involved in Situational Awareness, which is part of the Detect step. This involves sensors, knowledge management, information processing, and some level of data fusion. The goal of Situational Awareness, not unlike the Military Intelligence cycle, is to gather *Data*, turn that Data into *Information*, and then convert that Information into *Knowledge*. (This "knowledge" is referred to in military circles as "Actionable Intelligence.") These concepts introduced here will be discussed in further detail in the ensuing paragraphs.

The Information Challenge

Typical Command and Control Centers have numerous management tools to obtain information and status. A major issue is fragmented situational awareness resulting from disparate applications on separate displays and non-integrated data sources, such as cameras, alarms, Incident Management Systems, classified and unclassified information, and myriad inputs from individual sensors. A key to enhance preparedness and improve threat management capabilities for emergency management is to provide robust situational awareness and incident management tools in a C^2 environment that can be used in managing incidents as well as in the planning and execution of exercises. These tools can act as what are known in military parlance as "force multipliers." They achieve this state by off-loading upfront analysis from operators.

In order to be effective, Situational Awareness tools should be event driven and provide alarms that can be specifically tailored to Command Center needs. At the same time, these operator interfaces need to be tailored from warfighter capabilities to those of civilian agencies supporting Transportation Security. Where the military has a motto of "train as you fight, fight as you train," Transportation system operators do not have this luxury. Transportation systems such as airlines are, after all, in business to produce revenue and profits. Security is, unfortunately, secondary. As a result, Transportation Security systems must be simple to operate and require minimal training.

When taking the short view, one might think that there is a lack of information regarding data that would be useful in the implementation of a Transportation Security system. In fact, the opposite may well be true. There is too much information, more than any human being can easily process in a timely manner. For example, consider the security guard on the night shift watching 12 or 16 security cameras. The ability to concentrate fades quickly. Tell-tale details or changes may go unrecognized in the blink of an eye, literally *because* of a blink of an eye. The very operation of the eye and the brain may be faulted as the cognitive process of seeing itself tends to replicate that which was "seen" the last time. As a result, a change might be noticed but the odds are high that the recognition will occur after the event or behavior has already taken place. One may "research" this premise by taking a trip to an airport and watching how long it takes a carry-on bag scanner to go about his or her business. News broadcasts indicate that tests of these methods indicate an alarming rate of "false positives," that is, a bag is declared as "OK" when in fact it contains banned items.

Situational Awareness and Decision Support

Situational Awareness and Decision Support tools can be configured to provide an effective understanding of anything associated with the global transportation environment that could impact the security, safety, economy, or environment of the United States or any other country. Situational Awareness tools discern between legitimate and illegitimate activities, support threat-based risk management of valuable, but limited resources, and support all transportation security and safety operations. Acute "awareness" is the key to an equally important goal of "prevention." Increasing the coverage area is tantamount to early decision making.

The Harris Corporation's SafeGuard™ Situational Awareness system is depicted in Figure 10.3. Harris SafeGuard™ is a Service-Oriented Architecture-based (SOA) Situational Awareness solution. Such technologies increase operational efficiency and assure communication. Used in a Transportation Security implementation, Situational Awareness systems can save lives and protect property while ensuring that commerce of the nation remains uninterrupted. To accomplish this mission, the system operates under the guiding principles of Detection, Deterrence, Resumption, and Recovery. The key to

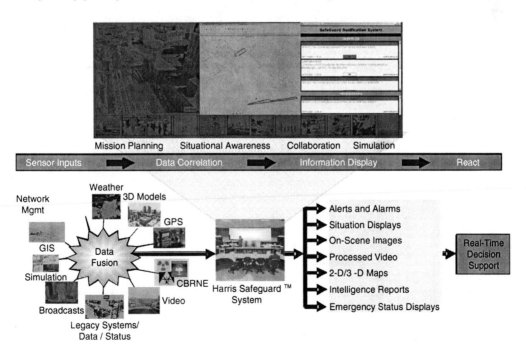

FIGURE 10.3 SafeGuard™ Situational Awareness and Incident Management Product

maintaining these principles is the provision of a User-defined Common Operating Picture (COP) that is provided anywhere there is network connectivity. (Network functions will be discussed further later in the chapter.) In order to effectively support Detection, Deterrence, Resumption, and Recovery, the COP should be capable of using a variety of presentation display devices from low-end desktop displays, laptops, Personal Data Assistants (PDAs), and large display walls.

The primary system functions are the monitoring of assets, mission planning, simulation management, situational awareness, collaboration, and integrated decision support capabilities. A fully capable system is used not only in operations for effective response to live incidents, but also for training and planning. A Situational Awareness system can be adapted to any number of Transportation Security threat scenarios from airport and coastal monitoring to vessel tracking and incident response. Sensor inputs are collected and the data are extracted from them and correlated to be displayed as information and (simultaneously) generate alerts, messages, and notifications of response personnel and decision makers.

The purpose of Situational Awareness systems is to provide the following:

- Capability to protect key assets
- Ability to implement security policies
- Collaboration
- Intrusion detection and access control integration
- Automatic protection upon alarm
- Remote permission validation
- Location restriction capability
- Tailorability of alarms and the Graphical User Interface (GUI) with special software
- Use of existing products and Computer Off-the-Shelf (COTS) hardware and software

Situational Awareness system features and benefits are described in Table 10.1 below.

Table 10.1 Situational Awareness Features and Benefits

Feature	Function	Benefit
Situational awareness	Asset tracking, intrusion detection, and Incident Command Center response to emergencies	Improves decision making and rapid response to emergencies to save lives and protect property
Common Operational Picture	Information sharing with multiple sites and levels	Allows coordinated and effective response
Flexible, Modular, Scalable architecture	Multi-level capability (from high-end systems at EOCs to first responders using laptops, tablets and PDAs)	Allows operational flexibility in providing and accessing data
Mission Planning	Design strategies, doctrine, and countermeasures	Mitigates vulnerabilities and reduces damage
Simulation	Forecast and prepare for events in advance	Allows detailed plans to be made and rehearsed
Scenario Generation	Support training and conducting exercises	Ensures effective preparedness
Filtering and abstraction of Situational Awareness Information	Delivery to multiple levels of stakeholders based upon their capability	Quickly disseminate the right data to the right stakeholder

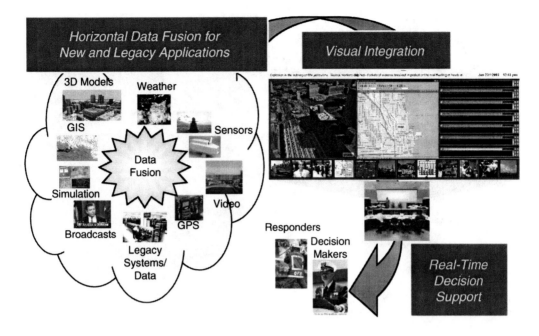

FIGURE 10.4 Sensor Data Correlation Concept of Operations for Situational Awareness

A Situational Awareness and Incident Management Product depicted in Figure 10.4 above provides the capability to rehearse, evaluate, and refine the incident management processes. This capability revolutionizes the means by which information is analyzed, viewed, and shared and dramatically improves the ability to do the following:

- Respond to incidents and maintain real-time situational awareness
- Track assets and maintain communication
- Plan mitigation and response activities
- Record events and benefit from lessons learned
- Perform Daily Operations, Prepare, and Train

The system provides mission planning, simulation management, situational awareness and collaboration capabilities integrated in a multi-media visualization environment. This environment is based upon proven commercial display technologies and proven processes for production of 3-D models such as Harris Corporation's RealSite®. The models are viewed using technology such as Harris Corporation's InReality™ Viewer or Google Earth, among others. The deployed system is shown in Figure 10.5 and consists of an equipment rack containing the integrated applications, control station interface equipment to drive the system, and multi-screen displays for effective multi-media presentation. The operational software performs the actual work of the system. Notice that some of the software items refer to "legacy" systems. These are systems that existed in the past and are integrated into the system using software encapsulation. The complete replacement of all systems by new systems is cost-prohibitive. Therefore, an effective Situational Awareness system should be capable of adapting old systems to new functions. (We will discuss this further in ensuing paragraphs.)

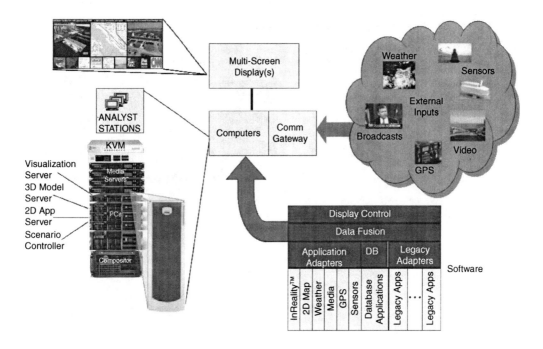

FIGURE 10.5 Situational Awareness system Components.

Transportation Security Management is achieved by the fusing or correlation of existing and on-scene multi-media data into decision-relevant information. When responding to a live incident, information is needed in real time and must be displayed in a manner that conveys the knowledge needed to act quickly and with confidence. Referring back to Figure 10.4, the Situational Awareness system synthesizes 3-D models, 2-D maps, dynamic data, GIS, video, broadcast information, and other data for effective knowledge presentation to make accurate decisions. GIS is the acronym for Geographic Information System and may also be referred to as a Geographical Information System or Geospatial Information System. GIS is any system that is used to capture, store, analyze, or manage data which are spatially referenced to Earth. Additionally, the Situational Awareness system consists of:

- MA multi-media presentation environment
- Control software that provides multi-media management of real-time view-ports (windows), including sizing and thumbnail pop-ups
- A photo-realistic, highly accurate 3-D model of all structures within the vicinity
- A 3-D model viewer that allows accurate 3-D navigation throughout the scene (including underwater bathymetry) and has mensuration line of sight, as well as GIS analysis capability. (Mensuration is the measurement of geometric quantities.)
- A 2-D map that is correlated with the 3-D model, providing all the benefits of flat map awareness and a linkage into the 3-D scene
- GIS mapping and attribution of critical infrastructure
- A geospatial display of near real-time dynamic information sourced from specialized ground, airborne, or underwater sensors

- Charting and timeline applications providing dynamic information to monitor the progression of an incident response
- GPS asset tracking within the 3-D model and 2-D map (meaning that tracked personnel or vehicles will be accurately correlated on both maps)
- Asset dispatching and tracking (where available) in support of asset allocation during a training exercise
- Predictive analysis capability (i.e., plume, chemical dispersion, and explosion impact)
- Collaboration (data ingest and dissemination) of situational awareness information to and from first responders and decision makers
- Display and control of standard office automation applications and Web sites
- A Simulation Manager, including logging and playback, to conduct training exercises
- Media servers for pre-recorded video providing event occurrences during training scenarios

Situational Awareness systems enable effective emergency response to incidents. Pre-event planning and exercises, as well as event and post-event response, are key to protecting occupants and infrastructure. Whether working to prevent man-made or natural disasters, or respond to one in real-time, an abundance of detail and information is needed, such as the following:

1. Where are the exits or escape routes near an incident location?
2. What is the distance between a reported fire and petroleum storage tanks or other hazards?
3. Are there any obstructions or infrastructures that would hinder dispatched response units?

Integrated Situational Awareness systems allows agencies to perform mission planning to design strategies, doctrine, and countermeasures for mitigating vulnerabilities and damage in addition to performing scenarios for training/exercises to ensure effective preparedness. They not only support these efforts, but also support response to and recovery from real emergencies and/or crisis situations. In addition, the flexible situation awareness design permits the simultaneous use of situation information in two or more modes (e.g., search and rescue, damage containment, evacuation management, and law enforcement). The end result is "just in time" incident management.

Situational Awareness systems allow collaboration between emergency response organizations. The Situational Awareness system ingests live situational informational such as dispatched emergency vehicles, placement of field resources, access blockages, availability of hospital capacities, and other information critical for effective response to the ongoing situation. In addition, local incident response information will be collaboratively shared with neighboring civilian authorities providing medical, fire rescue, and law enforcement support. The Situational Awareness system will enable the continuous and dynamic update of this information so that all response organizations are operating with the most accurate available information.

Situational Awareness systems are easy to deploy and complement existing systems. These systems will not necessarily replace existing equipment or applications in use at a

Command Center. The flexible hardware architecture takes advantage of existing site infrastructure (no throwaways), inserts into the existing infrastructure without impact, allows for view-port (window display) growth, and supports both distributed and concentrated processing. Additionally, many commercial (COTS) products are available for use in the solution as well. Situational Awareness systems must provide the ability to incorporate and interface to other products through their well-defined interfaces and adapter development kits. The Situational Awareness system should provide a suite of user interface capabilities for use in Transportation Security solutions. Most commercial software products that support Situational Awareness systems have established Application Program Interfaces (APIs) that allow disparate programs to share data with each other through the writing of an Extensible Markup Language (XML)–based messaging protocol. This protocol is extremely flexible when integrating existing and planned command and control systems. This allows the software components to interface and interoperate with existing or new commercial products. Those products that do not have established APIs or allow the development of XML-based interfaces are not recommended for ease of integration.

To be effective for Transportation Security, Situational Awareness systems must support the core technologies required for an Emergency Operations Center Situational Awareness, Incident Management and Training Tool. While many technologies exist that support individual visual and GIS functionality, they are not integrated to address situational awareness for incident management as a system. The following paragraphs briefly describe some of the technologies that compose the core of an effective Situational Awareness system.

Visualization and Display Tools

Situational Awareness systems provide powerful management control of multimedia real-time view-ports. Effective incident management for training or response requires personnel to have presentation of multimedia data including video, text, 3-D models, 2-D maps, charts, etc. Situational Awareness systems must have powerful graphical control software that provides the management of real-time digital graphics and data view-ports (windows) for effective knowledge presentation. Managing the components or view-ports on a large screen allows users to visualize and interact with the data and information they need in an emergency and collaborate for a cohesive response across agencies or commands. View-ports can come in and out of view, text information can continually scroll, application windows for video teleconferencing and other applications can be accessed, and so on. Figure 10.6 shows a multimedia layout consisting of three large display view-ports across the top. The top displays consist of a 3-D model, a correlated 2-D map, and live video. The nine smaller display thumbnail view-ports along the bottom represent other inputs to quickly replace the displays along the top of the screen should it be desirable to view them. Displays from other systems and sensors are also available in the background.

View-ports can span display screens and multiple view-ports can be displayed on a single screen. Interaction with the system alters the layout of the display windows depending on the state of the event. Each display view-port can be enlarged and consume greater real estate, and even consume a single screen. Selecting a thumbnail results in its display

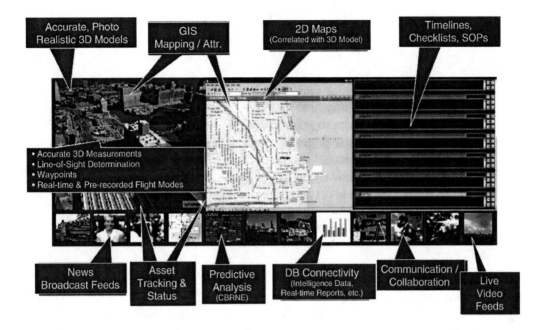

FIGURE 10.6 Situational Awareness system Visual Integration = Quick Action

within one of the three large view-ports. Geospatially accurate 3-D visualization provides a significant integrated situational awareness capability. Dependent upon the transportation system, city, port, and/or harbor 3-D models can be provided. Geospatially accurate 3-D models of a city or port and its surrounding area, with realistic texturing, are created from imagery from satellites, or other data sources. Once a 3-D model is created, the powerful and easy-to-use visualization software (such as Harris Corporation's InReality™ viewer) allows users to navigate with accuracy throughout the 3-D scene. The user can view scenes from either a first-person or north-up cartographic perspective while having continuously updated geo-position and heading information.

Using a combination of the keyboard, mouse, and/or joystick, users can navigate and place themselves at any location in the scene (e.g., on a pier). In addition to navigation and cinemagraphic controls, analytical tasks such as mensuration between any two points and range finding from the first-person viewer to any location can be performed to a high degree of accuracy. The software allows for quick real-time calculation of rooftop elevations and distances between any two points in 3-D, features that are beneficial to such efforts as distance-over-time calculations, strategic positioning (e.g., plotting positions of potential sharpshooters), and computing ranges of assets to targets within the viewed area.

The visualization tool also provides the capability to quickly zoom in and out of the scene and query specific locations in the scene with latitude, longitude, and height accuracies of 1 to 2 meters. These geospatial queries are performed by merely clicking on items such as buildings, vehicle garages, police stations, etc. within the model. Examples of other uses include finding access to a building or structure when roads are blocked, the length of hose runs from hydrants, height of access points to rescue people, or where to land helicopters for evacuation. The tool by itself can be used for security planning, scene familiarization, and mission rehearsal. A 2-D map application is also integrated in Safeguard™ and correlated with the 3-D model to focus in on areas or track assets. As one navigates through the 3-D scene, a location indicator and view direction are shown on the 2-D map.

Sensor readings, weather, alarms, transportation queues, etc., can be overlaid within the 3-D model or 2-D map. For example, population density within an area is important to help establish where to allocate resources throughout the day and provides critical knowledge for evacuation preparedness and planning (e.g., for hurricanes). Population density within the 3D model can be used to illustrate this effect. These density volumes could be color coded and translucent to indicate different density levels without obstructing the view of the 3D model or 2D map. Sensor data or weather data can be displayed in the same manner.

Access Control Systems

One technology that may be used to control the access of Transportation workers to transportation facilities is the Access Control Systems (ACS). The military has used such ID systems for years to ensure that access to controlled areas is only provided to cleared personnel. A 2007 incident involving gun and drug smuggling by Orlando International Airport baggage handlers indicates the need to control access at all facility areas to provide effective Transportation Security. The typical ACS consists of turnstiles, doors, access gates, vehicle access gates, badge readers, and (in some cases) biometrics interfaces that communicate with the centralized Situational Awareness system via a message queue. The request data are sent to the Access Control Database (ACD) to facilitate tiered access. Results from the tiered access controller and its associated database will determine the access request results. These results are then returned to the access point using the message queue.

The ACS would typically have three levels of access. Level One Access grants access to general restricted areas, and is limited to authorized personnel. Level Two Access grants access to sensitive restricted areas. Level Two access is limited to personnel in possession of a valid Transportation Security Clearance. Level Three Access grants access to very sensitive restricted areas, as designated by the supervisory authority and is limited to designated persons of trust who are in possession of a valid clearance.

The system would support two types of employee identification cards by interfacing with the associated credentialing system for access update card–registration info. Credentialing information is read from the Access Control Device (ACD) by the Situational Awareness system. The information is then associated with the current security level. This information assists in determining access request results, which are then forwarded to the ACS. Dependent upon the facility security status, the employee would be granted or denied access, dependent upon their security clearance level. The system permits flexibility in preparation for changing circumstances and security levels for all access points protected by physical barriers and card–biometric readers. All access requests pass to the Correlation Engine and ACD. This information is then correlated against current circumstances and security levels for determining access request results, which are then sent to the ACS.

To assist Situational Awareness system operators and security personnel, the Situational Awareness system would contain a reporting tool that queries the ACD for statistics such as: the number of persons that the system deems on the premises at any given time, number of invalid access attempts by access point, total number of accesses by access point for a given period, and other reports as needed. These reports would display, at a minimum, the name of the cardholder, the access point name/number, and the requested statistics that go with that person's identification and/or the access point. The system would also haves single-press "lockdown" to shut off all access in the event of an emergency. This feature is implemented by sending a message queue alert to the Situational Awareness system to use operator selectable access levels. A subsequent single-press will send a message queue message to the Situational Awareness system to use database stored access levels.

To enhance the security of the system, in addition to the standard password login, the system could optionally include a fingerprint (or other biometric device) reader that could be used as needed. The default authentication mechanism authenticates a user name and password. The system authenticates any requests for viewing live video, recorded video, and camera control against a Windows Server domain controller. The system provides access management to all system functions, such that types of actions performed by a user are limited by the user's access level.

At the heart of the ACS is the ACD. The ACS could support Transportation Worker ID cards and temporary worker–visitor cards by integrating the card readers into the Situational Awareness system via the message queue. The Situational Awareness system software will read credentialing information from the ACD that is populated by the credentialing system. In the event that an access point assigned Level One Access status receives an access request denied message, the appropriate security personnel will be alerted with a text message and audible alert via a notification Web page or via another designated notification system.

Another element of the ACS is vehicle access control. This portion of the ACS is integrated with the Video Surveillance System (VSS). In a typical application, the VSS is integrated with the ACS and provides specialized license plate recognition (LPR) software. The general ACS philosophy for vehicle access control can be described as follows. At each vehicle access gate, vehicles that are within an operator configurable virtual zone will have their license plates imaged by a dedicated camera or cameras that are supported by the video analytic software. When the vehicle's driver introduces Transportation Worker ID card or temporary enrollment pass to the reader at the gate, the system queries the ACD, which will have a data record with a license plate number field for each vehicle permitted access. For a correct query–database match for both the worker and the vehicle, the ACS will signal approval of the query, and ACS will activate the relevant vehicle gate access control hardware. For incorrect query–database matches, the ACS will signal denial of the access attempt. The ACS will then generate an alert for the relevant security personnel to investigate the vehicle.

A recommended architectural network topology is illustrated in Figure 10.7. A commercially available secure wireless networking technology is utilized throughout the system where the utmost communication security is required. This layer functions as the secure wireless data transport mechanism within the architecture, assuring data security and integrity. The use of 802.11 wireless technologies provides a network that minimizes space and remains independent of the existing infrastructure. In this scenario, sensor solutions can be deployed anywhere within the airport where power and wireless connectivity are available.

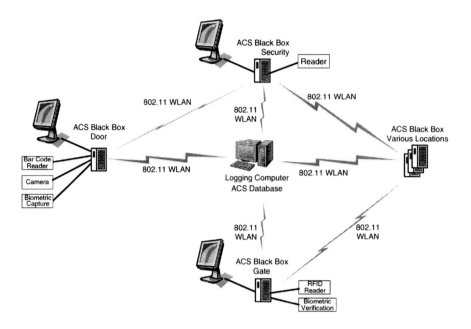

FIGURE 10.7 Access Control System Architectural Topology

Video Camera Systems

A key technology in Transportation Security is the integrated Video Surveillance System (VSS). Unlike the Security Guard discussed earlier who relies solely on his or her attentiveness and eyes, the VSS uses what is known as Automated Scene Understanding (ASU) to review incoming pictures for items of interest. Existing facility cameras may be easily integrated into the Situational Awareness system COP. For example, additional video from State and Federal Department of Transportation (DOT) camera system could be ingested and displayed in the system if so desired.

The Video Surveillance System (VSS) is typically installed to control CCTV cameras and provide automated alerts and alarms based on video analytics. Monitoring, configuration, and controlling of CCTV cameras are provided by the VSS. The VSS is composed of the CCTV camera suite, video analytics applied to the video feeds from those cameras, and the Perimeter Monitoring System (PMS). In addition, a Building Monitoring System (BMS) can be provided as well. The VSS processes video information and uses a Correlation Engine (CE) and an Alert Management Engine (AME). The processing results indicate either normal behavior or an incident event that requires one or all of the following actions:

- Notify the security station operator of the event.
- Send a cueing command to a camera.
- Request imagery to be retrieved and displayed from the VSS database.
- Request auxiliary control of a camera.

These Video Surveillance Systems differ greatly from typical DOT or Building Monitoring Systems in that they are software and hardware driven and depend on computers, rather than human operators, to perform their duties efficiently. VSSs of this type are readily available from many vendors, including, but not limited to ObjectVideo,

Vistascape (Siemens), PureTech Systems (ICx Technologies), Guardian Solutions, Vidient, and many others. (It should be noted that these systems are not an end in themselves. Effective Transportation Security Situational Awareness systems require far more than VSS-based alarms to be truly effective.) The VSS software provides data including but not limited to the following:

- Video analytic software alarms such as object left behind and unexpected zone entries
- Imagery of generated alarms
- Imagery requested by an operator
- Tracking data where an identified object is tracked
- A table that includes the locations of all VSS-connected equipment
- License plate data that will be referenced by other subsystems

As mentioned previously, the video surveillance system may use ASU software. This type of software is readily available from the companies mentioned previously and is frequently referred to as "smart video" image processing software. This software examines and recognizes programmed objects within view of the camera. Additionally, objects within the camera's view may be tracked by the system. These object behaviors, given proper camera placement, include but are not limited to the following:

- Object removal from an area
- Abandoned objects (also known as "object left behind")
- Tailgating (two people gaining entry to an access point using one badge swipe)
- Stopped vehicles
- Loitering
- Object counting and statistics
- Crowd formation

This type of system provides a user interface to define rules for generating alarms based on objects detected through the video analysis. Users may define one or more possibly overlapping detection zones for each fixed camera and Pan-Tilt-Zoom (PTZ) camera preset position. This camera setup selects which alarms can be generated based on detected object size, speed, direction, and behavior. During normal operation, CCTV cameras are automatically cued by inputs from subsystems, or manually controlled as a part of operator-invoked event investigations. The VSS (CCTV) system is capable of storing several preset locations for each fixed and PTZ camera without requiring or utilizing preset capabilities in the cameras themselves. Although fixed cameras cannot be moved, the "smart" video software has the ability to store fixed camera presets allowing for different video analytic settings to be activated under various activation criteria (i.e., during a security-level change). The activation criteria may be based on either time of day or as a response to an external input such as inclement weather, diurnal cycles, and national and local security threat levels.

For automatic cueing, the VSS adapter routes camera selection and positioning data to the video control software which directly communicates with the camera. The camera control software simultaneously directs multiple PTZ cameras to geographic coordinate positioning cues. Position data are processed by a Rules Engine and forwarded as Object Detected Events to the Events Engine. The Events Engine generates and forwards Camera

Control messages containing the position of the detected object to the Video Surveillance System which cues the appropriate camera(s) to view the location of the detected object. For manual camera control by an authorized operator, the VSS software provides a user-friendly Graphical User Interface (GUI) allowing pan, tilt, and zoom control of PTZ cameras through "point-and-click," "control-stick," "click to center," and "drag zoom" positioning interfaces.

The operator video displays allow security personnel to view video imagery of any location of interest using surveillance cameras in proximity. Operators may view imagery by selecting locations on 2D (or optional 3D) maps, or through the workstation GUI. Such interactions with the maps generate camera selection messages that direct the VSS to provide the video imagery of cameras covering the selected location. The security workstation GUI supports the same method of video viewing. Rapid configuration of selected surveillance and access control assets in response to changing circumstances is available through the system messaging infrastructure. All surveillance and access control assets subscribed to a given messaging topic will be configured simultaneously by reading the published commands to that topic. Should the need arise in the future to integrate any newly acquired or legacy systems, adapters using the protocol independent messaging infrastructure can be rapidly developed for those systems.

The distributed camera cueing elements of the VSS can be used by other integrated subsystems to direct PTZ cameras to geographic coordinate positioning cues. An example would be when a person attempts to enter an entrance using false credentials or an individual is tailgating at an access control point. In these instances, camera cueing is performed based on subsystem messages from another sensor or system. In this case, the correlation engine generates an event that delivers a camera control command to the video surveillance system to point the camera at a predetermined location specified by the operator. Once the investigation is complete, the PTZ camera in use is returned to normal operation through one of two methods. The security station operator can command the VSS to return cameras to normal operation through messaging, or a default timeout can be defined which will return the PTZ camera to normal operation when no operator input has been received for the specified amount of time.

As mentioned previously, the conventional way to view live video feeds is through the use of Closed-Circuit Television (CCTV) screens. This approach requires excessive real estate and is frequently ineffective. Using the Situational Awareness system's multiple large-screen approach, video outputs can be directed to various positions or view-ports within the large-screen display. The screen layout is adaptable. Each video view-port is selectable for larger display and can be moved or swapped with other video windows. Network video feeds ingested from cable or satellite TV are also addressed to video view-ports on the large screen. Newscasts can provide valuable information to aid in incident management. In addition, video cameras positioned throughout the area of interest are used to fill video window view-ports.

Actual video feeds may be recorded for later playback. Other methods include using a professional video studio to generate the video necessary for use in training situations. With either method, a video library is created for use in training events. The video is usually converted into a digital form, if not already in that form, so that it can be timed and controlled via the simulation controller. Simulated video is used in place of and in conjunction with private feed video during simulated events. SafeGuard™ contains integrated media servers to provide simulated video capability.

In addition, if a camera is stationary, its video output can also be projected onto the exact location in the 3-D model. For example, if a camera is focused on the front of a specific warehouse, the imagery or video from that camera can be projected onto the front of the same warehouse within the 3-D model, providing context as to where the video is being captured. By integrating COTS motion-detection surveillance software, movement throughout the incident area of interest can be reported within the 3-D model and 2D map.

Predictive Analysis Tools

Predictive analysis tools are specialized software tools that support Transportation Security by predicting chemical dispersion, explosion impact, transportation congestion, hurricane path evacuation planning, and floodplain predictions. This is an important capability that supports crisis managers by providing relevant data to support decision making and prioritize resources during unplanned events, whether natural or man-made. For example, with the recent threat of Chemical, Biological, Radiological, Nuclear, or Explosive (CBRNE) attack, it is difficult to estimate what effect such an attack would have within the surrounding community. It is also difficult to predict how chemicals would disperse based on the weather conditions. Forecasting with a plume model provides for real-time prediction of areas that would be affected. Plume models integrated into the 3D visualization provide views of different threat levels throughout the adjacent community. Sensor information from weather sources and chemical detectors can be integrated into a single application environment for chemical hazard predictions and continuous hazard monitoring. The system monitoring function alerts personnel (via audible or visual alarms) when hazardous conditions are detected and can initiate a series of actions in response to threat detection.

Information Sharing

Incident Response Information is shared between stakeholders, decision makers, and first responders for effective crisis management. Decisions are not made in isolation, and working with and sharing information with other agencies and authorities is critical. SafeGuard™ filters and abstracts situational awareness information for delivery to multiple levels of stakeholders based upon their capability. In addition, secure video teleconferencing is critical and allows remote decision makers to communicate and collaborate on managing the incident. It enables decision makers to work together on their systems at the same time.

Situational Awareness and Incident Management Systems ingest critical information from civilian authorities and emergency agencies (e.g., hospital bed status and dynamic data), and redistribute that information to all participating organizations, both fixed and mobile. Abstracted situational awareness information can be delivered to first responders using low-end platforms such as a Personal Digital Assistant (PDA). First responders will not only be able to receive situational awareness data but also will be able to send back data for automatic integration and display within the SafeGuard™ System (e.g., location of injured).

Scalable and Flexible Architecture

Situational Awareness systems should have a Scalable and Flexible Architecture to allow for easy integration into any Transportation Security facility. The technologies and capabilities described previously require a flexible and scalable architecture to be successfully implemented. The Situational Awareness system architectural components previously illustrated in Figure 10.5 can be packaged in a variety of configurations. The ideal system should consist entirely of COTS products. The computer platforms are standard PCs, some of which are fitted with high-end video graphics cards. The system should have no unique power requirements.

A *flexible* hardware architecture:

- Takes advantage of a site's existing equipment by using any resource that duplicates a SafeGuard™ component
- Leverages off site existing infrastructure capabilities (e.g., live video)
- Allows presentation display devices ranging from low-end desktop displays to large display walls, dependent upon the customer's configuration requirements
- Supports both distributed (compute-intensive applications on separate platforms) and concentrated (multiple applications on a single platform) processing
- Accommodates a desktop configuration or packaged rack configuration

A *scalable* hardware architecture:

- Contains built-in growth for both RGB (Red-Green-Blue) and composite video inputs to take advantage of planned video and computer resource additions or to accommodate remote connections in a multi-port configuration
- Permits base Situational Awareness system capability growth (new integrated applications) by simply adding PCs or other platforms for compute-intensive applications
- Supports single to multiple (e.g., 64) graphic wall displays and monitors

The Situational Awareness system software architecture should also be flexible and scalable. The following describes the three levels of integration required for new and legacy applications:

- *No integration is required*: The application is displayed on the large screen in a view-port and managed like all other multimedia view-ports in the system. Examples of non-integrated applications include asset management tools, video surveillance systems, etc.
- *Loose integration is required*: An application can be manipulated from a central command station that is controlling all applications in the system versus being controlled from a specific platform.
- *Tight integration is required*: The tool is fully integrated into the software infrastructure by developing adapters that publish/ subscribe data. The tool can then communicate with other integrated applications in the system (e.g., with the 3-D model application).

The Situational Awareness system architecture should also allow new sources of data (e.g., from police–fire dispatching systems, USCG search–rescue, and inputs from other sensors and tools) to be easily integrated into the system. In addition, software templates have been created to allow for quick interface development for access into a site's local database using standards such as SQL (Structured Query Language) for databases or XML for applications.

Additionally, the system can be packaged for safety, reliability, and redundancy. It should:

- Contain built-in growth for both RGB and composite video inputs to take advantage of planned video and computer resource additions or to accommodate remote connections in a multi-port configuration
- Permit basic capability growth (new integrated applications) by simply adding PCs or other platforms for compute-intensive applications
- Support multiple monitors and graphic wall displays
- Allow resources to be distributed anywhere on the network
- Allow control and execution to be assigned to single or multiple nodes for redundancy
- Provide network-based redundancy for 100% assured operation 24/7/365

Using a Situational Awareness System to Manage an Incident

An example of how a Situational Awareness system would be used to enhance preparedness and provide consequence management for a Transportation Security terrorism incident is summarized below:

- A bomb goes off in front of a railroad passenger terminal at peak travel time.
- Incident management and stakeholder personnel gather in the EOC.

The Situational Awareness system allows the incident response team to:

- View the incident scene with many types of data views available and allow the data to be selected, moved, and sized to provide information for decision making to the incident management team
- Automatically notify appropriate response personnel based upon predetermined rules and notifications
- Alert related agencies, such as fire and police, automatically
- Alert other personnel as necessary
- Pull up prerecorded video from the nearest surveillance camera to view the scene as the incident unfolds
- Use the 3-D model to look at the scene from all different views, evaluate look angles, distances, and evacuation alternatives to decide how resources should be deployed
- Point at locations on the 3-D model and pull attribute data for items, such as buildings, buoys, fire and police stations, and so on, and use the data in making decisions
- Use the asset-dispatching capability to dispatch an asset such as a police car or fire truck to get from the station to the scene on a preprogrammed route in the appropriate amount of time
- Use the 3-D model's accurate modeling data provided for critical infrastructure elements, such as fuel storage facilities and pipelines, to address potential risks and put in place actions to protect these resources
- Activate a site's existing surveillance cameras for that critical infrastructure by a thumbnail popup for live monitoring

- Use the perimeter feature on the 2-D map to outline the disaster area to initiate evacuation planning
- Use the drawing feature on the 2-D map to identify an evacuation route and ensure that it is clear of risk areas and rescue vehicle routes
- Use the 3-D map to navigate through the scene and retrieve accurate geospatial information
- Use the high-quality 2-D map to obtain a flat view of the scene and the unfolding events and correlate the data with the 3-D model providing awareness and linkage to the 3-D scene
- Pull up broadcast media such as CNN as well as Web pages to obtain relevant information such as weather and news
- Use the charting capability provided by SafeGuard™ to provide real-time updates in chart format for information within a predetermined or manually entered radius such as:
  - Police assets
  - Fire stations with appropriate equipment
  - Hospital bed availability
  - Number of blood units available, etc.
- Pull up incident and/or personnel management applications–status as required
- Use the timeline flow of events to monitor the progression of events in the operational or training scenario
- Provide playback of all or a portion of actual events at a later date for training purposes

Situational Awareness System Summary

Situational Awareness systems provide a capability that complements and improves existing security capabilities for managing incidents. They provide a flexible environment to quickly prepare for and respond to any Transportation Security incident including natural disasters, accidents, and acts of terrorism. SafeGuard™ Situational Awareness systems facilitate decision making by providing clear and easily understood information to the management team and stakeholders. The incident management team can easily visualize the whole scene and all of the available information, make good decisions and watch the result unfold on the SafeGuard™ displays. The technology providing these capabilities can be applied at any Transportation Security Command Center through the adaptation and integration of the Situational Awareness system to the specific installation and mode of transportation. Situational Awareness systems increase operational efficiency, assure communication, save lives, and protect property. Situational Awareness systems:

- Provide solution flexibility
- Provide application data fusion and multimedia data correlation
- Event–Driven Systems provide a relevant Common Operating Picture
- Easily integrate new sources of infrastructure data
- Provide domain independent data fusion (sensor agnostic)

- Provide open architectures that utilize commercially available components for rapid deployment
- Provide a flexible, modular, scalable architecture that:
  - Allows variety of presentation display devices from desktop displays to large display walls
  - Allows for easy integration of new or existing applications
  - Easily integrates existing infrastructure data sources
  - Does not replace existing site systems and tools
- Provides a Common Operating Picture anywhere there is network connectivity

Communications Network Management for Transportation Security

Military systems are currently focused on battlefield or situational awareness. While Napoleon may have said, "An Army travels on its stomach," today's warfighter depends heavily upon communications as well. The technology required to support these entities can be massive, such as the FAA's Federal Aviation Administration Telecommunications Infrastructure (FTI), developed for the FAA by the Harris Corporation. This communications network connects over 5,000 FAA Facilities. The network technology required to interconnect so many locations is extensive, yet may be done more efficiently when modern network technologies are applied. The design goal of the new network was to provide secure communications that are more reliable and less expensive to maintain and operate. The topology of the old FAA network was typical of most existing transportation support networks in that it was a mesh of unrelated leased and owned telecommunications lines, as illustrated to the left in Figure 10.8. The new network, on the right, is much simpler in layout, but requires more efficient technology, not unlike every other transportation support system in the United States of America today. The telecommunications required to support the nation's air activity is extremely complex, especially when one considers that this network is only for telecommunications and does not include the actual control of the aircraft by the air traffic control system. As such, network management is a key technology in transportation security.

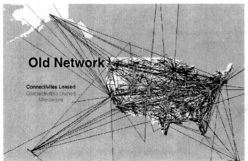 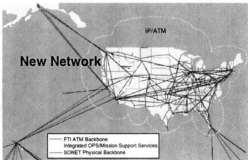

FIGURE 10.8 Old and Modern Network Topologies

Network Operations Centers

In order to provide an effective and responsive command and control capability, the decision makers must have access to key infrastructure elements and critical information about the elements and their interconnecting network(s). In today's telecommunications-rich environment, managing the network is divided into five functional areas: fault, configuration, account, performance, and security management (FCAPS). A network management system (NMS) uses these functions to manage information about the elements, control the software that keeps tabs on remote objects within the elements, and provide applications to display management data, monitor and control assets, and communicate with remote elements. A network operations center (NOC) is the hub that concentrates all of this activity into a designated facility for command center operations. The command center approach offers the ability to handle crisis management, incident response, and emergency preparedness in a variety of modern-day scenarios. NOC design and Integrated Communications Manager products can be configured to operate as a small, initial operational capability (IOC) and expand to a much larger nationwide final operational capability (FOC) as funding and time permit. An overview of the Harris NetBoss® Network Operations Center is illustrated in Figure 10.9 below.

Network Operations Center (NOC) for Situational Awareness Operations

While command centers and Network Operations Center (NOC) may be implemented at many levels, consider the implementation of a NOC at the airport level. The principles are the same as those applied to national scale military networks, although the amount of the

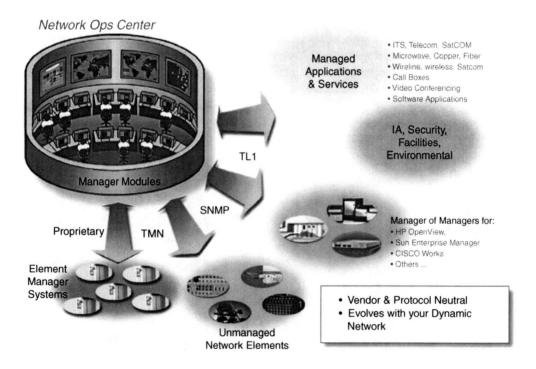

FIGURE 10.9 Harris NetBoss® Total Network Management Concept

equipment to be networked is adjusted, dependent upon the scale of the implementation. Having sensors located within the airport is important, but the control and monitoring of these sensors should be centralized to support the Situational Awareness system. For example, as part of the system, a centralized capability might be designed to pinpoint a carry-on bag left behind in a specific area of the airport. Using the Situational Awareness system, the ensuing response could be isolated to that area alone. This would avoid shutting down the entire terminal or concourse in order to respond to the incident. This sort of centralized decision support requires networking of the various sensors throughout the airport to provide a data network to all the edge devices within the solution.

An Airport NOC allows operational awareness of assets and access to vital information regarding events and incidents. The NOC may be defined as one or more locations from which control is exercised over a network of computers or telecommunications systems. Large organizations may operate more than one NOC, either to manage different networks or to provide geographic redundancy in the event of one site being unavailable or offline. The physical housing of the NOC may also contain many or all of the primary servers, routers, and other equipment essential to running the network, although it is not uncommon for a single NOC to monitor and control a number of geographically dispersed sites. Figure 10.10 shows a nominal approach for an airport infrastructure scenario.

The NOC coordinates network troubles, provides problem management and router configuration services, manages network changes, allocates and manages domain names

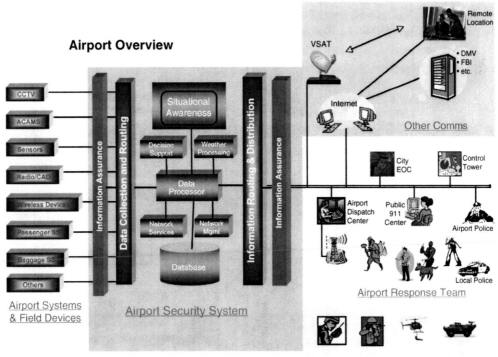

FIGURE 10.10 Airport Situational Awareness Command Center with Joint NOC

and IP addresses, monitors routers, switches, hubs and power systems that keep the network operating smoothly, manages the distribution and updating of software, and coordinates with affiliated networks. NOCs also provide network accessibility to users connecting to the network from outside of the physical space or area. NOC is normally used when referring to telecommunications providers, although a growing number of other organizations such as public utilities and private companies have also adopted the concept of a NOC to better manage their internal networks and provide pervasive monitoring services.

A military Command and Control (C^2) Center or a civilian agency Emergency Management NOC is usually composed of a Command segment (decision makers) and an Operations segment (implementers). The same would apply to all Transportation Security implementations. Figure 10.11 shows a nominal C^2 Center configuration.

A NOC provides a structured environment that effectively coordinates operational activities with all commanders, staff, responders, and participants related to the function of the mission. NOC technicians and engineers supporting the operation typically provide support 24 hours a day, 7 days a week. Typical tasks include the following:

- Monitoring operations of all backbone links and network devices
- Ensuring continuous operation of servers, routers, and services
- Troubleshooting of all network and system-related problems
- Opening tickets to track and document resolution of problems
- Supervised operating by highly skilled personnel 24 hours a day, 7 days a week

The Network Operations Center (NOC) enhances the continuity of command, enables controlled supervision of network assets, and prevents most service disruptions by providing around-the-clock proactive monitoring of the network and enterprise applications and services. The health and status of new and existing legacy communications

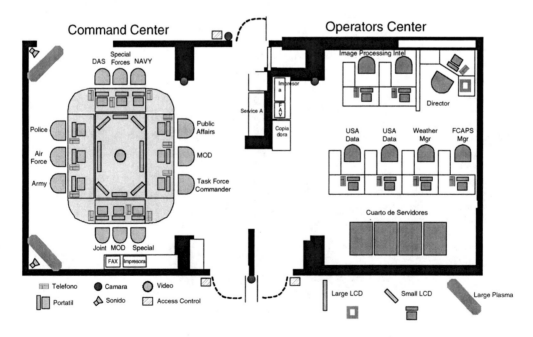

FIGURE 10.11 Typical Command Center Configuration

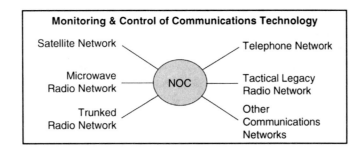

FIGURE 10.12 Health and Status Monitoring of Communications Media

networks can be easily monitored and controlled from a central location as shown in Figure 10.12.

The NOC provides the capability for effective service-based monitoring by clearly identifying services and their related infrastructure, as well as the effect of problems with these services on the mission goals and success criteria. At the heart of the NOC is usually a multi-switch gigabit network backbone and processing capability that can potentially join hundreds of local Ethernets into a large, geographically distributed network supporting thousands of networked devices and thousands of users. These devices include file servers and workstations, along with data visualization tools, critical computer-based surveillance and monitoring equipment, and responder control and dispatch systems. Implementation of the NOC will centralize responsibility for monitoring and responding to network and system performance issues and ensure a consistent level of access to the information that is generated by them.

The telecommunications industry has embraced the Telecommunications Management Network (TMN) model as a way to think logically about how the business of a service provider is managed. The TMN model, as shown in Figure 10.13 below, introduces the concept of Logical Layered Architecture which can be used to design the Transportation Security C^2 center. It consists of four layers, usually arranged in a triangle or pyramid, with business (or mission) management at the apex, service management the second layer, network management the third layer, and element management at the

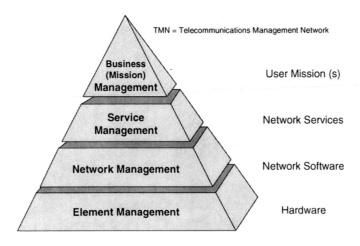

FIGURE 10.13 Basic TMN Logical Layering Architecture Model

bottom. The idea is that management decisions at each layer are different but interrelated. For example, detailed information is needed to keep a switch or a router operating (at the element management layer), but only a subset of that information is needed to keep the network operating (e.g., is the switch or router operating at full capacity?). Working from the top down, each layer imposes requirements on the layer below. Working from the bottom up, each layer provides a capability to the layer above.

Implementation of a network operations center requires careful consideration of the structure of the network, the purposes for which it is used, and similarly, the number and types of servers and applications used to provide the wide array of networked services that are not only presently in place but also planned for the future. A wide range of operating systems, database architectures, and an extensive range of server and workstation application software packages are supported within a distributed environment. Adherence to industry standards facilitates COTS use and proven upgrade paths as needs grow, or as change occurs, or as product improvements present opportunities for technology insertion. The NOC design is usually based on an industry standard OSI Network Management Model for fault, configuration, accounting, performance, and security (FCAPS) for monitoring, classification, diagnosis, resolution, and management of faults in applications, services, and the network.

FCAPS Requirements Summary

Fault management:

- Detect, log, notify users of, and automatically control, monitor, and fix network problems to keep the network infrastructure running effectively.
- Detect and correct network and network device problems.
- Implement repair–recovery procedures.
- Implement error-detection processes and diagnostic testing.
- Implement reporting mechanism.
- Implement concept of "trouble ticket system" to the responder entity (ground base fix, repair depot, etc.)

    Configuration and control management:

- Monitor network and system configuration information so that the effects on network operation of various versions of hardware and software elements can be tracked and managed.
- Control policy and rules.
- Control handling of detailed network configuration parameters.
- Configure and control RF links.
- Configure and control voice systems.
- Control network state and topology information.
- Re-configure when necessary.
- Control domain name service (DNS).
- Configure addresses to network hosts and devices.
- Configure IP encryptors (configuration of keys and addresses).
- Control inventory management.

Accounting management:

- Measure network utilization so that individual or group users on the network can be regulated and appropriately recorded and logged.
- Record use of management objects by users and/or groups.
- Provide and control user access.

Performance management:

- Measure and make available various aspects of network performance so that internetwork performance can be maintained at an acceptable and required level.
- Determine state of the system.
- Optimize network operation.
- Monitor performance of radio links.
- Monitor performance of LAN links.
- Monitor behavior of network elements:
  - Statistics
  - Interface statistics
  - Throughput
  - Error rates
  - Software statistics
  - Usage
  - System load
  - Disk space
  - Response time

Security management:

- Control access to network resources according to local guidelines so that the network cannot be sabotaged and so that sensitive information cannot be accessed by those without appropriate authorization.
- Manage security required for operating network.
- Protect information and data (information assurance).
- Control access to routers, communications servers, and services.
- Provide remote access to network elements for diagnostics.
- Distribute security-related information.
- Help in resolving a network issue versus a host issue.

NOC Framework Model

The framework of a Network Management System is usually based on SNMP v.1/v.3 and includes the following.

- Simple network management protocol (SNMP)
- Ability to be configured for Internet Protocol (IP) encryptors
- Structure of Management Information (SMI): [RFC 1155/RFC 1212/RFC2578-2580]

- Management Information Base (MIB): [RFC 1213/RFC 2011-2013/RFC 2700]
- Management Application Protocol (MAP): [RFC 1157/RFCs 2570-2573]

A "manager of managers" (MOM) tool, such as the Harris Corporation's NetBoss®, is usually included in the NOC design. The MOM tool consolidates and logically presents information captured from the network, applications, and services. It also provides work-flow automation to enhance the response of NOC personnel. The MOM is custom-scripted to present information from the point of view of the services affected and resources available in order to ensure appropriate impact analysis and prioritization of response. Multiple functions are constantly exercised to manage resources and provision the network, and multiple data sets are tracked, stored, and monitored, as shown in Figure 10.14.

Network management means different things to different people. In general, network management is a service that employs a variety of tools, applications, and devices to assist and maintain monitoring, controlling, and maintaining network devices. In a NOC architecture, network management will involve a distributed database containing network policies, automatic polling of network devices, and configuration of network addresses, as well as controlling Quality of Service (QoS) configurations generating real-time command and control of network devices. The network management platform will be the "Configuration Data Translator," a function that transforms configuration management data (policy-based management) into device local configurations (low-level policies) based on the generic capabilities of the devices within the network infrastructure. In general, network management will be a global service that employs a variety of tools, applications, and

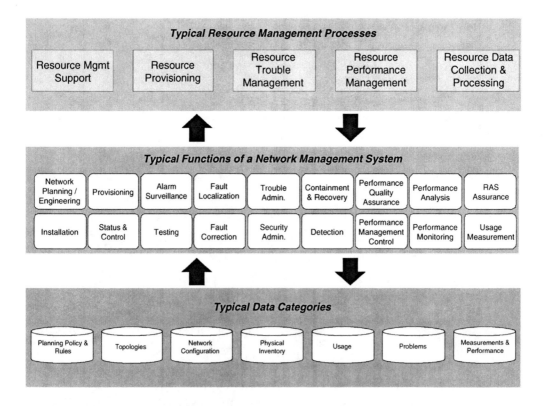

FIGURE 10.14 Hierarchical Handling of Management Data, NMS Functions, and Processes

devices to assist network personnel in monitoring, maintaining, and controlling devices within the network infrastructure.

Under the direction of the NOC director, a typical network management staffing model establishes four types of positions: network monitor, network engineer, application and server engineer, and network management system (NMS) engineer. Network monitors review faults detected by the manager of managers (MOM) and problem reports referred by the Help Desk. Whenever possible, network monitors resolve the fault or problem utilizing the troubleshooting and other tools available to them. Problems or faults that cannot be resolved by the network monitors are referred either to network engineers or to system engineers depending on the nature of the issue. The application server and system engineers are experienced in troubleshooting and resolving complex network and application or server issues. NMS engineers are responsible for development and maintenance of the MOM and all other management tools utilized by the NOC.

Benefits of an Integrated Communications Manager

The chief goal of the Network Operations Center is to proactively identify and resolve problems that lead to critical outages before such outages occur—and whenever possible, before any user is aware that a problem has occurred. Because NetBoss views its environment from a service perspective rather than solely on individual components, the operations staff has an immediate view of what will be affected when a problem occurs and can take timely and appropriate steps to prevent users from being affected. The NOC provides the capability of observing faults resulting from these kinds of issues that permit a much more rapid escalation to the appropriate service provider and constant monitoring to ensure that the provider resolves the problem promptly. In the event that an outage occurs that is the responsibility of the NOC, the immediate availability of expert staff around the clock results in much shorter resolution times than were ever before possible.

Importance of an Integrated Communications Manager

The key to quickly determining the degree to which server, network, or application faults have an effect on the usability of a service lies with the asset database that has been incorporated into the NOC suite of tools. This database should be custom-designed for the mission implementation. Asset discovery is done weekly following the weekly system maintenance period to ensure that the database reflects the true state of the installed system. When a fault is detected by the system, it is correlated to the asset database which groups assets by services so that NOC engineers immediately know which services, and subsequently which users, are affected by the problem. The NOC itself incorporates, but is not limited to, a state-of-the-art design utilizing COTS laptops, console workstations, high-end video cube wall displays, and ancillary LCD video display screens. The incorporation of a Web-access interface provides numerous high-level and detail displays showing the current status of the network resources, applications, and services, detailed fault status information, current news and weather information, and much more.

The NOC provides the network monitors and network engineers the ability to virtually and physically monitor enterprise applications and servers, the facility's electrical power feeds, HVAC system status, and the status of other infrastructure systems. The technology in use provides the 24-hour help desk with timely information on system status so that help desk analysts can knowledgeably assist end users. The call center management system used in the help desk design is tightly integrated with the NOC's MOM allowing problem tickets to be automatically generated as faults occur. In order to prevent the NOC staff from bombardment with fault alarms generated by multiple systems when problems occur, the event correlation and "de-duplication" capabilities of the system ensure that the system reports only the faults related to the underlying cause of the failure.

Network Management Summary

Without an effectively implemented NOC, the Situational Awareness and Command and Control (C^2) System would be like a brain without a nervous system. Network operations centers have long been a corporate standard for telecommunication service providers and other large business operations. The selection of the manager of managers (MOM) is an important discriminator for those seeking proven performance and scalability. The ideal MOM will have the ability to integrate information from other applications and a Situational Awareness tool.

Data Fusion and Data Correlation for Transportation Security Situational Awareness Systems

In a C^2 environment, operators are frequently inundated by a plethora of disparate information that cannot be analyzed quickly enough to provide a thorough threat assessment in a timely manner. Further, various data elements when reviewed and analyzed and combined properly can indicate a threat situation requiring observation or action that could not be gleaned from the individual elements. Lack of experience and frequent rotations frequently remove the most experienced operators from the field at the time when they become most effective.

Multi-sensor data fusion (also known as distributed sensing) is an engineering methodology that can be used to combine data from multiple and diverse sensors and sources in order to make inferences about events, activities, and situations. These systems are often compared with the human cognitive process where the brain fuses sensory information from the various sensory organs, evaluates situations, makes decisions, and directs action. In these instances, data from multiple sources are associated, correlated, and combined in order to generate useful information that leads to threat identification for the notification of operators in order to generate timely responses. In return, this information can then be applied to a situation to support the achievement of timely assessments, efficient threat predictability, and other significant information. These sources may be similar, such as radars, or dissimilar, such as electro-optic, acoustic, or passive electronic emissions measurement. This use of multiple sensors to increase the capabilities of intelligence has received considerable attention in recent years.

As sensor technology becomes more robust and more abundant, the need to filter, correlate, and glean new data from otherwise unintelligent sensor feeds becomes critical. The data fusion engine will be able to fuse these multiple sensors into useful information. Transforming this information into knowledge is the key concept behind acquiring an

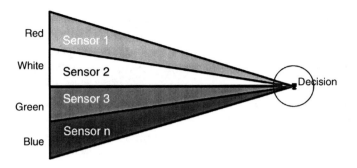

FIGURE 10.15 Multiple Sensor Data Fusion

understanding of this fused sensor data. Figure 10.15 shows the combination of multiple sensor data in order to produce a decision-based system. As a result, data fusion provides for the user information of tactical value with a very low rate of false positives.

The following are the advantages of Data Fusion and Correlation as applied to Transportation Security Situational Awareness systems:

- Robustness and reliability. The system is operational even if one or several sources of information are missing or malfunctioning.
- Extended coverage in space and time.
- Increased dimensionality of the data space. It increases the quality of the deduced information; it also reduces the vulnerability of the system.
- Reduced ambiguity. More complete information provides better discrimination between available hypotheses, providing a solution to the explosion of the information that is available today.

The military and intelligence communities over the last few years have discussed Data Fusion and Correlation capabilities in terms of "Levels" that were defined several years ago by the National Aeronautics and Space Administration's (NASA) Jet Propulsion Laboratory (JPL). Data Fusion can be broken down into several "levels" of fusion, as described below:

- Level 1—*Object Refinement:* fusing data to determine the identity and other attributes of entities and also to build tracks to represent their behavior.
- Level 2—*Situation Refinement:* fusing relationships between entities to group them together and form an interpretation of the patterns in the order of battle data.
- Level 3—*Impact Assessment:* an iterative process of fusing the combined activity and capability of enemy forces to infer their intentions and assess the threat that they pose.
- Level 4—*Process Refinement:* an ongoing monitoring and assessment of the fusion process to refine the process itself and to regulate the acquisition of data to achieve optimal results.

Level 4 interacts with all other levels. Level 1 is referred to as "low-level" fusion while 2 and 3 are referred to as "high-level" fusion. Data fusion is more commonly referenced to low-level fusion (level 1) while information fusion is referenced to high-level (levels 2 and 3) fusion. There are many techniques which one could use to perform the process of data fusion, namely rule-based systems, fuzzy logic, Bayesian networks, and other statistical and mathematical processes. Many commercial Fusion Engine products are available that are de facto Fusion Engines that ingest sensor inputs and perform decision-making. Other products such as Ilog or the Fair Isaac Blaze rules engine allow fusion algorithms to be dynamically evaluated and adjusted over time to achieve the best performance based on a trade-off. Based on the particular implementation, the operator may make decisions and take action based upon data content or the system may react automatically based upon preprogrammed responses to various stimuli. Within this section, some of these techniques will be discussed in order to draw inference from what design(s) best suit Situational Awareness system architectures.

Situational Awareness system solutions will typically include an inference engine and database as part of the information fusion component. One of the major challenges associated with developing a real time system that captures and processes the real time data associated with Transportation Security is the development of an infrastructure that is capable of handling massive volumes of data from sensors while performing real time correlation and object evaluation for the generation of alarms. The enormous volume of sensor data that is produced during critical periods coupled with the mass storage of data used for analytics and correlations of operational data requires a specialized system. The system must be capable of scaling to meet peak event rates while adapting and prioritizing tasks to ensure critical data is properly processed during peak levels of activity. In order to support the cueing of appropriate assets to potential threats, the Situational Awareness system architecture must support the correlations of operational data and the ability for responders to reach back and obtain current updated information. Finally, because of the public safety implication associated with the implementation of such a system, the system must always be available and support processing continuity in the event of disaster or facility failure. As a result of these requirements, the key characteristics of such a solution are:

- The system must always be available to process potential risk or emergency situations
- The ability to scale up and support both peak and future workload
- Support for new and emerging open standards
- Extensible architecture that can evolve and support new features and requirements

In any case, the term *fusion* or *Data Fusion* tends to be used rather loosely by laypersons where in fact, the term can have many meanings. For the purposes of Transportation Security Situational Awareness systems, however, we will define the term to mean the association of data from multiple sensors to derive conclusions that are not intuitively obvious to system operators. For our purposes, the derivation of information (whether from fusion, correlation, or data mining) is secondary to the desired results, namely the identification of a potentially dangerous scenario. The following example will serve to illustrate how this may be achieved in a harbor scenario, but is equally applicable to any Transportation Security scenario.

Data Correlation and Control

The heart of the Situational Awareness system is the Data Correlation and Control capability illustrated in Figure 10.16 below. These Data Correlation and Control capabilities

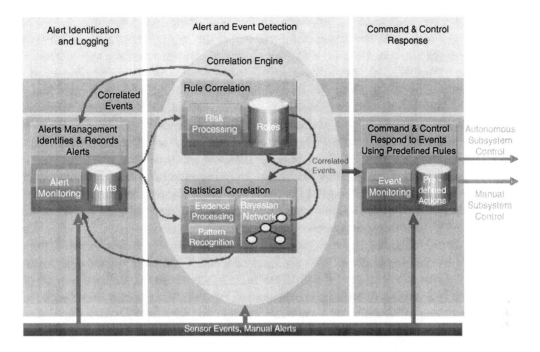

FIGURE 10.16 Situational Awareness system Data Correlation and Control

will ensure that the Situational Awareness system Vessel Tracking System (VTS) subcomponent will effectively provide Transportation Security Management. As previously mentioned, where conventional VMS typically pass data and information horizontally in an uncoordinated manner between vertically integrated subsystems, the Situational Awareness system Data Correlation capabilities turn *data* into *information* that is extrapolated into *knowledge*. Rather than generating simple alerts, the system associates concurrent events by using the techniques discussed earlier to extrapolate knowledge of concurrencies that are not detectable by conventional table-driven association-based systems. The result is faster detection, a higher probability that all events of interest do not pass unnoticed, and overall increase in the number of detected anomalies.

The rules engine tests events against the rules database and, when rules have been satisfied, it generates second-order events or alerts. The system's administrators, domain experts, and security personnel using a rules GUI provided by the system configure rules. The data correlation engine associates multiple sources of data and correlates them to generate useful information. This information is then applied to a situation to achieve timely employment of resources. The data correlation adds probabilities to event occurrences and uses these probabilities to send the right information at the right time to resources, which results in knowledge that can prevent and/or aid in an alert situation. Again, all techniques discussed previously come into play in this model.

One of the major functions of the information fusion component of Situational Awareness systems is to collect position object-tracking data from multiple sensors to combine or fuse duplicate data received from different sensors about the same object. The challenge of sensor fusion is finding a set of unique elements that specifically determine if the data can be fused. Different sensors transmit location data elements at different points in time that require the system to approximate the position of objects based on the time, direction, and velocity of the object being tracked. If the approximation algorithms are too stringent (or not correlated on a time axis), the objects are not fused and appear as two objects. Conversely, if the algorithms are too loose, the two objects may be incorrectly fused.

Unlike data fusion, data correlation attempts to augment or integrate information from additional and possibly external data sources. An example of this may be crew or manifest information derived from data received from other maritime databases. Data correlation is accomplished by developing XML-based adapters that interface with a common message layer that pushes or pulls data from the external system. The messaging layer inserts or updates the information in the object being tracked. As this new information is collected the system invokes the rules engine that determines what if any action should be taken as a result of the new correlated data.

The ability to use both human-generated and algorithmically derived knowledge to make risk predictions associated with the operation of an individual vessel, flag suspicious behavior, or detect abnormal operations is extremely powerful. Rule support takes advantage of the domain knowledge of operational experts and applies relevant doctrine and policies. Data-mining techniques can be used to fill any remaining gaps, by analyzing large data sets and finding the subtle and/or complex patterns that are extremely difficult for humans to spot. Data-mining techniques use sophisticated algorithms and models that are capable of detecting and representing complex patterns.

The Data Correlation and Control component provides the ability to trigger events based upon data item values or attributes contained within the incoming data. For example, if the system has a sensor interface to a hospital, the XML adapter will poll the hospital's database for available beds and report the information, allowing the system to "watch" the number of available beds and generate an event when the bed count reaches some preset value or threshold. To achieve this capability, the system uses a set of user-defined rules in order to determine when to trigger an event. The system parses each configured message and compares the information with a set of rules. Each rule has a corresponding action to be taken if the rule evaluates to "true." Returning to the hospital example, when the watched hospital reaches a certain capacity, the action taken may be a simple notification message or it could be as complex as generating a request to transfer equipment, manpower, or other assets to or from the hospital.

Because a Situational Awareness system is completely domain independent, another example may provide more insight and understanding into the system's capabilities. Using a network operations environment such as the one that was discussed earlier in the chapter as the domain, it is fairly commonplace for the system to read statistical data from routers, firewalls, and other network devices. With an SNMP traffic adapter implemented and a configured threshold established, the system can identify a denial-of-service attack while it is occurring, enhancing Information Assurance. In this example the SNMP traffic adapter reports firewall status including the number of hits or access attempts while the system monitors this value. When the access attempts per second reaches or exceeds some preconfigured value, the system will generate an event in response. The actions performed by the event are user configurable, but could include a system notification message or could even reconfigure the network, taking the firewall out of service, thereby isolating the network from the outside world.

Operational Example of a Transportation Security Situational Awareness System in a Seaport Scenario

As a result of the Maritime Security Act of 2002, improved seaport security policies and technologies are being implemented that secure ports while maintaining the flow of commerce. Ports must be prepared to respond to broader and more complex response scenarios, including natural disasters, accidents, and terrorism. To complicate matters, terrorist incidents typically involve multiple events. The ability to prepare and react to incidents is predicated on the ability to have timely, complete, and accurate situational awareness, derived from existing applications, data sources, and remote sensors, and depicted in a robust and flexible visualization tool. Many ports have built, or are in the process of building, Situational Awareness systems. Additionally, seaports have numerous management tools to obtain port information and status such as personnel, asset, and consequent management applications, along with surveillance equipment. In the following paragraphs, we will examine how an implementation of a Situational Awareness system might use data and information derived from sensors to protect a seaport. Keep in mind that all data sources are considered to be "sensors" for the purposes of Transportation Security Situational Awareness system discussions. Figure 10.17 below will be referenced throughout the following discussion.

In the illustration below, the Correlation Engine accepts inputs from disparate data sources. The vessel being tracked has its picture to the upper right of the figure. The Safe-Guard 2D display provides an icon that represents every vessel within an area of interest that is equipped with an Automated Identification System (AIS) transponder. AIS is a system used by ships and Vessel Traffic Services (VTS) for tracking, identifying, and locating vessels. The International Maritime Organization (IMO) requires AIS to be fitted aboard international voyaging ships of 300 or more gross tons, and all passenger ships. Each AIS-equipped vessel has a unique nine digit Maritime Mobile Service Identifier (MMSI) that is associated with the transponder mounted in the vessel. AIS uses this transponder to provide the electronic exchange of extensive data about the vessel. This data can include the MMSI identification, vessel name, position, heading, and speed. The data are transmitted with the MMSI on a regular basis. If the harbor is equipped with an AIS Base Station, the

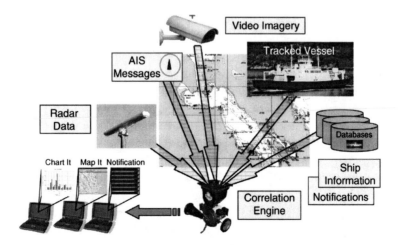

FIGURE 10.17 Seaport Situational Awareness system Data Correlation and Anomaly Detection

Base Station receives AIS messages from all vessels within range of its antenna. The Situational Awareness system can then display icons for each vessel in the area. The Situational Awareness system viewing software can then interpret these data for the operator.

If the operator overlays the Operator Station's cursor on a ship's icon and "clicks," the MMSI, ship name, course and speed, classification, call sign, registration number, and other information is displayed. This information is available to the Situational Awareness system server automatically and serves as an input to the Data Correlation Engine. The Situational Awareness system rules engine uses specific rules to determine ship proximity and speed based on AIS information. A Proximity Rule, for example, would analyze AIS transponder data and, based on the heading and speed of vessels in the area, could generates alarms when potential collisions are detected. An alarm could also be generated when ships are moving too fast in the harbor's channels. Any vessel or vehicle, or person equipped with a Global Positioning System (GPS) transponder can also be displayed on the 2D map. If the harbor radar is integrated into the Situational Awareness system, a corresponding position icon is displayed on the 2D map for each target in the area detected by the radar. Along with the icon, a velocity vector and information indicating speed and heading would also be provided.

As mentioned earlier, the Situational Awareness Security System Data Correlation Engine receives inputs from multiple sources. In this example, there are AIS message inputs as well as information from the radar system. This information might consist of location, speed, track, target size, etc. The harbor camera system also serves as a sensor. If the camera is a PTZ camera, it will track vessels either automatically or manually. Notification and other information is also available to the Correlation Engine. The Situational Awareness system Database could contain ship data from Lloyd's or Jane's ships registry database. This information frequently includes photos of the vessel. It is also required in the United States of America that an electronic notification of arrival be sent for all vessels 96 hours prior to arrival. This information includes the ship's name, call sign, tonnage, type and amount of cargo, including dangerous and polluting cargo, last port of call, etc. This information, along with the other sensor information, is provided to the Situational Awareness Data Correlation Engine as well. All of this information is used to provide transportation security.

For example, suppose that a ship is picked up via AIS entering the harbor. The ship's AIS MMSI indicates that the ship is a super-tanker christened the *Kobayashi Maru*. To ensure that this is true, other sensors are brought to bear to determine if the vessel is indeed the *Kobayashi Maru*. The Data Correlation Engine could detect if the radar target data indicate that the ship target size is not consistent with that of a supertanker, and if so an alert is generated. Radars simply track targets; they provide data rather than interpret them. The AIS system merely "hears" what the transponder transmits. (Bear in mind that AIS was developed as a safety tool, not a security tool; its intent is to prevent accidents.) In the event that the radar and AIS inputs "match," other specific actions may be taken as well depending upon the programming of the Correlation Engine Rules. The correlation engine may examine the notification information in the system database and determine that a supertanker is not scheduled for entry into the area under surveillance. This would generate an alert. Assuming that there is a notification for the *Kobayashi Maru,* the system could point the nearest long-range camera at the vessel and take a still shot. The correlation engine will examine the local database or a Jane's or Lloyd's database to get an illustration of the vessel. The system may then use video analysis software to compare the photo to the still shot, or both may be provided to an operator for manual comparison. This concept is illustrated in Figure 10.18 below.

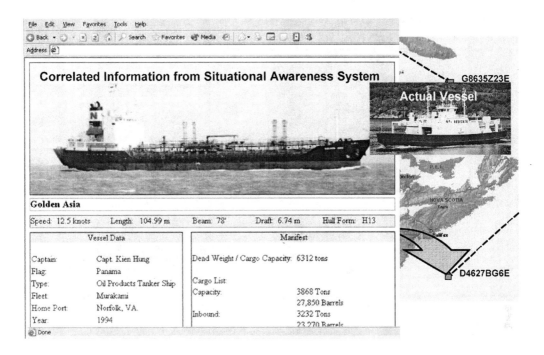

FIGURE 10.18 AIS Tracking and Registry Correlation

The Situational Awareness system also allows the operator to delineate a "GeoFence" in an area of a 2D map. A GeoFence is an imaginary boundary around an area of interest. The system tracks locations and generates alerts whenever a boundary is "crossed." In Figure 10.19, two GeoFences have been depicted in a harbor map. GeoFences may be "drawn" on the computer screen to delineate any area of interest. Specific Correlation Engine rules may be associated with particular GeoFences. The green dot indicates the position of the vessel of interest as it moves through the harbor channel. In the example in the illustration, ships that are traveling within the harbor speed mandates are indicated as green dots. The white square icons with the red borders represent response teams.

A close-up of the uppermost GeoFence is illustrated in Figure 10.20. In this instance, the icon has turned red and alerts have been generated since the vessel has entered a restricted area. Alerts and warnings generated by the Correlation Engine flagged the vessel as speeding and provided warnings when the vessel turned toward the restricted area. Response teams have been dispatched to intercept the vessel.

While this example has been associated with a harbor, a Situational Awareness system can be equally useful in other transportation security modalities as well, particularly airports and roadway systems. State Department of Transportation (DOT) system cameras can be used for more than mere traffic monitoring. If used as part of a DOT Situational Awareness system that will provide an end-to-end solution with:

- Asset tracking for improved efficiency
- Hazmat (hazardous materials) tracking
- GeoFencing for Hazmat route adherence
- Remote Web access with GIS mapping for DOT managers.

In an airport environment, Situational Awareness systems can go a long way toward ensuring transportation security. While having security sensors within the

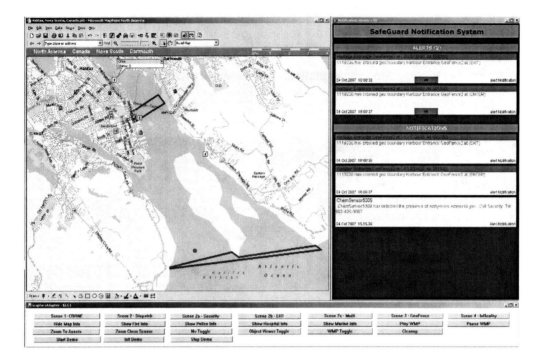

FIGURE 10.19 Depiction of Situational Awareness system GeoFence Capability

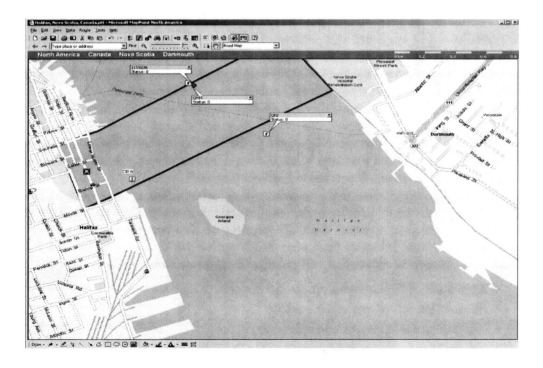

FIGURE 10.20 Situational Awareness system Response to GeoFence Penetration

airport is important, the control and monitoring of these sensors need to be centralized in order to effectively support situational awareness. As stated previously, the mission elements of Situational Awareness are to prepare and prevent, detect, respond, and recover. The existence of such an asset in an airport could save millions of dollars in lost operating expenses to airlines when a shutdown occurs due to a security incident. A single concourse could be shut down in the event of an incident, rather than an entire airport, as is typically done today. The technology exists today to provide a centralized monitoring capability that can pinpoint a carry-on bag left behind to a specific area of the airport and isolates a response to that area alone to avoid shutting down the entire terminal or concourse. In addition to incident response isolation, this sort of centralized, decision support system can be used for simulation and playback and supports the training of TSA personnel to respond to specific scenarios.

Summary

Network-based Situational Awareness systems are third generation product suites that provide extensive capabilities for transportation security Situational Awareness and Command and Control Systems. Situational Awareness systems have matured in the military and may be easily adapted to peaceful uses. Event detection can be performed by various types of sensors, such as SafeGuard™. SafeGuard™ is an event-driven system built on an XML-based messaging platform for use in transportation security. Situational Awareness systems provide alerts and information anywhere there is network connectivity. As such, a robust, well-designed, fast network is highly desirable. Key to the derivation of timely information and knowledge from available data are the data correlation and fusion capabilities. The Data Correlation Engine monitors messages transmitted throughout the enterprise, reads the data values from the messages, and generates events when the threshold conditions are met. Event generation in Situational Awareness systems sets in motion a set of user-definable actions that can control applications, generate data messages, and transform messages providing a virtually endless array of visualization control and data capabilities. The visual capabilities and applications included in Situational Awareness systems provide 2D and 3D mapping, video ingest and display, real-time charting, sequence of events logging and display, and Web-based display. Additionally, modern Situational Awareness systems provide a well-defined messaging protocol that is extremely flexible when integrating with existing and planned command and control systems. These systems are able to interface and interoperate with a variety of existing or third-party products.

References

Alberts, David S, & Hayes, Richard E., "Understanding Command and Control," Command and Control Research Program (CCRP), http://www.dodccrp.org/files/Alberts_UC2.pdf, 2006.

Canny, J., The future of human-computer interaction. *ACM Queue, 4,* 24–32, 2006.

Castro, Larry, "Information Sharing and Collaboration Challenges," Homeland Defense Training Conference, Washington, D.C., 6 November 2003.

Committee on Assessment of Security Technologies for Transportation, National Research Council, "Fusion of Security System Data to Improve Airport Security," http://www.nap.edu/catalog/11913.html 2007.

Dasarathy, B., "Sensor Fusion Potential Exploitation-Innovative Architectures and Illustrative Applications," *IEEE Proceedings, Vol. 85,* No. 1, 1997.

Delay, John, "An Integrated Approach to Digital Asset Management and Content Distribution: An Overview for U.S. Civilian, Defense, and Intelligence Agencies," Harris Corporation White Paper, 2004.

Duwadi, S. R., & Chase S. B., US Department of Transportation multiyear plan for bridge and tunnel security research, development, and deployment. Washington D.C.: USDOT, Publication FHWA-HRT-06-072, 2006.

Endsley, Mica R., & Garland, Daniel J. *Situation Awareness, Analysis, and Measurement*, Lawrence Eribaum Associates, Mahwah, New Jersey, 2000.

FEMA, "IS-700 National Incident Management System (NIMS), An Introduction," http://training .fema.gov/EMIWeb/IS/is700.asp, 2003.

Fox, Kevin, "FuzzyFusion™: Taking Information Assurance to the Next Level," Harris Corporation White Paper, 2001.

Ham, D. B., & Lockwood S. (PB) and SAIC, *National Needs Assessment for Ensuring Transportation Infrastructure Security,* The American Association of State Highway and Transportation Officials' Transportation Security Task Force, October 2002.

Harris Corporation. "Harris SafeGuard™ and Homeland Security," Harris Corporation Product Data Sheet, 2004.

Harris Corporation, "Processing Systems Core Capabilities Overview," Harris Corporation Product Data Sheet, 2007.

Harris Corporation, "RealSite™ and SafeGuard™ Support the FBI During Super Bowl XL," Harris Corporation Product Data 06–37, 2006.

Harris Stratex Networks, "NetBoss® Service Manager Overview," Harris Stratex Networks Product Data Sheet, 2007–2008.

Hecker, Robert, "Hurricane Katrina: Security Challenges and Response," AAPA Port Security Seminar, Seattle, July 2006.

Kielman, Joseph, "Threat and Vulnerability Testing and Assessment (TVTA)," Science and Technology Directorate. DHS, Washington, D.C., 2005.

Pepper, William, "Effectively Preparing for and Managing Port Incidents with the Harris SafeGuard™ System," Harris Corporation White Paper, 2005.

Pepper, William, "Emergency Management Communication Interoperability," Harris Corporation White Paper, 2003.

Pepper, William "Enterprise Data Center Security," BiometriTech 2003. New York, 2006.

Pepper, William, "Operator-Centric Multi-Sensor Information Data Fusion, Correlation, and Anomaly Notification for Decision Support in a Situational Awareness Environment," Transpo 2006, Tampa, 2006.

Owen, Todd, "Customs and Border Protection and Container Security Initiatives at Seaports," AAPA Port Security Seminar, Seattle, July 2006.

Seppa, S. A., "NOC For Command Center Operations," Harris Corporation White Paper, June, 2007.

Steinberg, A. N., Bowman, C. L., and White F. E., "Revisions to the JDL Model", *Joint NATO/ IRIS Conference Proceedings*, Quebec, October, 1998.

Steinberg, A. N., Bowman C. L., and White F. E., Sensor Fusion: Architectures, Algorithms, and Applications, *Proceedings of the SPIE, Vol. 3719,* 1999.

Ullman, David G, "OO-OO-OO!" The Sound of a Broken OODA Loop, Crosstalk, April, 2007, http://;www.stsc.hill.af.mil/CrossTalk/2007/04/0704Ullman.html

United States Coast Guard, "What is the Automatic Identification System (AIS)?," http://www
.navcen.uscg.gov/enav/ais/default.htm 2008.

U.S. Congress, "Maritime Transportation Security Act of 2002," 116 STAT. 2064 PUBLIC LAW
107–295, NOV. 25, 2002.

White, F. E., "A Model for Data Fusion," *Proc. 1st National Symposium on Sensor Fusion,* 1988.

Transportation Security Solutions

11

Automatic Identification and Data Capture (AIDC): The Foundation of Military Logistics

Corey A. Cook and Thomas A. Bruno

Objectives of the Chapter:

- Describe an automatic identification and data capture Automatic Identification and Data Capture (AIDC) system
- Explain the role of automated identification technology (AIT) in logistics
- Discuss the logistics process and functional enhancement
- Discuss the evolution and use of bar coding
- Discuss contact memory buttons (CMBs) as an electronic seal
- Discuss radio frequency (RF) communication
- Discuss satellite-tracking systems (STSs)

Introduction

What is the significance of the number 65,354? The number 65,354 reflects the number of steel shipping containers lost in transit to Iraq and Afghanistan while in the U.S. Central Command logistics transit system as of November 2007, according to the current Speaker of the House's fact sheet. While the loss of any one container shows negligence, the loss of high-dollar material, munitions, and medical supplies is absolutely criminal. Within the medical arena, the reality is that the loss of even one medical supply container can mean the difference between life and death. In terms of munitions and weapons, the loss in transit can have even greater consequences.

Container loss in transit, while a crucial factor in material accountability, is only one aspect facing the military transportation arena. In the ever-changing and complex military support environment, the identification, storage, tracking, and reconstitution of material assets have become increasingly difficult. These elements are vital in supporting transit ability and accountability. Simply stated, you need to know what is in a box before you ship a box.

A combination of automatic identification technologies (AITs) are required to satisfy the asset visibility needs of industry. No single technology will solve all information exchange requirements, even as no one tool will support all home repair tasks. A toolbox approach, wherein one or more technologies are employed for each situation to provide requisite asset visibility effectively and efficiently, is the recommended approach to providing asset visibility. This chapter will focus on the definition, types, and use of AIT utilized to enhance the medical expeditionary arena.

Military Logistics Technology
Automated Identification Technology

AIT is defined as a suite of technologies that enable the automatic capture of source data, thereby enhancing the ability to identify, track, document, and control deploying and redeploying forces, equipment, personnel, and sustainment cargo.[1]

AIDC is a family of technologies that, if employed properly, improves the accuracy, efficiency, and timeliness of material identification, marking, and data collection. AIDC media and devices include linear and two-dimensional bar code symbols and their readers, magnetic stripe cards, integrated circuit chip (ICC) cards (e.g., smart cards), optical memory cards (OMCs), active and passive radio frequency identification (RFID), contact memory (button memory) devices, magnetic storage media, and wireless devices.[2]

The use of AIDC enables the automatic capture of source data for transfer between automated information systems (AISs) to identify, track, and document supplies throughout the Department of Defense (DoD) supply chain. The use of AIDC assists in the control of deploying and redeploying forces, equipment, personnel, and sustainment cargo to reduce processing times, improve data accuracy, and logistics business processes.

AIDC is a part of information technology. This rapidly developing science requires coordinated effort to support Joint Chiefs of Staff initiatives for standardized systems. AIDC uses cost-effective and operationally efficient technologies supporting logistics business processes. The use of AIDC requires standardization to ensure business systems are compatible, interchangeable, maintainable, and interoperable.

AIDC is a key enabler to gain timely visibility of all logistics assets within business processes to meet the goals of total asset visibility—regardless of the location of the materiel: in process (being procured or repaired), in storage (being stored as inventory), in use, or in transit (being shipped to another location). System requirements define the design, analysis, fielding, installation, training, and implementation critical to a standardized and interoperable framework, architecture, and data structure.

Asset visibility is critically important as operational commitments continue to grow far faster than resources with which to execute missions. Automatic identification and data capture (AIDC) media, also commonly referred to as automatic identification technology (AIT), provide the ability to create machine-readable data more easily, quickly, and with much greater accuracy than manually keying data into information systems or handwriting data on paper forms. Increasingly heavier burdens on human infrastructures require the necessity to work smarter, not harder. AIDC creates tremendous value to asset visibility efforts when implemented wisely and with forethought.

---

[1] March 2008 DoD AIT Implementation Plan (TRANSCOM).

[2] Department of the Army, Total Asset Visibility Action Plan, October 1993.

AIDC is designed and implemented to create, capture, store, and pass accurate and timely information on the status of assets—whether in storage, in process (during manufacturing or repair), in use, or in transit. AIDC facilitates the creation of machine-readable data, capturing data with a device (bar code scanner, contact memory button reader–writer, RFID interrogator, etc.) and transferring the AIDC data so it can be aggregated and viewed throughout the enterprise via networks. The use of AIDC requires minimal human intervention, enabling the rapid capture of detailed information for interface with automated information systems. The goal of any AIDC implementation is to assist the user in selecting the "right" technology for the required task.

AISs are repositories of the information that describes an organization's activities. AISs must have accurate and timely data in order to compile, analyze, and distribute data converted into information. AIS managers must have strict configuration control of their data models, database tables, and business rules, as well as the authority to change any of these.

Logisticians place special emphasis on transactions used to configure, modify, move, or utilize materials, components, and products throughout the supply chain. AIDC media in several forms are utilized to ensure that item identification is properly captured throughout the process. AISs provide asset and material status for maintenance-critical, serialized components through the use of a closed-loop, life-cycle tracking system that is enabled by AIDC. In addition to providing asset visibility, the AIDC-enabled AIS allows access to maintenance and usage history.

A toolbox approach to AIDC includes a variety of media similar to a home tool or fishing tackle box. This approach provides a combination of technologies for a component, a product, a package, or a transportation label, which enables asset visibility with a substantial increase in speed and effectiveness of data retrieval. AIDC media include bar codes, magnetic stripes, integrated circuit or "smart" cards, optical memory cards (OMCs), and contact memory buttons (CMBs). AIDC is often used with radio frequency data communications (RFDC), radio frequency identification (RFID) tags, and satellite-tracking systems (STSs). The use of wireless communications allows more timely transmission of data from point of collection to the system of record. This offers the best value support to provide logistics data and information when, where, and how it is needed.

The use of AIDC enables asset management with real-time, accurate data that can be viewed by many parties from functionally specific perspectives to support each of the diverse uses. It provides an AIS with the ability to cross-reference critical acquisition, maintenance, operation, transportation, and supply data elements, such as part numbers, document numbers, and job control numbers. The need for a strong data management program must complement and enable the use of AIDC. Without strong data management, data may be accurately captured at the source, but data integrity can be lost as they move between AISs. All AIDC data can be created, collected, and passed to an AIS to provide visibility along the supply chain from manufacturer to supply to maintenance, and while in transportation.

In a fully integrated automatic identification and data capture (AIDC)–collection environment, U.S. Navy configuration management (CM), damage control, financial, food service, hazardous material (Hazmat), maintenance, medical, ordnance, personnel, safety, security, ship service, supply, or transportation functions create and use accurate, timely information about their activities and the products they maintain. Near real-time information is available on demand and is updated easily by the maintainer in this automated environment. The information provides the logistician specific product genealogy, supply, transport, repair, and performance histories, as well as links to appropriate technical data and troubleshooting guides, and tracks comprehensive configuration information.

Focused logistics, the Joint Chiefs' systematic approach for developing full-spectrum supportability across the range of possible missions envisioned in support of Joint Vision 2020, defines four tenets. The tenets are the framework for designing a logistics template in joint war-fighting technology. One of the four tenets is information fusion: the timely and accurate accessing and integrating of logistics data across units and combat support agencies throughout the world that provides reliable asset visibility and access to logistics resources in support of the war fighter. Focused logistics specifically identify AIDC as a vital component of information fusion because it ensures the capturing of current and accurate source data for existing and future services, agencies, and commanders at all levels of access in a number of automated information systems.

The Office of the Secretary of Defense (OSD) published the *Logistics Automatic Identification Technology Concept of Operations* (CONOPS), established a DoD Logistics AIT Office, assigned a program manager for AIDC, activated the Maintenance Technology Senior Steering Group, and developed a DoD Logistics AIT Implementation Plan. These actions, in turn, have placed a number of requirements on the Services, as well as the Joint Chiefs and Defense agencies. These requirements include developing and adopting AIDC standards, ensuring interoperability and compatibility, improving business practices, and developing a strategy for funding AIDC applications.

Logistics Processes

According to several studies, at least one error must be expected in every 200 to 400 keystrokes. If a typical computer entry involves 15 keystrokes, the result is a 5% error from this source alone. As a consequence, AIS data accuracy is far less effective than it could be simply because the original source data from manual input will be flawed if not corrected. The more centralized the data, the greater the risk of the data being corrupted. Here again, a strong data management program must be implemented in conjunction with AIDC technology in order to maximize data integrity.

Functional applications and business processes must be modified to accept AIDC data, which will have the most immediate impact. Cost savings will be more closely associated with cost avoidance than cost reduction since it is the responsibility of combatant commanders, platform–weapon system managers, etc. to effect cost reductions.

The DoD has three overarching requirements for AIDC devices to support supply, transportation, and maintenance functions. They should be (1) integrated into the AISs supporting their logistics functions, (2) used to maximize the use of pre-positioned data in these AISs, and (3) compatible throughout.

For materiel received at a storage site, maintenance facility, shipping activity, port, destination receiving activity, or consignee, AIDC devices should be capable of transferring item identification data (such as manufacturer ID, part number, traceability code, quantity, SKU) and transaction controlling data (such as document numbers) to the supporting AIS. The data transfer process should facilitate the receipt, disposition, and documentation processes.

If prepositioned data are not available when a shipment is received, AIDC devices should be capable of transferring the stored information to the AIS that supports the particular logistics process. The AIS can use this information to create a record for the item and complete the required transaction.

AIDC devices should be used to convey item or transaction identification information, access and update specific records through AIS interfaces, and permit logistics personnel to enter only new or updated information. Examples of specific functions and processes that AIDC can enhance include the following:

- *Storage sites* are responsible for receiving materiel from vendors, reconciling purchase orders and contract commitments, stocking materiel in bins or on shelves, conducting physical inventories, and issuing materiel from storage. AIDC support should facilitate the transfer of information to AISs and confirm the accuracy of specific item data with information in the AISs.

- *Maintenance facilities* are responsible for tracking materiel through work-in-progress by collecting identification data, time, and location of materiel in every step of the repair process. The AIDC devices should collect information on materiel consumed in the repair process and carry comprehensive maintenance history data for sequential logistics processes to use.

- *Shipping and receiving activities and ports* are responsible for confirming materiel has been received, preparing shipment documentation, combining items into consolidated shipment units, dividing shipment units into individual items, and updating shipment status.

- *Logistics management* is responsible for assisting in locating and identifying materiel, shipments, and containers. AIDC devices should be capable of supporting the transfer of items among maintenance facilities, storage sites, and shipping activities. AIDC devices, in concert with associated AISs, should be capable of supporting the reconstitution of DoD shipments.

The effective use of AIDC depends on assured and reliable communications. Success will be constrained by the quality and capacity of communications lines. The use of AIDC necessitates the integration of data, AISs, standards, and communications networks. This can facilitate the integration of functional, operational, and business processes that have operated independently.

Maximizing capabilities requires supporting a wide range of information exchange needs. To support the needs, several security issues need to be addressed. The exchange of unclassified information from highly classified systems to ones with lower classifications, access controls, encryption, and data aggregation is a key security issue that needs to be resolved.

The implementation of AIDC depends on the quality and timeliness of the data provided to users. Data availability and data accuracy are constrained by quality controls of source system data. Because data are readily available and consequently widely used, data providers should feel a heightened sense of responsibility to provide data that are as timely and accurate as possible. Strict controls are required as to who may create, update, or delete data values. This necessitates a strong data management program.

The following sections are referenced in the U.S. DoD, Army, Navy, and Air Force AIT implementation plans.

Bar Coding

Bar codes have infiltrated every facet of our lives since their inception in 1974. They have become an accepted and crucial part of our society, but little is known about what they represent. Different industries have their own standards for bar code content and format. There are many issues to consider when using bar codes in a management information system.

A bar code is machine-readable graphic representation of data that may be alphabetical or numeric or both. Bar codes are a way of encoding numbers or letters by using a combination of bars and spaces of varying widths, both of which are read. The appropriate usage of bar codes can reduce inefficiencies and improve productivity. Simply put, bar codes are a fast, easy, accurate way of capturing and entering data.

Actions Upon Reading a Bar Code   The pattern is read with an optical scanner that transfers the information contained in the bar code to a computer. Just as your Social Security number does not contain your name or address, a bar code references an associated record that contains descriptive data and other pertinent information. For example, a bar code found on a loaf of bread does not contain the product name, type of bread, or price; it contains a 12-digit product number. When the cashier scans this number at the checkout, it is transmitted to the store's computer, which finds the record associated with that item number in its database.

The matching item record contains a description of the product, vendor name, price, quantity on hand, etc. The computer does a price lookup, displays the price on the cash register, and subtracts the quantity purchased from the quantity on hand. The entire transaction is done instantly as opposed to the time it takes the cashier to key in a 12-digit number for every item you are purchasing. Similarly, the AIS in the issue room uses bar code data to access the descriptive data such as the product name, vendor, national stock numbers, and item price.

The most common type of bar code used in the business sector is the Universal Product Code (UPC), present on most general merchandise. Bar coding allows individual items to be uniquely identified, accurately tracked, and automatically inventoried. Bar coding is fast becoming a requirement for the delivery of merchandise to many purchasers, including governments and military installations. Bar coding is also being used in other areas that require high speed or that deal with large inventories. For businesses, bar code systems can save time, increase productivity, improve product quality, reduce paper documentation, improve reporting accuracy, and minimize inventory overhead. Bar coding also provides an accurate, digital record-keeping system that tracks and stores information about products and processes and can be easily accessed and transmitted electronically.

Linear Bar Codes (LBCs)   The most basic form of AIDC identification uses a linear bar code. Linear bar codes are a series of bars and spaces to create a bar code symbol. Linear bar codes have limited data storage capacity and rely on the existence of a flat surface. Linear bar codes are one-dimensional bar codes. This means the information is carried on one axis, either like the pickets in a fence or the rungs of a stepladder, and represents a limited group of characters. LBC applications require the existence and availability of an external database to support business processes. The LBC usually identifies a data element, such as a document number or Transportation Control Number (TCN), and serves as an automated key to information prepositioned in an AIS database.

The DoD recognizes two standard LBCs. Code 128 was developed in 1981 and is becoming increasingly popular with commercial users. While similar to a Code 39 bar code, there are significant differences. Code 39 is a double-density code, meaning it encodes two numeric digits as a single character, allowing the encoding of more data in the same space. Numeric Code 128 is 47% smaller than Code 39, while alphanumeric Code 128 is 23% smaller. Although the DoD accepts Code 128 as standard, there are currently few DoD applications that use it.

Linear Bar Code Symbol Requirements   The linear symbologies referenced in this chapter are Code 39, U.P.C. versions A and E, EAN-13, UCC/EAN Code 128, and the EAN. UCC Reduced Space Symbology (RSS) family. Users contemplating applications of Code 128 with Data Identifiers must be familiar with the issues related to their use. In the future, all purchased equipment should be capable of printing, reading, and verifying all linear symbologies identified in this standard.

Reduced Space Symbology (RSS)   The EAN.UCC RSS is a family containing three linear symbologies and three stacked variants.

- RSS-14 encodes the full 14-digit EAN.UCC item identification in a linear symbol that can be scanned in any direction. RSS-Limited encodes item identification suitable for use on small items but not for use at point of sale.
- RSS-Expanded encodes the 14-digit EAN.UCC item identification plus supplementary AI elements.
- RSS-14 Stacked is a variant that is stacked in two rows, either as a truncated version (above) used for small item marking, or as an omni-directional version (below) designed to be read by omni-directional scanners.

LBCs are the standard for all packaging entering the DoD system to enable electronic scanning of the item. Usually, a small scanning "gun" or reader is used to scan the label, and that starts the update process. When a DoD user scans an LBC, it must be scanned in sequence for the information to go seamlessly into the AIS. The LBCs used in the DoD do not have embedded data identification or Data Identifiers (DIs) to tell what the LBC represents. DoD bar codes are commonly referred to as "naked" because they cannot tell if the 17 characters in the LBC represent a TCN, NSN, or serial number.

Two-Dimensional Bar Codes (2DBCs)   Developed by AIT corporations, and tested and adopted by standards organizations such as ANSI–ISO, two-dimensional (2-D) high-capacity symbols get their name from the fact that they store data in two directions—vertically and horizontally. They store significantly more data than LBCs, thus facilitating more complex applications. While most LBC applications require the existence availability of an external database to support business processes, 2-D symbols function as a portable data file (PDF) that travels with a shipment.

2DBCs have error detection and may include error-correction features. They can sustain considerable damage and still be read. The most commonly used 2-D symbology within the DoD is the PDF417 symbol. It is composed of a stack of bar code rows, ranging from a minimum of 3 rows to a maximum of 90 rows. A PDF417 symbol can encode approximately 1850 numeric characters and approximately 960 alphanumeric characters with up to eight levels of quality assurance. This symbology was initially used by the

Defense Logistics Agency (DLA) in 1998 to store and transfer Transportation Control and Movement Document (TCMD) and Issue Receipt Release Document (IRRD) data on MSLs for individual shipments and for consolidated shipments involving eight or less line items. All DoD shippers were required to use 2-D MSLs by the end of the second quarter of FY 2002. The PDF417 symbol requires a 2-D scanner.

Data Matrix is a high-capacity matrix symbology popular for marking small items, such as integrated circuits, printed circuit boards, and parts. A Data Matrix symbol can encode up to 2000 characters. UID policy and MILSTD130L mandate the use of Data Matrix symbology. In mid-2001, the consumer electronics and automotive industries started using this symbology to identify specific vendor Commercial Activity–G (CAGE), Data Universal Numbering System (DUNS) Uniform Code Council–European Article Number (UCC–EAN)), part number, and traceability codes (see number, lot and batch numbers). The Data Matrix symbology is used by the Consumer Electronics Association (CEA)–Electronic Industries Alliance (EIA), Telecommunication Industry Forum (TCIF), Automotive Industry Action Group (AIAG), and Air Transport Association (ATA). The Data Matrix bar code requires a special image scanner.

Data Matrix is a two-dimensional array of square or round cells arranged in contiguous rows and columns. It is a binary code in which the dark data cells are given a value of "1" and the light cells a value of "0." The data cells are read from left to right, top to bottom. Each human-readable character encoded into the symbol is typically assigned eight data cells. The character "M," for example, would be 01001101.

MaxiCode, a 2-D symbol, is a medium-capacity matrix symbology especially designed for high-speed scanning, package sorting, and tracking. A MaxiCode symbol can encode up to 93 characters. The United Parcel Service (UPS) initially introduced it in 1992. Although the DoD accepts MaxiCode as the standard for logistics sorting applications, there are currently no DoD applications using MaxiCode, other than DLA for UPS shipping. The MaxiCode bar code requires a special image scanner.

InfoDot™ is an application of 2DBC technology for use on small items that may be otherwise difficult to identify, such as surgical instruments and small, high-value tools. Each InfoDot™ label is laser engraved with a unique 2-D Data Matrix bar code that can be applied directly to an instrument allowing it to be tracked throughout the complete surgical instrument sterilization cycle.[3]

The InfoDot™ works like a license plate. It "links" a specific instrument to a data file that contains a wealth of information about that particular instrument, including (but not limited to) purchase date, manufacturer part number, generic description, cost, maintenance–repair history, loaner status, sterilization cycles, and instrument utilization by case. Every time the instrument is scanned, a date–time stamp tracks the instrument at each step in the cycle. Instruments marked with the InfoDot™ are scanned using a CCD reader.

Specifications for the InfoDot™ technology include the following:

- Material: .003-inch black destructible acrylate bar codes.
- All alphanumeric bar codes are laser engraved.
- Guaranteed to have no skips or duplicate numbers in the label sequence.

---

[3] Michael Meeks, APTEC Technology.

- Standard symbology is 2-D Data Matrix ECC-200, which includes the full set of ASCII characters.
- Data Matrix is a highly redundant bar code. Up to 60% of the bar code can be destroyed and still readable.
- Along with the bar code, human-readable equivalents can be included on the 1/4-inch and 3/8-inch InfoDot™ labels.
- Black background with white copy.
- Standard diameter sizes: 3/16 inch, 1/4 inch, and 3/8 inch.
- Pressure-sensitive acrylic adhesive.

InfoDot™ labels are supplied 100 per sheet. The Foil InfoDot™ is an ideal identification solution for mall parts tracking. Whereas the standard InfoDot™ performs best in acidic or caustic environments, the new Foil InfoDot™ specializes in more extreme conditions where abrasion and high temperatures may be an issue. The 3/8-inch diameter label uses a Data Matrix ECC-200 bar code to encode as much data as a traditional linear bar code, while taking up only a 10th of the space and maintaining bar code readability if as much as 60% of the label is damaged.

Specific information is as follows:

- Applications: Tool and instrument control, fixed assets, and calibration.
- Material: .003-inch-thick anodized aluminum.
- Affixing Methods: Pressure-sensitive adhesive—very high-bond, high-performance adhesive; excellent resistance to heat and chemicals.
- Data Matrix ECC-200.[4]
- QR Code[5]: For product packaging, PDF417 shall be used as the default 2-D symbology. With specific mutual agreement between trading partners, either Data Matrix ECC-200 or QR Code may be used. This is relevant where DoD clients make direct vendor purchases or commercial over-the-counter purchases.
- Either 2-D–capable imaging or 2-D–capable laser scanning technologies can read both LBC symbols and PDF417 symbology. Data Matrix ECC-200 and QR Code require 2-D–capable image-scanning technology. Users should ensure that the scanning technology they select is capable of reading the symbols they choose to read.
- The minimum narrow element dimension "X" for the PDF417, Data Matrix ECC-200, and QR Code symbologies shall be 0.254 mm. The recommended "X" dimensions for each symbology are as follows:
  - 0.254 mm for PDF417
  - 0.38 mm (cell size) for Data Matrix ECC-200
  - 0.38 mm (cell size) for QR Code. The X dimension shall be determined by the printing capability of the supplier–printer of the label.

---

[4] ISO–IEC 16022.

[5] ISO–IEC 18004.

Product Marking

The evolution of product marking began within the retail sector. The retail sector has based its entire logistics and point-of-sale (POS) systems on a simple Universal Product Code (UPC). The key to the UPC success is the fact that it represents unique product commodity identification. The industrial sector must take unique product identification one step further than the commodity. Unlike the retail sector, the industrial sector requires serialized control of each individual product item–component that is unique across manufacturers. Each product item–component requires a unique machine-readable data plate that identifies it throughout its entire life cycle—during manufacturing and through the acquisition process, field maintenance, and disposal.

Technical Components

Symbology is the physical layout of the dark and light areas that make up the symbol. Several large industrial sectors have agreed on Data Matrix as a common 2-D symbology for product marking, which is an ISO standard 2-D symbol capable of storing up to 2000 characters of machine-readable data. It was specifically designed to benefit from ISO 15434 syntax, including macros to enable the compression of the symbol's footprint. The enveloping structure separators are encoded in specific "code words" within the symbology specification. In the case of part marking, less than 100 characters of data are currently encoded in the Data Matrix symbol.

Three mandatory *data elements* have been selected by several large industry sectors including automotive, electronics, telecommunications, chemical, and aviation to uniquely identify a product item or component. These elements include manufacturer's identification (ID), part number, and serial number. The aviation industry has levied the requirement that serial numbers be unique within a manufacturer's ID.

Syntax is the method of combining data elements into a message string. An example would be the combination of manufacturer ID, part number, and traceability code (serial number) into a single data string. The work of the ANSI MH10 committee has been endorsed and adopted by ISO 15434 for use with all high-capacity AIDC media. High-capacity media includes 2-D symbols, CMBs, RFID, and OMCs. This standard was designed with the flexibility to accommodate future high-capacity AIDC media. ATA has deviated from the ISO syntax standard, incorporating a proprietary use of "/" characters to separate data elements, as well as not encoding the unique header and trailer.

Semantics describes or gives meaning to the data that is encoded in a symbol or AIDC device by establishing a "DI" to precede each element. An example of embedding a DI in the element is "1P12345678." The "1P" preceding the number identifies the following string of characters as a part number. Another example is "S239495032." The "S" preceding the number identifies the following string of characters as a serial number. Programmers can use intelligent DIs to automatically parse the data as the symbol is read. ISO has established a common set of DIs in ISO 15418. These DIs are used by national and international organizations across many functional areas including automotive, electronics, and telecommunications. The ATA, representing commercial airlines, has chosen not to adhere to this international high-capacity automatic data capture (ADC) semantics standard. The ATA established a proprietary semantics standard within ATA SPEC 2000, which is based on an alternative set of text element identifiers (TEIs) instead of the ISO 15418 set of DIs. An example of ATA semantics is "PNR 12345678." The "PNR" serves the same purpose as the ISO 15418 "1P." The commercial airlines are the only industry using ATA SPEC 2000 semantics for bar coding.

The DoD's Position

The DoD is firmly committed to the standardization of product marking, emulating commercial industries. The ability to use common markings across the entire life cycle of a product item–component offers benefits to all. The same markings that are affixed by the manufacturers of an item can enable efficiencies in their own manufacturing processes, facilitate aggregation of components into end items, facilitate the purchaser's acquisition processes, and track the item through a life cycle of use and maintenance, without added cost for additional labeling, to meet unique requirements by the various industries that may play a part in the item's life cycle.

In support of this commitment, the DoD Logistics AIT Office and Air Force and Navy AIT Offices all participate in the MH10 machine-readable committee and the EIA bar code subcommittee. This involvement has enabled the DoD to infuse its requirements into the ANSI–ISO semantics and syntax standards.

The DoD part marking standard, MILSTD 130, currently requires a data plate containing specified information about the part with linear (Code 39) machine-readable markings. The Air Force is the executive agent for the maintenance of MILSTD 130. The DoD Logistics AIT Office has been working on a modification to MILSTD 130 that commercializes the DoD part marking standard. Industries that follow ISO encoding methods will be allowed to use the same part marking standard on items sold to the DoD that is used for their commercial sales. This supports acquisition reform and is further strengthened by the insertion of commercial standards into the DoD's acquisition regulations.

Bar Code Technology Summary

Bar codes are the simplest media to use but require the correct technology standards, data standard, conformance (print quality), and application standards to take full advantage of the low-cost technology.

Even though bar codes are somewhat prone to damage, they are successfully used in ·outdoor environments. One example of such use is the fuel farm at Patuxent River, MD. Bar codes have been placed on aircraft fuel access panels, fuel trucks, and storage tanks to keep track of fuel inventory, distribution, and billing. In over 4 years, only two of the 180 bar codes used had to be replaced due to loss or damage. In the retail industry, there are approximately 8 billion bar codes scanned per day. Federal Express (FedEx) estimates they scan a shipping package 23 times from origin to destination.

The 2DBC greatly enhances the supply and transportation data that can be electronically loaded into the AIS. This bar code contains the TCMD, Transportation Control Number (TCN), and source of supply information, making receipt and validation faster and more automated.

Contact memory buttons (CMBs) are not an ANSI or ISO standard; CMB technology was introduced in the commercial market in 1991. CMBs have a memory capacity range from 128 bytes to 256 megabytes. They are truly portable data carriers that are designed for both hostile operating environments and surface areas where limited space is available to mark and track serial-numbered components. If a finger can touch the component, then a CMB can be read or written to, thus allowing use in tight and dark spaces.

CMBs can be affixed by epoxy and/or mechanical methods. CMBs are available in non-reflective surfaces for discrete applications such as on weapons.

Use of CMBs as an Electronic Seal   The Naval Air Systems Command (NAVAIR) has qualified CMBs for use on aviation components utilizing strict guidelines for flight safety. Much of the strength, durability, and environmental testing for the CMBs has been accomplished by NAVAIR. For naval aviation applications, most components will be marked with a CMB. This holds true for commodities in all functional areas of this organization. This device meets Navy information space requirements, as well as survivability requirements in most operating environments.

The CMB is a relatively new AIDC medium, and there is no International Standards Organization (ISO) or American National Standards Institute (ANSI) standard. The DoD has de facto standards. The CMBs were tested in accordance with DoD MILSTD 810E by NAVAIR and by the Navy AIT Support Facility at St. Inigoes, MD, and should be used for internal Navy standardization. Additionally, Oregon State University exhaustively tested the CMBs in accordance with DoD MILSTD 810F with great success.

Over 300,000 CMBs have been installed by the Navy Air Warfare Center, Aircraft Division, Patuxent, MD, to capture data on aviation components, calibration data on electronic components, helicopter logbooks, and technical manuals carried with aircraft. They are used in both military and commercial applications. The Navy AIT Project Office is working with the DoD Logistics AIT Office on the formal process to have the CMB made an ANSI and an ISO standard. This process began in October 2000. The U.S. Post Office utilizes buttons in its collection and delivery process. Ford Motor Company uses them to record production data during the manufacturing of electronic components. Boeing Aerospace uses CMBs at the Boeing Aircraft Maintenance Facility, Wichita, KS, to track high-value facilities-maintenance items throughout their life cycle, recording and maintaining inventory as well as maintenance management information.

CMBs are high-capacity storage devices that are made to be extremely durable. They have passive read and active read–write capability and range in physical size and memory capacity. There are a number of products that can be used to attach a CMB to a component. One is HYSOL: EA 9394 (Navy helicopters) LOCTITE Corporation, Ultra Copper RTV Silicone Gasket Maker, an approved epoxy to bond the CMB to the surface part. The surface must be properly prepared and cleaned to the bare metal surface.

For planning purposes, it will take approximately one-half hour to prep the surface area and mix the epoxy or glue. It will take approximately one-half hour to install the CMB. The CMB ambient temperature should be near 70° Fahrenheit to ensure correct bonding. Under normal conditions, it takes about 24 hours (80%) to 5 days (100%) to properly cure; however, the use of a heat lamp for approximately 1 and one-half hours will significantly accelerate the process.

CMB advantages and disadvantages:

- Ideal for maintenance applications
- Supports full life-cycle management
- High data capacity AIT medium
- Offers expansive read–write capability
- Stand-alone data file survivable
- Needs to touch for reading

Current RFID Applications

Transportation	Manufacturing	Security	Financial	Other
Airline Transponder	AGV Control	Access Control	Electronic Cash	Animal Identification
Container ID	Assembly Line ID	Auto Immobilizer	Automated Fueling	Finish Line
Global Positioning	Configuration Management	Baggage Tag	Payphone Token	Gambling Token
Pallet ID	Factory Automation	Boarding Pass	Ski Tickets	Gas Cylinder ID
Parking Control	Forklift Positioning	EAS	University Cards	Laundry Tracking
Toll Collection	Inventory Control	Electronic Keys		
Traffic Management	Maintenance Logs	Fleet Management		
Truck Fleet Tracking	Paint Shop	People Locating		
	Process Control	Security Areas		
		Theft Prevention		
		Vehicle Access Control		Rail Car Identification
		Vehicle Movement		Time & Attendance

FIGURE 11.1 Radio Frequency (RF) Communications

- Limited vendor selection
- No existing standards

RF communications encompasses two major uses: radio frequency data capture (RFDC) and radio frequency identification (RFID). Each technology has its role and may be used in combination or separately as best fits the organizational situation and needs.

Commercial Applications There is an opportunity to create tremendous cost savings within the supply chain by using RFID. This is due to reduced inventory, better information timing, and better distribution to retail outlets or customers. There is, potentially, less storage space required. Not only can you have improved inventory control, but you also can drastically reduce or eliminate inventory check-ins.

RFID can assist with theft detection and reduction, and it should help minimize product counterfeiting. It is less prone to human errors, provided it works. However, there are some problems with RF systems that have to be addressed. One problem is matching of tags to readers. If a tag and reader are not well matched, then reads can be compromised or missed altogether. If a read is missed, the product ceases to exist (unless there is a built-in technological redundancy, i.e., bar code, and that is counterproductive to using a new technology in the first place).

Tags and readers coming from different sources need to be minimized (or at least qualified) until international standards for performance and architecture are final.

Of course, this point also applies to mixing tags of different frequencies. Obviously, training is required and battery life continues to be an issue with active RFID tags. Additionally, expended batteries must be treated as hazardous material.

RFID technology is currently used throughout numerous industries. Here are just a few applications:

- Asset identification and tracking: The ability to identify and track assets is critical, whether for a retail store, wholesale distributor, manufacturer, hospital, law firm, or government agency. RFID technology is used for real-time inventory control systems, item synchronization, security, and throughout the supply chain.

- Animal identification: Ear tags with transponders encased in a tough plastic allow farmers to identify and track their farm animals quickly and easily. A similar offering is available for labeling household pets. In the latter solution, a tag is actually embedded under the animal's skin.

- Electronic toll collection: RFID tags allow vehicles passing a toll station at high speed to debit their accounts automatically. There are already more than 2 million road transponders in use today.

- Fuel-dispensing loyalty programs: Exxon/Mobil uses RFID in the Speedpass payment system. Customers can pay at the pump with a wave of their key tag. Over 3 million Americans already use this program.

- Automobile security: Transponders are also embedded into keys. A reader in the vehicle ignition system transmits a signal to the transponder that answers back with a unique code allowing the vehicle to start. More than 16 million vehicles around the world have this feature today.

Radio Frequency Data Communications   RFDC essentially provides mobile communications between the point of activity (POA) and the AIS through the network. RFDC involves wireless-enabled devices, such as laptops and personal digital assistants (PDAs), specifically designed to incorporate AIDC media readers and an integrated capability to transmit data over wireless local area networks (WLANs), including BlueTooth, 802.11 a-g, and ultra-wideband into the AIS for the business process.

RFDC greatly expands the number of portable data capture applications by allowing users to be truly mobile while online to update the host computer with real-time data from each transaction through the WLAN connection rather than at a desk or workstation.

The basic question to ask is whether RFDC is a cost-effective solution for one's application and environment, or can batch data capture terminals meet the requirements equally well at lesser cost? One must evaluate the organizational requirements to answer this and other pertinent questions.

The benefits of RFDC include real-time transaction updates, bi-directional communications, the elimination of cables, and reduced costs.

The operating environment requires a technical assessment to ensure RF communications are secure and that the network will provide sufficient RF signal strength to support the business processes. When data capture gets interrupted, so does business.

The radio frequency site survey is an essential element for proper operation of an RFDC system and supports use of RFID as well. Site surveys are fundamental tools for those who deploy WLANs. Understanding the RF environment in which the WLAN will exist and the level at which access points are "visible" throughout the facility is critical to a solid deployment. This service includes a comprehensive test of the site, which will analyze the environment and check for any potential interference. A well-executed site survey aids implementation, security, functionality, and performance tasks. Upon completion, the study will determine the quantity and location of antennas and data transmission devices needed for optimal coverage.

While it is difficult to quantify the benefits that RFDC capabilities bring to overall investment, the following are some common benefits across multiple industries:

- Increased productivity: Information is readily accessible in offices, conference rooms, hallways, and laboratories—anywhere employees require it. This enables rapid decisions to be made that are based upon accurate information. Impromptu meetings can become as effective as planned ones. In addition, less time can be spent on administrative tasks prior to meetings (e.g., printing notes and handouts) when all the necessary information for a meeting will be at the attendees' fingertips.

- Better decision making: Anytime, anywhere access to real-time corporate information means that employees are always "in the know." Outdated information no longer prevents employees from making accurate decisions.

- Improved communications: WLANs ensure that employees always remain accessible, via e-mail, instant messaging, or even voice services. This dramatically improves both internal and external communications, ensuring that the right resources are available when required. For example, field personnel can reach the appropriate product manager when a question arises during a customer presentation. Customers can reach managers and support personnel if product-related issues arise.

- Reduced capital expenditures: WLANs alleviate the need for expensive network cabling. Of particular note is the ability to use the Voice Over Internet Protocol (VOIP) to provide voice in addition to data communications over the same device. This will be especially valuable for personnel working in remote or dangerous areas who need emergency communications available to them.

Classes of RFID tags   RFID is used to identify, categorize, and locate people and material automatically within relatively short distances (a few inches to 300 feet). RFID capabilities—particularly those provided by *active* RF tags—are beneficial when a user needs to locate and redirect individual containers or have standoff, in-the-box visibility of container contents. RFID also may be used to support personnel in a forward area with an inadequate systems or communications infrastructure and to facilitate the AIS capture of asset data within the time criteria established. The active RFID capability offers significant capabilities for yard management, port operations, and in-transit visibility (ITV) that cannot be provided by passive RF tags.

There are several classes of RFID tags that address both active and passive tags. The challenges with RFID vary based on the frequency used. Generally speaking, the higher frequencies provide faster data transfer rates. Power limitations associated with the FCC have an effect on the range achievable at a frequency.

Active RFID Tags   The RFID transponders are known as tags. They contain information that can range from a permanent ID number programmed into the tag by the manufacturer to a variable 128-kilobyte memory that can be programmed by a controller using RF energy. The controller is referred to as a *reader* or an *interrogator*.

Typical Read Range Passive Tags (unlicensed operation)	125-135 kHz	13.56 MHz	862-928 MHz	2450 MHz
United States	1 meter	0.7 meter 0.9 meter 0.75 meter	3 meter (915) 3-5 meter	1-2 meter
Europe	1 meter	1 meter	N/A 1.5-2.5 meter	40-70 cm .5W 1-2 Meter 4W
Asia Japan	1 meter	0.7 meter 0.8 meter	N/A	0.5-1 meter
Typical Write Range Passive Tags (unlicensed operation)				
United States	1 meter	0.35 meter 0.9 meter 0.75 meter	3 meter (915) 2-3.5 meter	0.7-1.4 meter
Europe	1 meter	0.5 meter 1 meter	N/A 1.0-1.8 meter	30-50 cm .5W 70-140 cm W
Asia Japan	1 meter	0.35 meter 0.8 meter	N/A	35-70 cm

FIGURE 11.2 Radio Frequency Identification Technology (RFID) Courtesy of Craig Harmon, QED Corporation

Performance

	125-135 kHz	13.56 MHz	862-928 MHz	2450 MHz
Typical Read/ Write Speed Through Field	15 - 20 tps	32 bit 16 ms R 22 ms W 50 tps 1.5 ms tag	32 bit 16 ms R 22 ms W 40 tps (ID) 12 ms/8 byte R 25 ms/byte W	40 tps (ID) 12 ms/8 byte R 25 ms/byte W
Susceptibility of RF Interference in a Manufacturing Environment	Low	Low / Medium	Low	Low
Typical Data Transfer Rate (kbps)	4	26.9 26.4 106 and 424	25 38	38
Memory Storage Capacities (bits)	2000	256-2000	1024	1024
Selective Data Block Locking?	Yes	Yes	Yes	Yes
# tags read successfully in FoV simultaneously (anti-collision)	2^{35}	100 2^{64} 8000	250+ 2^{256}	2^{256}
Selectively Write to One Tag When Multiple Tags are in the FoV?	Yes	Yes	Yes	Yes

FIGURE 11.3 Performance

An interrogator and a tag use RF energy to communicate with each other. The interrogator sends an RF signal that "wakes up" the tag, and the tag transmits information to the interrogator. In addition to reading the tag, the interrogator can write new information on the tag, thus permitting a user to alter the tag's information within the effective range. Interrogators can be networked to provide extensive coverage for a system.

The DoD uses active RF tags for pallet shipments, CCP-prepared sea van containers, ammunition and unit-cargo shipments moving in containers, movements of unit equipment, and pre-positioned cargo. Air pallets are attached at the CCPs to enable ITV and TAV. Shipping activities, depots, ammunition plants, and activities that support prepositioned ships and units prepare and attach RF tags. The tags contain supply and transportation data, as well as some limited data to identify the tag (known as "license plate" information).

The tags update local AISs and support a number of logistics processes, including container management, port operations, receipt, and distribution. The Army is the most extensive user and employs data-rich, active, omni-directional RF tags that accommodate line-item detail information to provide ITV and standoff, in-the-box visibility of container contents. Other DoD components are incorporating the use of RF tags in other logistics applications. There is a DoD-mandated data format for active tags, commonly referred to as JTAV 2.0 format, which should be used to ensure interoperability.

Passive RFID Tags   Other, less capable RF tags—*passive* tags—operate similarly to *active* RFID tags. The data capability of passive tags is much less than that of active tags (up to 256 bits compared with the 128KB capacity of active tags), and interrogation is generally constrained to shorter distances and line of sight. Commercial transportation applications of passive RFID technology are limited, particularly to the rail industry.

Such capability necessitates the most intensive installation of all AIDC media. Installation of interrogators and readers, as well as the ability to write to a tag, get frequency clearance, extra batteries, and infrastructure modifications makes this type of capability the most costly of all AIDC types. However, it also enables the most hands-free data exchange because the interrogators read the information on the tags as they pass through; no one has to use a scanner or physically read a tag up close. The major advantages of RFID application are speed and range (non-contact). RFID is approved for use and is used to remotely identify materials automatically while in transit.

The ease of using RF tags is apropos only if the tags, readers, computers, and software requirements match with a given situation. Difficulties can arise from any of the components in an RF system and, if you mix and match components without proper testing, the difficulties compound. For that reason, the way to make RF friendly in a given situation is to match the tags to the readers and make sure that they cover the areas you expect. Matching tags to readers may seem obvious and, with the industry in its infancy, it might be when components are purchased from a single source. However, as the industry grows and prices start to drive decisions, the relevance of a price break may make companies look to alternative vendors for their components. This decision, while financially beneficial, could create problems costing more money than that of the price break.

There are many reasons why an organization should evaluate RF applications. Some of the more important ones are track and trace capability, product authentication, and security issues.

Some applications are proving to be financially beneficial, and others are being reported as proving that the technology can work in a given environment. Obviously, these reasons are not the same, and one must remember that being able to work and being able to work *economically* in the real world are two very different statements.

Regardless, the reasons for examining an RF system typically center on its being able to work economically in a given environment or distribution system. Since distribution systems vary, so too will the reasons for looking at the RF system components and their ability to work well. Whatever the reasons, common business goals are to economically improve the production and distribution cycles while having access to better and faster information.

Track and Trace   The use of RF-enabled smart labels for the purpose of tracking and tracing product location is one of the primary reasons for the interest in this technology. When decision makers knows where a product has been, or where it currently is, they then have the ability to modify the production, storage, shipping, and, perhaps, sales requirements to best meet the needs of their system. For this to work, the RF systems have to be accurate and reliable throughout the entire supply chain. Unfortunately, when a tag, reader, computer, or software component fails, so too does the information and supply chain. Regardless, location tracking, inventory control and work-in-process are realistic goals for this technology in its current state.

Product Authentication   The growth of counterfeit products can threaten the financial strength of a company. Being able to quickly discern if products in the supply chain are counterfeit helps ensure the removal of the fake products. This is particularly true for luxury goods and pharmaceuticals, two areas that will derive immediate benefit from the implementation of RF tagging. Other authentication factors that are important and should be considered are physical tampering detection, chain-of-custody issues, component identification, and warranty and repair work. However, the recent debacle at Benetton clothing suggests that the public still have misconceptions about using RF for product authentication.

Security   Security enhancements will benefit a company by knowing the location of a particular asset (given that assets can be either physical items or personnel), protecting against theft due to the tracking of movement, and assisting with access control within facilities or at borders.

The ultimate benefit to a company will come when all these issues can be accomplished with the same tag used for everything else in the supply chain, including sales information. However, don't hold your breath on this being a standard part of the 5-cent tag's ability. And, in fact, the old adage still applies in that there is no single tag that will work for all applications.

Components of an RF System   This section will not cover all component possibilities because the market continues to grow, and a full discussion would take far too much time and space. The short version for this section is that components consist of the tags, readers, computers, and accompanying software that a company must utilize. Each of these components comes in multiple "flavors," and a large part of what you will want to utilize depends upon where and how the components will be used.

With so many options available in a relatively new market, interoperability is a key concept in making the systems approach work. Toward this end, standardization is both a good thing and a thing of concern. Like EDI, it is still open to requirements of uniqueness by competing companies and industries. Further, the chosen standard for a system may not be the best standard because the chosen standard may not work in many applications, leaving some companies or industries out of luck in meshing with the rest of the supply chain. Lastly, updates to equipment and technology have a tendency to render workable situations obsolete, particularly when there is a competitive advantage to be gained.

RF Technology Summary

It is important to match RF readers with appropriate tags. In doing so, there are many questions to consider for optimizing the system. You *must* take the systems approach to installing and running a viable RFID system, or it will be overly expensive and unlikely to succeed. The use of the RF airways requires coordination, typically through the communications organization. This is especially important when deploying to OCONUS locations that are subject to host nation regulations.

Spectrum users will do the following:

- Obtain frequency use authorization for each use of the electromagnetic spectrum through their appropriate joint force component
- Use frequencies as assigned and operate systems according to parameters authorized by the frequency certification and assignment processes
- Coordinate any need to exceed or operate outside the parameters authorized through the appropriate joint force component
- Ensure the emitting equipment is properly maintained to preclude unintentional violation of authorized spectrum use parameters
- Report incidents of unacceptable EMI to the appropriate joint force component or to the joint force JFMO

For installing an RFDC or RFID system, reliability (first and foremost), performance, and then the price should be considered. If you are not achieving reads well in excess of 3 sigma, then you should ask yourself, "Why am I spending all this money?"

Satellite-Tracking System (STS)

An STS provides the ability to track the exact location of vehicles and convoys. The latitude and longitude locations of trucks, trains, and other transportation assets equipped with a transceiver are transmitted periodically via a satellite to a ground station. Some systems also provide two-way communications between a vehicle operator and a ground station for safety, security, and rerouting. This technology is in the commercial motor carrier industry. However, this capability is easily adapted to rail, bus, barge, aircraft, military organic, and other forms of continental modes.

The U.S. European Command (EUCOM) is using satellites to track convoys and critical shipments moving to Bosnia. Likewise, the U.S. Central Command (CENTCOM) is using satellites to track convoys and critical shipments moving within the Iraq area of operations.

A system has five components: a subscriber unit, satellite, Earth station, network control center (NCC), and logistics managers. A subscriber unit is installed on the conveyance being tracked. The unit exchanges information with an Earth station via satellite. The Earth station is connected to an NCC that stores information in electronic mailboxes. Logistics managers access their mailboxes to receive information from subscriber units and return information to them.

Satellite tracking enables visibility of convoys and equipment by use of a small device that interfaces with satellites in orbit. This information is forwarded to national systems so that visibility of the location and status of the vehicle–convoy can be achieved. There are several commercial satellite systems available, each of which has advantages and disadvantages that must be considered in selecting mobile satellite services for voice and/or data. Of particular interest is the use of Iridium satellites that provide worldwide coverage using the equipment without changing frequencies, antennas, or configuration. Iridium is the only truly global service provider of voice and data services; others promote their "worldwide service" but omit to define the conditions under which worldwide service can be provided.

The Iridium satellite system provides complete coverage of the Earth, including oceans, airways, and polar regions. Iridium delivers essential communications services to and from remote areas where terrestrial communications are not available. Service is ideally suited for industrial applications such as heavy construction, defense–military, emergency services, maritime, mining, forestry, oil and gas, and aviation. Iridium currently provides services to the DoD. Iridium launched commercial global satellites and services. Service enhancements include improved voice quality and simplified pricing plans, and Iridium expanded the service portfolio to include data services. Iridium Satellite LLC is focused on providing affordable, dependable, long-term global communications solutions.

The Iridium system is a satellite-based, wireless personal communications network providing a robust suite of features to virtually any destination anywhere on Earth. The Iridium system is a satellite-based mobile satellite services (MSS) system supporting global wireless digital communications. Iridium provides voice, messaging, and data services to mobile subscribers using handheld user terminals.

Iridium Satellite LLC has operations in Leesburg, VA, where the Satellite Network Operations Center is located, and gateway facilities in Tempe, AZ, and Oahu, HI. The DoD relies on Iridium for global communications capabilities through its own gateway in Hawaii.

The Iridium system composes three principal components: (1) the satellite network, (2) the ground network, and (3) the Iridium subscriber products. The design of the Iridium network allows voice and data to be routed virtually anywhere in the world. Voice and data calls are relayed from one satellite to another until they reach the satellite above the Iridium subscriber unit (handset or modem) and the signal is relayed back to Earth.

The Iridium constellation consists of 66 operational low-Earth orbiting (LEO) satellites and 14 spares orbiting in a constellation of six polar planes. Each plane has 11 mission satellites performing as nodes in the network. The 14 additional satellites orbit as spares ready to replace any unserviceable satellite. This constellation ensures that every region on the globe is covered by at least one satellite at all times. The satellites are in a near-polar orbit at an altitude of 485 miles (780 km) where they circle the Earth once every 100 minutes at 16,832 miles per hour. Each satellite is cross-linked to 4 other satellites, 2 satellites in the same orbital plane and 2 in an adjacent plane.

The ground network is comprised of the System Control Segment (SCS) and telephony gateways used to connect into the terrestrial telephone system for central management. It provides global operational support and control services for the satellite constellation, delivers satellite-tracking data to the gateways, and performs the termination control function of messaging services. The SCS consists of three main components:

(1) four Telemetry Tracking and Control sites, (2) the Operational Support Network, and (3) the Satellite Network Operation Center. The primary linkage between the SCS, the satellites, and the gateways is via K-band feeder links and cross-links throughout the satellite constellation. Gateways are the terrestrial infrastructure that provides telephony services, messaging, and support to the network operations. The key features of gateways are their support and management of mobile subscribers and the interconnection of the Iridium network to the terrestrial phone system. Gateways also provide network management functions for their own network elements and links.

Enhanced Mobile Satellite Services (EMSS) has the capability to support low-rate voice and data services from mobile, lightweight terminals. EMSS is commercially provided with modifications to allow for unique DoD features, such as end-to-end encryption and protection of sensitive user information.

The DoD has a dedicated government EMSS gateway in Wahiawa, HI, for government use through the Defense Switched Network (DSN). EMSS subscribers have direct connection into Defense Information Systems Network (DISN) capable of providing secure services and nonsecure access to commercial telephone services. A user terminal will support secure communications by adding a removable National Security Agency (NSA)–approved Type I Communications Security (COMSEC) sleeve that fits onto the commercial user terminal.

The satellite modem functions globally without modification using a single approved frequency spectrum; it can be commanded via satellite to activate, deactivate, or change reporting intervals. The terminal functions as a wireless modem via a laptop or handheld device to allow users to relay critical supply and logistics data from remote locations globally. The device enables two-way communications anywhere in the world without changing equipment, frequencies, or processes. It requires no infrastructure beyond the constellation of LEO satellites that support both voice and data communications.

The model 9505 Iridium Tracking Terminal provides accurate location data to within 4 feet without human intervention. The Global Positioning System (GPS) component of the terminal provides location data via the Iridium modem to a designated receiving station. The interval between reports can be set from 1 minute to many hours and changed remotely via the embedded microprocessor.

The data from the Iridium Tracking Terminal can be received at the download site, recorded, and displayed in both tabular and via Geographic Information System (GIS) for map-based presentation. The displays can be accessed via the Internet with user name and password protection.

The Iridium model 9505 Tracking Terminal has been installed and tested on motor vehicles, river barges, and a private aircraft, which it successfully tracked for nearly 2 hours, from takeoff to landing. Data from the tracking terminal were downloaded and automatically displayed via a Web-based map display. Data include time, latitude, longitude, and altitude using GPS satellites. Vehicle or battery power may be used. GPS and Iridium antennas are commercially available in various configurations.

Satellite Data Modem Model 9505   The Ruggedized Iridium Modem 9505 is designed to operate in a ruggedized environment and is manufactured to MIL Standard 810 requirements. NAL Research Corporation of Manassas, VA, is the sole manufacturer of Iridium modems. The standard configuration connects through the standard RS 232 port of a laptop or desktop computer. Through utilization of HyperTerminal protocol and an expanded listing of AT commands, a user can send files or conduct chat sessions with any other type of modem. The user can determine signal strength with this model of modem. Transfer rates of 2,400 bits per second are standard and increased through standard compression. Multiple types of antennas are available for use depending on the mode of operation and user requirements. The modem provides a global data transfer capability.

Satellite Tracking and Data Terminal 9505   The satellite-tracking and data terminal functions globally, to include the polar regions and all ocean areas using a single approved frequency spectrum. This terminal, developed for the Department of Defense, can operate independently, or as a modem the terminal contains a micro-controller for two-way command and control and a Global Positioning System chip. The terminal reports the time, date, latitude, longitude, altitude, and the unique terminal identification number at preset intervals determined by the user. Operation of the terminal can be accomplished through a variety of power sources to include an ultra-high-capacity battery pack. The standard configuration connects using the standard RS 232 port of a laptop or desktop computer or stands alone if used strictly as a terminal. Through utilization of HyperTerminal protocol, and an expanded listing of AT commands, a user can send files or conduct chat sessions with any other type of modem. Transfer rates of 2,400 bits per second are standard and increased through standard compression. Multiple types of antennas are available for use depending on the mode of operation and user requirements. An active GPS antenna is part of the overall configuration with the terminal accuracy tested at less than 4 meters.

Electronic Data Interchange (EDI)   EDI is the computer-to-computer exchange of routine business information in a standard format. The federal government, as part of the president's reinventing government program, began a major initiative for "streamlining procurement through electronic commerce."

In recent years, many private-sector companies have reaped substantial benefits from automating their internal operations, such as accounting, order entry, purchasing, scheduling, and material processing. Those same companies are now focusing on automating their external operations using EDI and, in doing so, are reporting significant economic rewards—between $2 and $10 or more in direct cost savings from every document that they transmit electronically to their trading partners.

In spite of the magnitude of direct cost savings achieved through EDI, many proponents note that the real benefits of EDI come from using it as a tool to simplify and improve business procedures—*business process re-engineering*, in the current parlance. As a consequence, they are reporting $4 to $5 in indirect cost savings for every $1 in direct cost savings from various business improvements made possible by EDI, such as reduced inventories, improved competitive pricing strategies, enhanced auditing procedures, and streamlined operations.

The government has long recognized the economic and strategic advantages of EDI, both in the defense and the civilian sectors. Those advantages are becoming even more important as federal budgets are constrained. Additionally, as a major player in world trade, the United States needs to improve its ability to compete. The adoption of common federal implementation conventions and the promotion of "better business practices" are expected to help in promoting our nation's commerce, electronically.

The terms *standards, conventions,* and *guideline* are defined as follows: Standards are the technical documentation approved by ANSI ASC X12; specifically, transaction sets, segments, data elements, code sets, and interchange control structure.

Standards provide the framework for how a specific EDI message will be formatted for transmission. Conventions are the common practices and/or interpretations of the use of ANSI ASC X12 standards. Conventions define how trading partners will use the standards for their mutual needs. The guideline contains instructions on the use of EDI. It provides additional information to assist in conducting EDI. The guideline is intended to provide assistance and should not be your sole source of information.

Implementation notes help explain how trading partners will use the standards for each convention. For convenience, they are clearly marked and laced throughout the convention at the appropriate point. Implementation notes are the bridge from the standards to the convention.

EDI is the standard that the DoD must use as it enables electronic data transfer from AIS to AIS. The Defense Logistics Management Standards Office (DLMSO) is responsible for DoD Logistics EDI transaction business rules and procedures (http://www.dla.mil/j-6/dlmso).

This will open data flows between AISs that are dissimilar, so long as the data are sent in a standard format. There are technical data mediators and/or translators required when data in AISs are dissimilar. The goal is for AISs to use standard data to the maximum degree possible to negate or greatly reduce the use of mediators–translators. This is especially beneficial when joint task forces work with coalition partners and when working with other governments or conducting business with industry.

Biometrics   There are many types of biometrics. The fingerprint is the most commonly employed type of biometric; however, others have value in different applications. Other biometrics include facial recognition, retinal scan, iris scan, face geometry, infrared, voice recognition, thermal imaging, and hand geometry.

The terms *biometrics* and *biometry* have been used since early in the 20th century to refer to the field of development of statistical and mathematical methods applicable to data analysis problems in the biological sciences. Statistical methods for the analysis of data from agricultural field experiments to compare the yields of different varieties of wheat, for the analysis of data from human clinical trials evaluating the relative effectiveness of competing therapies for disease, or for the analysis of data from environmental studies on the effects of air or water pollution on the appearance of human disease in a region or country are all examples of problems that would fall under the umbrella of biometrics, as the term has been historically used. The journal *Biometrics* is a scholarly publication sponsored by a non-profit professional society (the International Biometric Society) devoted to the dissemination of accounts of the development of such methods and their application in real scientific contexts.

Recently, the term *biometrics* has also been used to refer to the emerging field of technology devoted to identification of individuals using biological traits, such as those based on retinal or iris scanning, fingerprints, or face recognition. Neither the journal *Biometrics* nor the International Biometric Society is engaged in research, marketing, or reporting related to this technology. Likewise, the editors and staff of the journal are not knowledgeable in this area.

Products are based on in-house developed biometric fingerprint authentication technology. The technology allows isolation of the characteristic features of a human fingerprint to match them with a stored template to secure identity. The security areas that this addresses are IT security, where secure and convenient login on computers and networks are replacing pin codes and passwords; physical access, where secure and convenient physical access to buildings and rooms is replacing keys, codes, and old card systems; and embedded solutions focused on embedding our technology in products provided by strategic partners, in applications ranging from handheld devices to cars.

AIDC Technologies Summary

The use of AIDC is rapidly advancing. Some media are more mature than others. A bar code requires direct line of sight from a scanner to a bar code. Typically, an LBC or a 2-D symbol has a limitation for curvature on a round surface to be no more than approximately 10 degrees of curvature. An LBC should be placed on the long axis of a round surface to allow for bar code scanning.

Bar code media are available with varied options such as printed on a label or direct part marking; depending on the type of label needed, they vary greatly in associated costs of labels, etchings, printers, and scanners. Size and type of bar codes selected are dependent on amount of information required. A major consideration for any type of bar code symbology is the real estate available on which to place the bar code. Bar codes can be printed on different types of material such as paper or plastic, with chemical etching, or micro-sandblasting on metal, rubber, or plastic.

The cost to implement any AIDC solution is dependent on the scope, architecture, and application of the integration; in the prototypes, careful examination of the AIDC technologies and selection based on the match of features to the requirement is needed. The implementation of a particular AIDC medium because of availability, familiarity with its capability, or other nonrequirements-based reasons has occasionally proven to be less than successful.

Automated Information Systems Integration with AISs

AIDC facilitates the collection, aggregation, and transfer of data but is only effective when integrated with logistics AISs. The strength of AIDC as an enabling technology is its ability to capture data rapidly and accurately and transfer the data to AISs with little or no human intervention. The use of AIDC supports the DoD strategy of capturing information once and making it readily available to all users.

The role of AIDC is an integral part of DoD logistics; AISs should be recognized. Therefore, any discussion of AIDC should also include a brief analysis of current AIS conditions. Some combat service support AISs are characterized by one or more of the following problems:

- Slow, inaccurate, and labor-intensive data input
- Inability to access the AIS because assured communications are not available
- Ineffective data transfer among AISs (AIDC can be part of the solution to these problems.)
- Ineffective, immature, or nonexistent functionality

However, no single AIDC device can support all DoD requirements and applications. A mix of AIDC capabilities is needed throughout the DoD logistics chain. If existing AISs can meet data timelines requirements, their capabilities should be used. A variety of AIDC can be employed to facilitate the capture of data to help the AIS meet the timeline requirements, as well as provide additional capabilities that cannot be provided by an AIS. Therefore, the DoD needs to apply a suite of AIDC that permits automatic capture, aggregation, and transfer of data to AISs to enhance logistics management. Strong data management must be implemented to protect its integrity throughout the process.

Collecting Initial Source Data

Bar codes should be the initial means to collect data about items moving in the logistics chain and provide the data to AISs. After "initial source" data have been provided to AISs, additional keystrokes or other manual means to input similar data can be avoided. Emphasis is placed on the use of prepositioned data to the greatest extent possible. Once source data reside in a logistics AIS, the first choice to meet user needs is to obtain the data from those AISs. If this approach is not feasible, a companion AIS that receives the data from the original AIS can be used. Although the use of prepositioned data is preferred, frequently AISs cannot meet the time criteria or provide sufficiently detailed asset information. In such cases, AIDC can be used to meet customer needs. Correct data will be available through the use of strong data management.

Improving Receipt Processing

OMCs have proven to be a useful tool for improving receipt processing. They document the contents of multipacks and containers as warranted and justified. For example, large installations daily receive many containers or multipacks at their central receiving points. Each container can have numerous orders; each order can be for a different end user. If analysis confirms that the volume of those shipments warrants the investment in equipment through savings in receipt processing time or increased accuracy, the OMC or the CMB could be the preferred AIDC alternative. While bar codes or RFID could possibly satisfy the need, OMCs and CMBs would be better, based on cost, capability, and user needs.

Providing asset visibility RFID may be the preferred alternative if, in addition to the large volume of containers, the existing AISs cannot provide asset visibility in the required time frames. In addition, if a user has to locate and redirect individual containers or needs stand-off, in-the-box visibility of container contents, RFID should be used. RFID can also facilitate the AIS capture of asset data within the proposed time criteria, especially in a forward area with an inadequate systems or communications infrastructure.

Near-term AIS enhancement or replacement will continue to improve asset visibility and data timeliness. No near-term AIS capabilities for yard or port management functions or stand-off, in-the-box visibility are anticipated, and these tools are needed in both normal DoD business processes and military operations. Likely trouble spots around the world will not have an established infrastructure, the absence of which may preclude immediate use of AISs, even if they could provide the data visibility required. RFID technology is being used in contingencies to provide timely and accurate shipment data not available in AISs and enabling in-theater management of large volumes of containers.

Tracking and Redirecting Surface Transportation Assets

When security, safety, or operational situations require rapid redirection of transportation assets and the cargo they contain, satellite-tracking capabilities should be used. Satellites are particularly effective to track ground movements. The DoD continues to develop the capability to use satellites to track and redirect surface transportation assets within theaters of operation. The DoD needs to be able to track and redirect movements of unit equipment and supplies in theaters of operation.

Sound Data Management Practices Are Mandatory for Implementation of AIDC   Poor data discipline and data management will be the Achilles' heel of any implementation. A great deal of money can be spent on AIDC technologies without realizing the full potential if the organization has poor data management practices.

Data collected correctly at the beginning of the process (i.e., data capture at the source by bar code, data matrix, etc.) are accurate only until they go into the first database. If the structured data within that first database do not comply with specified standards for data, or are inconsistent with the captured data, there is room for error when the data are passed. Additionally, if the captured data are inconsistent with the structured data internal to other AISs to which they are transferred, such as financial, item management, transportation, supply, or disposal systems, then data accuracy is lost.

Strict accountability throughout the business process must be maintained to ensure that unauthorized people do not change or manipulate the data. Business goals of total asset visibility, supply-chain management, and reduction in customer wait time all depend totally on the accuracy, validity, and integrity of the data in order to analyze and make business changes to support the customer. AIDC implementation requires data and data management discipline as part of a systems approach.

Historically, very little management attention or resources have been spent to actively manage the core resource of the data that are collected, transported, moved, and relied upon to make core business or operational decisions.

Data management has been designated as a technical responsibility and relegated to the IT part of the organization. Data are fundamental to running a business enterprise or executing military operations and verify whether corporate policies and procedures are being implemented. Logisticians and other functional business process owners must become actively involved in data management to ensure their business interests are being served.

Data Basics   A data element is the most elementary unit of data. However, a data element is not data—it is the "name" that the business gives to the definition and specifications so that the data values can be structured and configured for storage and manipulation in computer information systems; thus, the term *structured data*.

Specific values assigned to a data element as part of a business process are known as "data." However, in order for the "data" to be meaningful, the business owner must understand the underlying data elements in terms that relate to the business process that is performed, the appropriate business rules, the relationships of the business process with other business owners, the states (status) of the information or data assigned, who will be allowed to create, retrieve, update, or delete specific "data values," and so on.

Metadata, or "data about data," describe the content, quality, condition, and other characteristics of data. Metadata describe how and when and by whom a particular set of data was collected, and how the data are formatted. Metadata are essential for understanding information stored in data warehouses and have become increasingly important in XML-based Web applications. OSD is migrating from MILS 80-card column transaction sets to the Defense Logistics Management Standards (DLMS) variable-length transaction sets.

Unique Identification (UID) of Tangible Items   The DoD vision for UID is to implement policy, regulations, and supporting processes that establish a strategic imperative for uniquely identifying tangible items. The policy relies, to the maximum extent practical, on commercial item markings and does not impose unique government data requirements. To that end, uniquely identified tangible items will facilitate item tracking in DoD business systems and provide reliable and accurate technical and financial data for management, financial accountability, and asset management.

On July 29, 2003, the acting undersecretary of Defense (Acquisition, Technology, and Logistics) signed the Policy for Unique Identification (UID) of Tangible Items— New Equipment, Major Modifications, and Procurements of Equipment and Spares.

Expected UID Outcomes   In setting forth a UID policy, the following strategic outcomes were defined:

- Data integration across DoD, government, and industry systems as envisioned by the DoD Business Enterprise Architecture (BEA)
- Incorporation into the Wide Area Work Flow (WAWF)
- Improved item management and accountability
- Improved asset visibility and life-cycle management
- Clean audit opinions on the property, plant, and equipment and operating materials and supplies portions of DoD financial statements

This equates to the following expected benefits:

- Engineering will provide for the seamless transfer of product data (specifications or bills of material) into the supply chain to allow for faster production ramp-up and to speed up engineering change processes.
- Acquisition will provide for establishment of requirements and the efficient capture of the UID data elements through the contracting process.
- Financial management will provide clean audit opinions on item portions of DoD financial statements.
- Property, plant, and equipment accountability will provide physical controls and accountability over tangible items to reduce the risk of undetected theft and loss, unexpected shortages of critical items, and unnecessary purchases of items already on hand.
- Logistics will provide improved asset visibility and life-cycle management.
- The industry supply chain will provide enhanced ability to supply innovative, tailored products and to strengthen customer relationships, fostering better buyer– vendor partnerships.

Additionally, one can expect to see greater simplicity, standardization, speed, and certainty in automated data capture and electronic information exchange throughout DoD and industry processes. Standard contract language has been provided for the marking and evaluation of items, to smooth the way for a project manager's implementation effort.

Summary

The philosophy has been to specify the minimum essential elements necessary to achieve the objectives for unique identification of the DoD's assets. To the maximum extent practical, current methods should be used among suppliers, including commercial practices, and there is a preference for international standards. This is in the best interest of the DoD, the coalition partners, and industry. The DoD has involved the international community and industry in the development of this policy and continues to collaborate with them for implementation. The need for the integration across the acquisition, financial, and logistics domains guides the DoD.

Though AIDC is not a system, the principles of data management must be applied for successful implementation to occur. AIDC *will not* succeed in a chaotic data environment in which data are not properly managed. The successful implementation of AIDC technology requires the talents and skills of multifunctional business experts at both the wholesale and retail levels, information technology specialists, data specialists, standards specialists, and multiple other talents.

References

Finkenzeller, Klaus. RFID Handbook: *Fundamentals and Applications in Contactless Smart Cards and Identification.* Chichester, England: John Wiley & Sons, Ltd., 2003.

Kleist, Robert A., Chapman, Theodore A., Sakai, David A., and Brad S. Jarvis. *RFID.*

Labeling: 2 Smart Labeling Concepts and Applications for the Consumer Packaged Goods and Supply Chain. Irvine, CA: Printronix, 2004.

12

Infrastructure Recovery Initiatives: A Retrospective Assessment

Ralph V. Locurcio, Brig. Gen. (Ret.), P.E.

Objectives of This Chapter:

- To propose a more responsive recovery system for the United States
- To investigate a system of "regional" response and recovery operations
- To analyze a major disaster recovery to obtain lessons learned
- To relate these lessons learned to transportation recovery operations
- To propose a system that is self-sustaining to include operations, training, and support elements

Introduction

Transportation systems have become such a vital part of modern urban society that any interruption of these systems represents a severe hardship at best, and a full-scale disaster at worst. During the past hundred years modern societies have moved from agrarian, subsistence economies, where the essentials of life support were generated locally, to highly interdependent urbanized organizations that rely on transportation networks to support the financial, commercial, and social needs of the population. As these modern industrial and postindustrial societies centralized around urban centers, the problem of population density stimulated the development of collective transportation systems to increase the efficiency of support operations. Economies of scale, enabled by the availability of these systems, allowed the concentration of many subsistence and support operations into regional systems. Virtually all modernized societies have now developed collective systems such as regional food supply and distribution, regional hospitals, regional schools, and so on. Inherent in the centralization of these systems, and their consequent dependence on transportation, is the unexpected consequence of transportation security. Indeed, the millions of people concentrated in these urban centers have now become completely dependent upon the full and complete operation of these integrated

FIGURE 12.1 Modern Transportation Systems Are Critical to Urban Life

transportation systems for their very survival as depicted in Fig. 12.1, above. Unfortunately, the security of these systems is not as well developed as one might expect. This chapter will seek to review recent examples of major infrastructure recovery operations and propose a regionally integrated system for the United States.

Recent Examples of Disaster Recovery Operations

There are several recent and poignant examples of how transportation security has threatened regional urban populations. On September 11, 2001, the attack on the World Trade Center in New York City was a horrific example of how a disaster can affect a highly concentrated and interdependent urban population. The World Trade Center was not only a vital center of commerce and business activity for New York City, and for the entire northeast region of the United States for that matter, but also a vital transportation hub through which thousands of citizens flowed on a daily basis. Immediately following the attack, this vital hub was removed from the transportation network supporting New York City, causing severe reductions to many vital operations. A secondary effect was the impact to other transportation systems in the region that were overloaded. Fortunately, due to good management operations on the part of the responsible agencies in the New York region, alternative systems were put in place to relieve the congestion and restore operations. As a result, the citizens of New York City experienced only a temporary reduction in support services. By contrast, when Hurricane Katrina slammed into the Gulf Coast region of the United States, the transportation management systems were not used effectively to relieve the stress caused by the disaster event. The result was complete gridlock, where virtually all urban support systems ground to a halt. Operations that supplied food, water, medicine, police, emergency services, supplies, fuel, recovery, and reconstruction operations were all completely halted, resulting in a severe threat to the survival of virtually millions of citizens trapped in New Orleans and urban centers within the entire region.

Regional Transportation Operations: The FIRST Concept

The failure of this most recent disaster response and the problem of integrating regional infrastructure management organizations demonstrate a critical need for an integrated, regional approach to disaster planning, preparation, training, and recovery operations, especially where transportation systems are involved. We need only review the findings of the House Committee on Armed Services, 98th Congress, and the resultant success of our military operations, to see the need for substantial change. It is fairly obvious that both the cause and the impact of most disasters extend beyond local political and state boundaries. Similarly, these problems generally exceed the response, financial, technical, and management capabilities of local political organizations. Given the forecast for increasing activity in natural disasters, coupled with the continuing global threat from terrorism, our federal and local governments need a more structured, regional methodology for response and recovery support of these events. The FIRST program, proposed by the University College at the Florida Institute of Technology (FIT) and developed by Dr. Clifford Bragdon jointly with the author, seeks to use new technology, better organization, and continuous training in disaster recovery operations to overcome these deficiencies. The FIRST team's goal is to design an organization that can effectively integrate the collective resources of a region in response to a crisis situation, such as the interruption of transportation networks. The team would draw upon the collective professions of planning, engineering, architecture, transportation, logistics, and medical and social sciences. Using virtual simulation, a 5-D simulation technology, they would produce a dynamic visualization of our urban-based society from which the team could build scenarios to develop practical organizational design options and real-time training and implementation tools.

Step 1: Response Cells

The first element of the FIT proposal is the establishment of a series of "regional" planning and response coordination cells funded and staffed by the federal government. These cells would ideally be operated and staffed by the Federal Emergency Management Agency (FEMA), or any subsequent or alternative agency assigned the responsibility for federal assistance for disaster recovery. These regional cells would, first and foremost, establish and maintain critical lines of communication and coordination between the federal response agency and the respective state and municipal recovery agencies. The establishment of these communications facilities would overcome several problems experienced in the Katrina recovery. First of all, the cells would, of necessity, have to develop, procure, and field the physical hardware and software necessary to establish a working level of communications in the region. This hardware should be "hardened" to withstand extremely severe storm events and similarly to resist disruption by a terrorist element. In addition, the hardware should be equipped with backup power supply to ensure operation in a power outage. Once a reliable network of communications hardware has been established, the personnel staffing the cell should conduct periodic communications checks with state and municipal agencies within the region. These tests will accomplish two important functions. Obviously, they will determine if the hardware is working properly and ensure that personnel assigned to the task are familiar with the operation of the system. More importantly, however, these frequent tests will establish a working

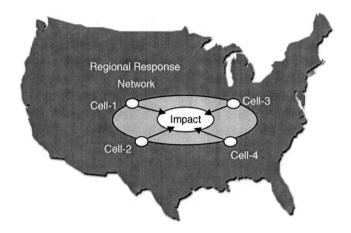

FIGURE 12.2 The FIRST Model

relationship between the federal, state, and municipal response personnel within the region, which will serve to blur or even eradicate the bureaucratic boundaries between these agencies. Of necessity, these "technical" communications should be followed by periodic "physical" communications between the response cell staff and the supported state and municipal staffs to establish familiarity and to cement working relationships between the staffers. We know from research that mere technical communications, such as telephone or e-mail messages, are not sufficient to bridge the interpersonal gaps between individuals who must interact on a common goal. When personnel meet on a personal level, they establish interpersonal bonds that pay big dividends in later operational scenarios that require a level of trust that cannot be established through technical communications alone.

Step 2: Response Organizations and Policies

Once a reliable means of physical interaction has been established between the response cell and the supported agencies, the next important step is for them to decide on how they will operate under the various scenarios that might be experienced following a natural disaster or terrorist event. In other words, they must decide what tasks will be required to accomplish the recovery and who will be responsible for accomplishing each specific task. Most probably the organization will develop a "matrix-type" organization, as depicted in the diagram in Figure 12.3, where project managers oversee the various response and recovery systems so output meets key goals.

These recovery tasks will most certainly be authorized and funded by enabling legislation passed by the federal, state, and municipal assemblies. However, interpretation of this legislation must always be tested with a simulation of various events, and "war gaming" of responses to potential scenarios within the region where the response cell is located. These response scenarios will be unique to each region due to differences in geography, demography, available resources, physical obstacles, and so on. For example, one region might have a cold weather constraint, where another might have a major river to contend with. From these exercises will flow a preferred organizational structure, with accepted lines of authority, communication, and responsibility for accomplishing specific

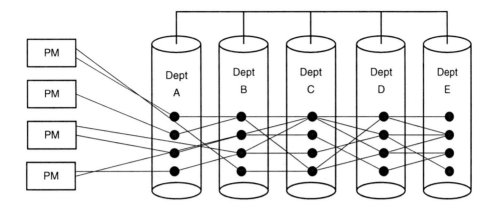

FIGURE 12.3 Classic "Matrix" Organization Structure

tasks under proposed situations. Again, perhaps more importantly, the staffs of the regional cell and the supported agencies will have worked together to establish these operating policies, and through the process of that interaction they will overcome any cultural or bureaucratic obstacles that might impede operations. Instead, these personnel will have established bonds of trust and cooperation that will serve them well in any future real crisis situation.

Step 3: Training

The final element of this proposal contains a recommendation that the staffs in these regional cells first be trained to ensure they have the requisite skills to accomplish whatever tasks might be required in an emergency situation. These skills can range all the way from knowing how to estimate the quantities of emergency water, ice, or food to supply to a stricken area, to how best to deliver and distribute those essential commodities, to understanding the dangers and techniques associated with handling explosive devices. Many of these skills will be unique to the region in question, so the training will have to be tailored to each regional response cell, and furthermore, these training exercises should be conducted in a "joint" mode with personnel from all political elements in the region attending together so that the working relationships that will ensure effective operations in a crisis are formed during these training exercises.

Finally, one could not expect these response cells, or the state or municipal governments, to develop all the required organizational, staffing, and training material necessary to field these response cells without some help from a qualified agency. The FIT proposal therefore recommends that this training and organization function be supported by a national disaster response center located at a university with the necessary faculty and supporting equipment to conduct research into past disaster response methodologies from which suggested policies, procedures, equipment, organizations, and training courses and exercises can be developed and tested. In addition, the establishment of a national center would ensure that common policies and procedures could be replicated throughout the country to ensure that resources were conserved and that the various regions would in the end have the ability to be mutually supportive. This last feature would be especially important in the event a disaster were of sufficient proportions to require that personnel from other regions be brought in to assist. In the Kuwait recovery, for example, personnel from all Corps of Engineers district offices were able to work together effectively without friction or confusion because they had all been developed in a common training and procedural base.

Step 4: Simulation

Once the regional cells have been established, equipped, and trained, the national disaster response center could perform a critical additional function that would enhance the ability of the cells to respond to any disaster in a timely and effective manner. Given the research base that would be needed to establish and support the organizational and training functions of the response cells, the national center would therefore have an outstanding skill, data, equipment, and personnel base that could be effectively utilized to provide real-time consulting services to the response cells to prepare for any potential or suspected event. Using sophisticated computers, the national center could simulate any number of possible disaster scenarios to give the staffs in the response cells a better understanding of the possible outcomes they might face as the reality of the disaster situation unfolded. In the Katrina event, for example, such a capability would have allowed the national center to very rapidly simulate the probable flooding of neighborhoods for several storm scenarios, which would have allowed the municipal, state, and federal officials to predict the numbers of personnel who would need to be evacuated, from which they could have preplanned for the use of transportation resources and road or rail networks. In the photo in Figure 12.4, a computer simulation of a railroad disaster in an urbanized area is shown that could be used to train response cell staffs in recovery operations. The environment, geography, and timing of this scenario could be very easily varied to suit any desired training objective. Military organizations have been using such training methods for many years to provide a wide variety of training opportunities at a very low operational cost.

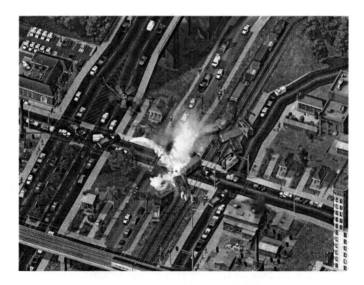

FIGURE 12.4 Computer Simulation of an Urban Railroad Disaster

Scientific Background and Approach

At Florida Tech a unique team of experts has been assembled to prepare this proposal. The FIRST team approach would be to utilize accurate and reliable databases and geospatial information to create a database and computer platform to perform effective modeling and simulation. Members of this team have been responsible for the rebuilding of Kuwait ($650 million); 5-dimensional simulation of operations for major ports, airports, and transportation systems; intermodal safety and security analysis and solutions for the U.S. Department of Transportation; safety and security training for federal, state, and local governments; hurricane and disaster recovery and reconstruction; development of training manuals and protocols; patents; and virtual distance learning and delivery mechanisms to maximize educational communication. Some of the specific events that these personnel have been responsible for include the following:

1. Rebuilding Kuwait following the first Gulf War
2. Disaster recovery from hurricanes Hugo, Iniki, Marilyn, and many others
3. Medical assistance for post-traumatic stress disorders
4. National certified training in safety and security
5. Simulation, visualization, and planning for all transport modes
6. Planning for the execution of the 2004 Olympics in Atlanta, GA
7. Risk assessment and situational analysis for major transportation events
8. Consultation with the Office of the President, the Department of Homeland Security (DHS), the Department of Transportation (DOT), and the U.S. Army Corps of Engineers

The FIRST team sees the following tasks as essential to the development of these regional response cells and a national disaster response center:

1. Inventory recent natural and/or man-made disasters and associated responses.
2. Develop a computerized regional response integration model.
3. Integrate planning, engineering, housing, medical, transportation, and human factors.
4. Validate the model using data from selected scenarios.
5. Test the model with the regional disaster planning agencies at all levels.
6. Develop functional methodologies and organizational policies for specific regions.
7. Conduct training for the personnel staffing the regional response cells.
8. Advise regional response cells on planning for suspected or actual disaster events.

An example of a successfully coordinated recovery operation can be seen in the reconstruction of Kuwait following the 1991 Gulf War. Examination of this operation can provide valuable insights into the project management principles that underlie any successful recovery operation and that would form the basis of training for these response and recovery cells. This chapter will examine the Kuwait operation in detail and conclude with recommendations for the employment of these principles in a regional transportation disaster recovery system for the United States or any similarly urbanized modern society.

FIGURE 12.5 Kuwait International Airport and oil fields as the KERO team drove into Kuwait to begin recovery operations on March 4, 1991

Project Management in the Kuwait Recovery Operation

In the aftermath of Desert Storm, the U.S. Army Corps of Engineers undertook the challenging task of reconstructing the basic infrastructure of the state of Kuwait to reestablish life support systems sufficient to guarantee the safety and security of the post-war population. Of course, these operations included the reconstitution of the transportation systems and network that supported virtually all social and survival systems in Kuwait. Kuwait is a small country with only one natural resource, oil. Therefore, virtually all other commodities necessary to support the population of 2.5 million people must be transported to Kuwait by land, air, or sea. The Kuwait Emergency Reconstruction Operation (KERO) was therefore dedicated to the reconstitution of these systems before any semblance of population security could be restored. Consequently, the Kuwait reconstruction may well be a model for future international aid and assistance to restore civil infrastructure, establish internal security, and enable political stability.

What follows is a description of the project management activities utilized in Kuwait, with implications for similar projects wherever they may be conducted. What is significant for the engineering profession today, considering the growth in the practice of project management, is an account of the major management decisions that governed operations in Kuwait, with a commentary on the effect these decisions had on the success of the overall recovery project. This chapter reviews the following subjects as they were applied during the KERO operation. These are considered to be core elements essential to the success of any project management process.

Principles of Disaster Recovery Construction

Based on an analysis of the experience from the Kuwait recovery operation, the following key items need to be addressed in any major disaster recovery operation. Although these factors do not address transportation aspects in particular, in the author's opinion they have been found to be essential elements of any major recovery operation.

Project Environment and Background

All projects exist within a context or a social environment that influences the management decisions of the project team. In truth, the project's ultimate purpose, in most cases, is to serve this context by providing design, construction, or manufacturing services that improve upon the condition of the environment. The Kuwait recovery was no exception, and we will, therefore, begin with a description of the situation and conditions that underlie the need for the KERO project and the desired outcomes.

During the Iraqi occupation and the resulting Gulf War, more than 90% of the Kuwaiti population of 1.2 million fled to Egypt, England, and other safe havens. Anticipating extensive devastation from the Iraqi occupation and the resulting Gulf War, the emir of Kuwait needed to rebuild the country as quickly as possible to reestablish normal government operations and to guarantee the safety and security of his citizens. Consequently, the emir formed a government committee known as the Kuwait Emergency Recovery Program (KERP) to manage reconstruction operations. With functions similar to the Federal Emergency Management Agency (FEMA) in the United States, this committee was headed by Dr. Ibrahim Al Shaheen, who was later named a national minister. Dr. Shaheen's task was to restore the civil infrastructure of Kuwait, that is, the municipal services and government functions, to a condition that would support the returning population of Kuwait in peace and safety after the war.

After considering several private contractual options, Dr. Shaheen advised the emir of Kuwait to ask President Bush for recovery assistance from the U.S. Army Corps of Engineers, with all costs to be reimbursed by the government of Kuwait. The Corps had extensive experience in natural disaster recovery operations and recent experience with several major disasters, such as Hurricane Hugo and the San Francisco earthquake. Additionally, the Corps had a thorough working knowledge of the Middle East and the region's construction environment and business culture. Dr. Shaheen and his advisors assumed that this experience, and the Corps professionals who had accomplished those recovery operations, could just as successfully apply their expertise to the devastation wrought by a military disaster in Kuwait. They were correct.

FIGURE 12.6 Critical Transportation Systems Destroyed in Kuwait

On January 14, 1991, as the air attack began and plans for the ground attack to liberate Kuwait were being finalized, the Corps signed a $46.5 million Foreign Military Sales (FMS) Case with Kuwait to begin the process of assistance. The Corps launched the recovery operation from its Transatlantic Division Office in Winchester, VA, formerly known as the Middle East/Africa Projects Office, which was to provide command, control, and logistical support throughout the operation. The recovery task force, dubbed the Kuwait Emergency Recovery Office (KERO), was organized in the Virginia office of the Corps' Transatlantic Division and later moved into Kuwait from a staging area in Dhahran, Saudi Arabia, following the liberation of Kuwait in February 1991. Once in Kuwait, KERO operated under the local direction of the Defense Reconstruction Assistance Office (DRAO), which, in turn, reported to the Department of Defense and the U.S. ambassador to Kuwait, for guidance and direction.

The KERO team entered Kuwait on March 4, 1991, and began operations almost immediately. In the 300 days following liberation, this team, which averaged 140 U.S. and 60 Kuwaiti professionals, placed over $550 million in repair work through contracts with major U.S. and foreign construction firms. Working 7 days a week and an average of 12–14 hours per day, they surveyed, repaired, and restored major infrastructure systems and facilities. The scope of operations, as shown in Figure 12.7, included the repair of the national network of 5000 km of 300-KV electrical distribution lines, substations, water mains, and pumping units; the highway network; sanitary mains; two seaports; the international airport; more than 150 public schools; and over 850 public buildings, including police, fire, medical service facilities, ministerial headquarters, and some defense facilities. The details of the scope and magnitude of this operation have been reported in several accounts of the national and international media and will not be repeated here. A complete historical account of the Kuwait recovery is provided in *After Desert Storm: The U.S. Army and the Reconstruction of Kuwait*, by Dr. Janet McDonald, published by the Department of the Army in 1999.

Planning for the Recovery Operation

Planning is one of the most important steps in conducting a successful project, no matter how large or small. A well-defined plan can save time, money, and team morale. No

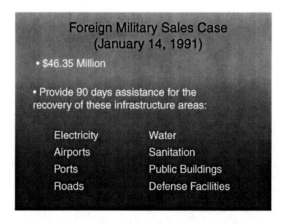

FIGURE 12.7 KERO Scope of Work

project can really succeed without a comprehensive plan to outline the key milestones and events that guide the day-to-day operations of the project team. In our daily lives, none of us would begin an expedition into unknown territory without a road map. That would be a foolish and costly adventure. Similarly, a project, by definition, is a journey into the unknown, and the project manager is the leader of that expedition. Therefore, it is the project manager's most fundamental responsibility to define the vision, or destination, for the project and to clearly identify the route the team must take to achieve success. Once the project is under way, the project manager must then make decisions to adjust the plan for unforeseen problems that arise along the way. This is a leadership rather than an engineering function, and although a project manager must be competent in the technical aspects of his project, his or her most important role is as the leader of the project team.

Project Management is first and foremost *a leadership activity!*

Unfortunately, although I do not have statistical proof, I would say that generally only about 25% of all projects begin with a documented plan prior to the start of operations. Lacking a formal directive to do so, many project managers launch into their duties without taking time to formulate and publish a written plan. Their reasons for this failure are numerous. Some of the most common are no time, no funding, no need, no information, and perhaps the only understandable reason, no training.

In general, project managers are action-oriented individuals who like to see results. Many feel that they don't need to slow down to plan out what they already know. Most would probably argue that upper management overworks them, and therefore, they are too busy to follow all the steps that textbook project management procedures recommend. They cut out seemingly unimportant steps, like publishing a plan, to meet deadlines and trim cost. Another frequent argument is that the project budget that was sold to the client by marketing or management did not include any billable hours for planning. Therefore, the project manager's only recourse is to somehow make time for planning through after-hours work or by reducing team budgets. Finally, some project managers would argue that there is too little information available at the beginning of

FIGURE 12.8 KERO Team Holds Its First Planning Session in Kuwait March 4, 1991

the project to formulate a definitive written plan, and once that information is available, the plan would be counterproductive because the project is too far along. Thus, planning is frequently viewed as a waste of time and energy.

None of these arguments is acceptable. Perhaps the only defensible, but still unacceptable, argument is no training. Many, in fact most, project managers come by their titles not because they have the requisite training and qualifications in project management but, rather, because they have survived the ordeal of completing a project without major incident. In reality, every poll that I conduct at the beginning of a project management training session produces the same result: Fewer than 10% of the attendees, most of them experienced managers, have had any formal training in project management. In short, many who wear the mantle of project manager simply do not know what the key elements of a sound project plan are, or that sound management operations always begin with a good plan.

In the case of the KERO project, the team had only 45 days to begin the operations, precious little time to prepare for such a monumental undertaking. Funds were extremely tight at the outset and risk was high. No one had asked for a plan prior to funding or supporting the operation, and lastly, there was absolutely no information upon which to base a definitive plan: no scope, no site plan, and no summary of damages to be repaired. No one could expect the team to develop a sound management plan with so little time and information.

Despite these impediments, planning for the Kuwait recovery began in Washington, D.C., on January 16, 1991. This was, coincidentally, the same day that the air campaign began, initiating military operations in the Gulf War. The first stage of planning dealt primarily with the task of gathering sufficient information to determine the scope and nature of the engineering tasks that would be required to restore the civil infrastructure in Kuwait, once military operations had ended. As noted earlier, the Kuwaiti government had signed a contract with the United States to provide funds in the amount of $46.5 million to accomplish the repairs outlined in Figure 12.7 within 90 days. The Kuwaiti

FIGURE 12.9 Little was known about damage to key transportations systems

government would then have the option to either extend or terminate the operation. No further information was available or given.

Given such a tremendous task and so little information, it would seem impossible to develop a plan for project operations. However, I would argue that a lack of information is no excuse for a lack of planning. In fact, it may be the most convincing argument for conducting a comprehensive planning cycle. When faced with a paucity of information about the future, we have only to rely on three sources for our inspiration: (1) our experience, (2) our training, and (3) our imagination. In the case of the Kuwait operation, we first looked to the only other definable recovery operation of a similar nature, the Marshall Plan for war-torn Europe following the Second World War. Unfortunately, the scope of the Marshall Plan was several orders of magnitude greater than the Kuwait recovery, and the time available to cull through the many volumes of historical information prevented any detailed study of that operation. On the other hand, the experience that the Corps of Engineers had recently had with recovery from several natural disasters in America was directly applicable to the scope and magnitude of the Kuwait operation. The experience gained by Corps personnel would prove invaluable. In the end, KERO team leaders would have to use their imagination to develop a "notional" scope of recovery operations for Kuwait, and it was against this "notional" scope that operations were planned. The results of these assumptions, as applied to the Kuwait operation, are summarized in Figure 12.10.

Initial planning was begun in Winchester, VA, with a team of approximately 20 individuals who had both overseas and natural disaster experience. This team made several crucial decisions early on, which are summarized in Figure 12.10. First, operations would be divided into an emergency phase and a recovery phase, as shown in the diagram. During the emergency phase, civil affairs military units and Kuwaiti government teams would address critical life support and security issues while the KERO team

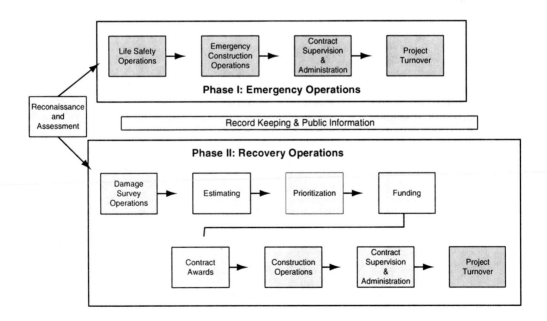

FIGURE 12.10 Summary of Recovery Operations

catalogued the damage to civil infrastructure. In addition, contractor teams would be mobilized, equipped, and moved to Kuwait to conduct specified emergency tasks, such as debris removal, to reopen the city's roads and remove hazards. During this emergency phase, teams of Corps volunteers from KERO would be organized to identify and define the scope of specific recovery projects.

Using Corps procedures, the KERO teams would prepare an individual damage survey report (DSR) and cost estimate for each potential project. The DSR would define specific elements to be repaired at each site and establish a budget for that project. These procedures are identical to those used routinely during Corps natural disaster recovery operations. The DSRs would then be aggregated into a repair program by damage assistance groups (DAGs), which were organized by function: roads, electricity, buildings, etc., and assigned to a KERO project manager for prioritization and direction. These project managers were functionally aligned with a Kuwaiti ministry that had responsibility for that function; i.e., the KERO project manager for roads worked directly with the Ministry of Roads as his client. During the subsequent recovery phase, contractors would be hired to execute the repairs defined by these individual DSRs. See Figure 12.11 for the project prioritization strategy developed for the Kuwait operation.

Similarly, a decision was made to model the KERO organizational structure on a typical Corps project office that might be used for natural disaster recovery operations, as depicted in Figure 12.12. Corps personnel were familiar with this organization and the associated lines of authority and procedures, all of which had proven to be effective in the past. Subsequently, a detailed logistical plan, which is described in Figure 12.12,

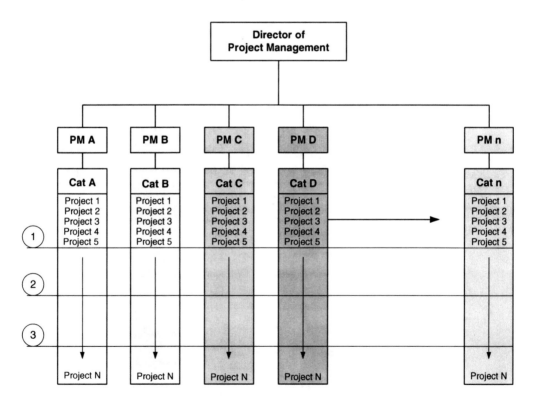

FIGURE 12.11 Project Prioritization Strategy

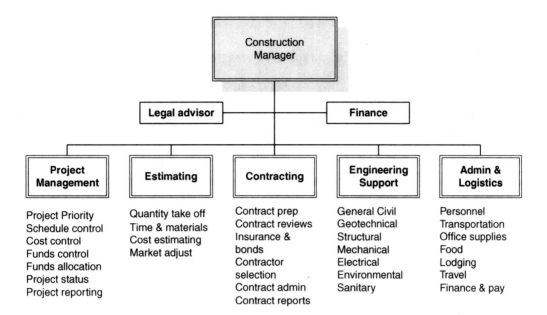

FIGURE 12.12 A Typical Corps of Engineers Project Office Structure

was developed based on this organizational model. At the same time, several contracting models were investigated. However, due to a lack of information, no firm decisions were reached about the specific contracting vehicles to be used. Rather, these decisions were postponed until further research could be accomplished.

During this same period, an advance party of eight experienced managers was deployed to a staging area in Dhahran, Saudi Arabia, where they co-located with the Corps engineering elements that were servicing the infrastructure needs of Operation Desert Shield, the buildup of combat forces prior to the ground attack into Iraq. This advance party was given the task of gathering information and formulating a specific plan of action for the recovery operations. After several weeks of meetings between Corps and Kuwaiti officials, planning efforts in Washington began to bog. The greatest reason for the slowdown was the voracious and frequent demand for detailed briefings by multileveled Washington bureaucracies. These agencies were demanding detailed information about operations that had not yet begun and second-guessing assumptions for those plans. The second reason for the slowdown was the fact that there simply wasn't enough accurate information in Washington to make meaningful plans. In effect, the team was trying to plan an operation by remote control and with second-hand information, a dangerous practice. The best information, scarce as it was, was to be had at the frontlines, as close to the area of operations as possible. In this case, that location was Dhahran, Saudi Arabia. And it was for this reason that the commander of the Corps office responsible for control of the project made an important decision: He sent the KERO commander, who was also the project manager for the overall operation, forward to Saudi Arabia to complete the plan. There he would be far away from the bureaucratic tangle and closest to the best sources of information about the project. On arrival in Saudi Arabia, the project manager met immediately with the advance party to assess the status of their planning efforts. To his surprise, no plan had been formulated, and very few decisions had been made. This raises a very important point that will be repeated over and over again in this, and any meaningful discussion of recovery management.

Project Management is first and foremost *a leadership activity!*

That is to say, project management is an activity that establishes the vision, justification, and parameters for the success of the operation and, most importantly, provides the decisions necessary to define and schedule the tasks necessary to move the project forward, and then to make adjustments for any obstacles encountered. Without this direction and focus, which can be provided only by a leader with the requisite competence and authority, no meaningful progress will be made. This is precisely what was happening in Saudi Arabia. There was no leadership on-site, no decisions had been made, and therefore no real planning had been accomplished.

On arrival, the project manager's first action was to assess the status of planning and meet with the client's representatives, in this case the U.S. ambassador and the officials of the Kuwaiti government, to clarify the scope and timing of KERO responsibilities. Once again, it was immediately apparent that there was very little information available about the actual damage to the infrastructure. Therefore, the advance party had to postulate a worst-case scenario and devise a flexible plan that could be implemented prior to the end of hostilities and then expanded or contracted depending on the actual situation once the team entered Kuwait.

To accomplish the recovery task the KERO team had to break down the overall scope, "Repair Kuwait," into manageable elements. First, the scope was divided into distinct work sectors, some functional and some geographical, as shown in Figure 12.13.

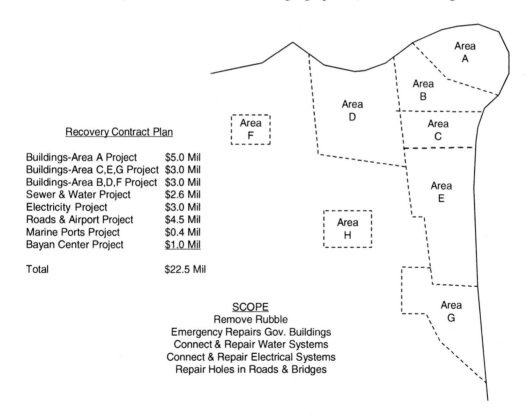

Recovery Contract Plan

Buildings-Area A Project	$5.0 Mil
Buildings-Area C,E,G Project	$3.0 Mil
Buildings-Area B,D,F Project	$3.0 Mil
Sewer & Water Project	$2.6 Mil
Electricity Project	$3.0 Mil
Roads & Airport Project	$4.5 Mil
Marine Ports Project	$0.4 Mil
Bayan Center Project	$1.0 Mil
Total	$22.5 Mil

SCOPE
Remove Rubble
Emergency Repairs Gov. Buildings
Connect & Repair Water Systems
Connect & Repair Electrical Systems
Repair Holes in Roads & Bridges

FIGURE 12.13 The KERO Project Plan

Although comprised of a series of related work items, each defined by an individual damage survey report (DSR), these sectors were each considered as a single project, which would be accomplished through a separate contract. Correspondingly, a project manager, whose only function was to ensure the success of that sector, would be assigned to manage each sector. For example, road repairs throughout the country were defined as a project. One project manager would direct this roads project, and the work would all be accomplished under one construction contract. Similarly, all domestic water and sewer lines, and related pumping equipment, were considered as one water and sewer project. All electrical work, power plants, distribution lines, and substations were defined as the electrical project, and so on for each component of the civil infrastructure. A separate budget was devised for each of these projects, and ultimately, a unique contract was written and a separate contractor was selected to accomplish all related construction. This very simplified procedure emphasizes the principle of clarity and unity in project operations. Only *one* project manager, *one* budget, *one* contract, and *one* contractor were assigned to each element of the civil infrastructure. Consequently, in a rapidly moving and changing environment, the number of variables had been reduced to a minimum and the chances for success thereby increased. See Figures 12.17 and 12.18.

As an added measure of control and management, each of these project sectors was aligned with the corresponding Kuwaiti government ministry responsible for the management and operation of that sector. For example, the KERO project manager for the roads project worked directly with the Kuwaiti Minister of Roads. Similarly, the KERO manager for the water and sewer project worked directly with the Minister of Water, and so on. These ministers were the clients who ultimately accepted the work delivered by the individual project managers, and therefore, each project manager responded to only *one* client, an ideal situation. Government buildings, on the other hand, were handled a bit differently. Because there were over 1,100 government buildings in all, spread out all over the country, a slightly different method of organization was needed to break the task down to manageable proportions and guarantee simplicity of operations. In the case of government buildings, the country was subdivided into geographical, rather than functional, sectors, and each of these sectors was designated a project. By this method, they could mobilize one construction contractor per geographical sector, thereby avoiding any potential problems associated with dividing the work among competing contractors in a given area.

Similarly, this would keep the contract administration and budget operations separate for each sector. For example, the Central Business District (CBD) was designated "Area A." All government buildings within that sector were assigned to one contractor and one project manager. Similarly, areas B, D, and E were grouped into one contract and assigned to one project manager. A single budget was developed for all work in each sector, and one contractor was hired for construction operations. In all there were five geographical sectors allocated to government buildings as shown on the project plan in Figure 12.17.

Essentially, the two diagrams in Figures 12.17 and 12.18 define the initial project plans for the reconstruction of Kuwait. By way of review, I will cover several notable features of these plans. First, the plans were simple and clear. The entire scope of operations, or vision, was well defined, and procedures were established to clarify details and overcome obstacles as the work progressed. This plan was published for all members of the KERO team, reviewed at progress meetings, and used to brief all incoming personnel throughout the life of the KERO operation.

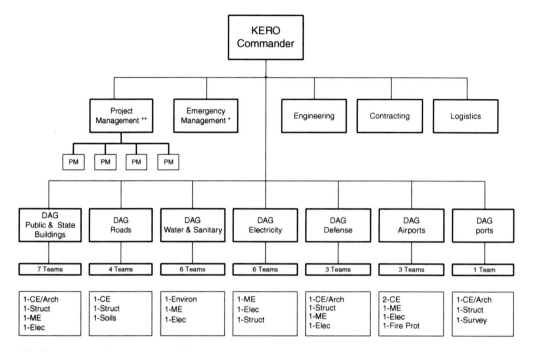

* Provides primary direction during "emergency" phase
** Provides primary direction during "recovery" phase

FIGURE 12.14 Damage Assessment Operations

FIGURE 12.15 KERO team members review reports prepared by damage assistance groups

Second, the plan broke the work down into individual tasks, each of which was to be managed by a single project manager whose responsibility it was to ensure the success of the project(s) assigned to his or her sector. This allowed each PM to focus his or her attention on a clear set of priorities and to serve as an advocate for the success of his or her sector. Finally, each project manager was assigned to one client, in this case a Kuwaiti ministry. This gave each client a single point of contact for information and control. Project managers were instructed to become completely familiar with the culture and mission of their ministry so that they could make informed decisions on project features, trade-offs, and schedule, just as if they were the clients themselves. By this means, the client was assured of a constant flow of information, complete understanding of the intended outcome for each project, and, therefore, complete control of the operation.

Organization and Staffing

In the case of the KERO operation, there was not much time to organize, staff, and train the initial task force of approximately 140 Corps employees, all of whom were experienced volunteers from various districts and divisions who would rotate to Kuwait on a 3-month cycle. To minimize the dysfunction and confusion caused by an unfamiliar organizational structure, a decision was made to create a small "district-like" office for the Kuwait recovery. In short, civilian or military volunteers from any Corps organization would be able to arrive in KERO and "recognize" the working environment with little training and only situational orientation. KERO used two variations of a district organizational structure, each suited to the particular operation at the time, but both employing a common headquarters structure that did not change.

During the initial emergency phase of the recovery, KERO field offices were set up according to Corps emergency management practice. Labeled *damage assistance groups* (DAGs), each group had a variable number of assigned damage assistance teams (DATs), depending on the mission of the group (see Figure 12.18). As mentioned earlier, DAGs were aligned with specific Kuwait ministry offices: buildings, roads, airport, electricity, water and sewer, ports, defense, etc., according to pre-war Kuwaiti management conventions. Adoption of these conventions simplified the interface with the host nation, as leaders and volunteers from these Kuwaiti ministry offices were readily available to work with KERO.

Additionally, this organization allowed Kuwaiti volunteers to work directly with their parent ministry. This structure was used for about 45 days while damage survey reports (DSRs) were being completed and contractors were mobilizing. Actual construction during this time was minimal; however, more than 1,200 DSRs and associated cost estimates were prepared. These reports later became the basis for all future budgets and recovery work orders.

During the subsequent recovery phase, when construction management became the dominant operational consideration, a more conventional Corps project management structure, with traditional resident offices to manage contractors in the field, was adopted to ensure that project delivery and contract administration were accomplished according to Corps procedures and quality standards (see Figure 12.10). Following the same practice used during the emergency phase, project managers (PMs) were assigned to coordinate work with the responsible Kuwait ministry officials. PMs were tasked with developing a work program for each ministry to control all projects and funds for that ministry from concept to turnover. For example, all roads projects constituted the "Roads ministry program," which was the responsibility of one PM. The PM established project priorities; developed project and program budgets; decided project scope, features, and quality standards; monitored progress through all technical phases (design, contracting, construction); reported on the progress of each project and the entire program; and supervised eventual project

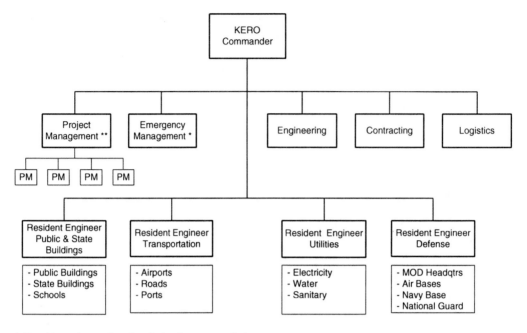

* Provides primary direction during "emergency" phase
** Provides primary direction during "recovery" phase

FIGURE 12.16 Construction Operations Plan

turnover. Once into the recovery phase, these ministry work programs were organized along functional lines under the direction of a resident engineer, as shown in Figure 12.16. Resident offices were responsible for construction management, contract administration, modifications, and claims resolution and budget control for all assigned programs.

The KERO technical divisions handled the more specialized areas of the project development cycle: The Engineering Services Department handled analysis, design, estimating, specifications, value engineering, and field consultation. Contracting handled contract preparation, solicitation, review, award, small business administration, and coordination. Other professional elements such as Legal, Safety, Audit, etc. were also present on the KERO team to round out the professional package and ensure timely project completion. These divisions reported directly to the commander to ensure complete independence of their operations and timely reporting if problems arose. Additional technical and administrative functions were managed in the United States at the Transatlantic Division Office (see Figure 12.14).

All these decisions proved to be correct and advisable for future operations, but there were a few problem areas. For example, the conversion from DAGs to resident offices revealed that new roles and responsibilities had to be sorted out in real time under intensive operational pressure. Also, different skills were needed to staff the resident offices. Whereas the DAGs used personnel with design experience, the resident offices needed personnel with construction and contract administration experience. Similarly, the decision to rotate field-level engineering professionals every 3 months was a difficult, but workable, staffing concept. On the other hand, the rotation of key leaders such as deputies, division chiefs, resident engineers, the chief of project management, the resource manager, or the property book officer on the same 3-month cycle proved extremely disruptive. For the future, these key personnel should be selected for the duration of the operation, or not less than a 6-month tour, with a guaranteed overlap of old and new managers to ensure continuity of operations.

The Project Management Process

In 1991, the Corps of Engineers did not have a formal system of project management as part of its operating procedures. Instead, projects were traditionally managed from within the functional discipline that had the largest share of the work. For example, if a project had a preponderance of mechanical engineering, a project manager from that department was selected to lead the project. Under this approach, the project management function was decentralized with little or no central control of project operations. As a result, clients were forced to deal with a variety of managers and processes to gain control of their projects.

This method of managing projects was an adaptation of older organizational models developed during the "industrial age" of manufacturing. As manufacturing management grew during the early 20th century, similar functions were grouped into departments for the sake of efficiency and ease of management. It made good economic sense to have one manager specialize in a given functional area. As such, each manager became an expert in his or her specialty and consequently was able to optimize both quality and productivity in that area. Quality was checked at the end of the process, and adjustments were made to

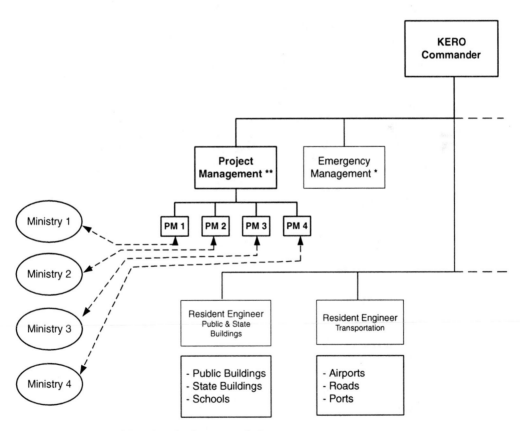

* Provides primary direction during "emergency" phase
** Provides primary direction during "recovery" phase

FIGURE 12.17 Project Management Operations

overcome errors and defects and improve the efficiency of the process as a whole. Over time repetitive errors were eliminated, and the manufacturing process guaranteed product quality. This seems to work well for mass production of similar products, such as automobiles, refrigerators, and the like. However, this form of management is not well suited to the design of "unique" products such as buildings or water systems, which are one-of-a-kind, very expensive, and not mass produced. The obvious problem with this decentralized management is that no one manager has responsibility for the entire project. Even worse, the customer has even less influence over the development of his or her project, and no one knows how it will turn out until the project is completed and all the funds are expended. When dealing with such large, one-of-a-kind products, this industrial form of management is disastrous. Errors cannot be corrected in the next item off the assembly line and often result in expensive legal battles between the customer, the designer, and the contractor.

Around the time of the Kuwait operation, the Corps of Engineers had been considering a new, more customer-oriented form of organization for project management. Under this new system, project managers would be given responsibility for the entire project

FIGURE 12.18 Project managers made transportation system repairs a top priority

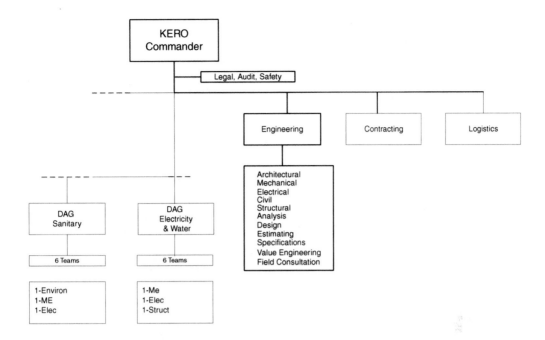

FIGURE 12.19 Engineering operations

development cycle, centralized and assigned to a separate department reporting directly to the leader of the organization. This new system of organization was expected to improve several important features, which were lacking in the former system. First and foremost, it would provide direct and unfiltered communications about project status to the commander or director who had the power to correct errors immediately. Second, elevating the project managers within the structure would give them greater visibility of the entire life cycle of project development and, therefore, the ability to detect errors along the way. Next, the new organization would give project managers authority equal to other department heads, which added considerable power to their opinions in discussions with functional managers. And finally, and most importantly, this new organization provided a "single point of contact" to the client. In other words, the PM would be the one person in the organization who was both authorized and responsible to interact with the client and make decisions about all of the projects under development. This would give the client unprecedented control over the project(s).

Although the Corps of Engineers had not formally adopted this new management system, KERO managers decided to implement it on a test basis, as shown in Figure 12.17. Under this arrangement, all project managers in KERO were assigned to a Department of Project Management, which reported directly to the KERO commander. Within the PM Department, individual project managers were assigned to each Kuwaiti ministry to provide a single point of contact for all projects in that ministry. Each PM was completely responsible for ensuring that the scope, cost, quality, and schedule of all projects met the needs of his or her assigned ministry. Consequently, the client had only one person to contact within KERO to gain complete knowledge of all projects. Similarly, all functional elements in KERO were directed to look to only one PM for decisions concerning projects for a given ministry. In addition, the PMs were instructed to serve as advocates for their client's projects. As such, they had to understand the mission, operations, and culture of the ministry well enough to know how each project served the needs of that ministry. With this insightful knowledge they were able to make intelligent decisions about project features or trade-offs throughout the life cycle of project development, thereby reducing errors and saving time and resources. Within KERO, they were empowered to control all variables associated with their projects to ensure that outcomes met their client's objectives.

Contracting

Speed, and later, control were the driving forces in all KERO contracting operations from the outset. KERO staffers planned for a 45-day competitive emergency contract award cycle to be executed from Saudi Arabia. They were eventually forced to utilize a 10-day contingency plan when the war ended more quickly than expected. Working nearly around the clock and without specific knowledge of the conditions in Kuwait, the KERO staff, with Kuwaiti counterparts, completed necessary project scoping, solicitation, pre-qualification, and proposal evaluation actions to award eight "cost-plus" letter contracts worth approximately $25.4 million. These letter contracts would be employed on a "task order" basis to execute individual DSRs as quickly as possible. Each task order would then be definitized after mobilization and converted to a fixed price, once the actual scope of the repair mission was known (see Figure 12.21).

To divide the work among the contractors, Kuwait projects were organized into either functional or geographic work areas according to pre-war work management conventions. These work areas became the geographical or functional scope of the contracts. Consequently, the eight contracts were divided as follows: general building repair (in three areas of the city), all road repairs, all sewer–water pipe repairs, all port surveys, and all electrical repairs. A target cost estimate was assigned to each contract, and the remainder of the funds was held in reserve. Later, as needs became known, several additional program areas, such as hospitals and communications facilities, were added. Funds for these areas were supplied from these reserves with the approval of the KERP. Eventually, additional funds were provided from the augmented 1991–1992 budget request. Since KERO contract scoping was aligned with pre-war ministry functional responsibilities, the KERO management structure was perfectly aligned to partner with Kuwaiti officials during recovery operations (see Figure 12.10).

These letter contracts were essentially cost-plus instruments. Therefore, a precise method of controlling and documenting the flow of work to the contractor was required once construction operations began. As described earlier, the original DAGs were converted to standard Corps resident offices for contract control and administration once mobilization was complete and construction had begun in earnest. For the purpose of passing specific work requirements to the contractors, the DSRs and associated cost estimates prepared during the emergency phase were converted into work orders. These documents, which resided in the project database, were reviewed by PMs and adjusted to define the exact scope of repair work. The work orders were then provided to the Engineering Services office, where the original cost estimate for the work order was reviewed, and corrected if needed, using the Corps' newly automated M-CACES process. This extremely valuable tool allowed detailed estimates to be developed quickly and accurately, sometimes overnight. It also allowed the Corps to adjust the unit cost of labor and materials in accordance with changes in the market values of these items as time

FIGURE 12.20 Airport Lobby Restored in 30 Days with "Task Order" Contract

passed. These work orders, detailed cost estimates, and records of negotiations were incorporated into a contract modification that served as the plans and specs during construction. Ultimately they, in conjunction with daily construction management logs for time and materials, became the basis for final negotiations with the contractor on the cost of each work order. As the KERO operation neared completion, these documents were audited to verify that over–under payments were made (see Figure 12.21).

As the operation matured, some larger projects were advertised individually to enhance competition and encourage participation of both U.S. and Kuwaiti small businesses. However, after several months of operations this conversion to individual competitive contracting was extremely difficult with the limited KERO staff. By this time, over $100 million had been spent in cost-plus contracts, with well over 150 modifications to the original eight contracts in progress. In addition, the time required to formulate, advertise, and award a competitive, fixed-price contract in the uncertain universe of bidders that was available in Kuwait was both risky and time consuming. Consequently, for

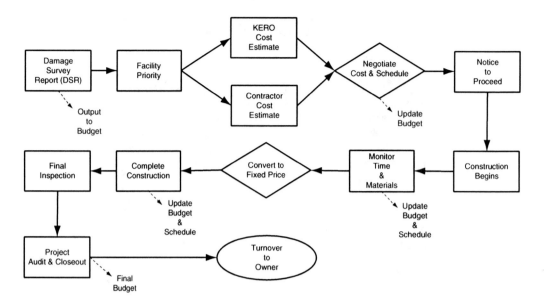

FIGURE 12.21 Contracting operations

speed and control, eight basic contracts, each covering a specific area or function, were increased with work orders added as negotiated modifications to the original contract.

The system of dividing the country into work areas (Figure 12.10) performed reasonably well during the early stages of the Kuwait recovery and served to keep the contractors separated geographically. However, in future operations, if time and staffing are sufficient, this geographical orientation could possibly be discarded after a brief period of construction to allow more competition within each area. Additionally, more general contractors should be employed during the mid- to later stages of the recovery, also to enhance competition. That is, if the projected work flow and funding are sufficient to justify the additional contractor mobilization costs, which must be borne by the host. If more contractors were available, competition for individual work orders would result in greater cost efficiency. As a rule of thumb, each general contractor operating in an international setting must foresee a potential workload of $50 million to $100 million to justify mobilization costs. In the case of Kuwait, the contracting operation was initially sized for a $50 million to $100 million horizon based on the best information available at the time guidance from the government of Kuwait was received. The eventual $300 million to $400 million workload could not have been forecast accurately enough to justify the risk of opting for additional mobilization. In fact, a plan with as many as 10 major contracts was considered at one point, but the option was discarded because the cost of mobilization would have consumed an inordinate percentage of the funds available for construction.

Another method of contracting worthy of consideration is "job order contracting." This method awards an indefinite delivery–type contract to a general construction contractor based on competitively bid fixed-unit prices for a list of desired construction activities. Later, actual quantities are specified in the field via delivery orders generated from damage survey reports or similar documents. The unit prices are fixed upon award, except for out-of-scope work that is negotiated as a modification. This is an attractive method of contracting in such situations because it offers fixed pricing as well as flexibility in scope and quantity of placement. The restrictions in this case are two-fold. First,

FIGURE 12.22 Reconstruction of the Kuwaiti Parliament building was under a separate "design-build" contract

the considerable up-front work involved in preparing the unit price contract specification for solicitation, award, and administration is time prohibitive unless these data are already available and computerized from prior work in the region.

Typically, a job order contract can have as many as 20,000 line items for unit price bidding. Therefore, it requires some previous experience with contracting in the area. Second, as with other fixed-price contracting, the method presumes at least general knowledge of scope and the cost and availability of materials, labor, and other factors that mitigate the risk of a fixed-price bid. In the case of Kuwait, even if the voluminous specification could have been prepared and distributed in time, it is doubtful that contractors would have accepted the considerable risk of a fixed-price contract given that virtually all pricing factors were unknown. In retrospect, it appears that some form of cost-plus contracting is inevitable in such an operation. The instrument must be flexible enough to shape the work to the scope as it becomes known, responsive enough to meet the urgent requirements of the crisis, and yet controllable to minimize risk and ensure

FIGURE 12.23 Fully Stocked KERO Supply Room

cost efficiency. Staffing plans must consider these factors and allow for adequate contract supervision, administration, and, most certainly, one or more audits. The role of auditors very early in the contract scenario cannot be overemphasized for any cost-type contract. While engineers are supervising contract execution, the auditors can work with the contractors to shape the allowable range of contract overhead to suit the work environment. They will also specify the level of cost and pricing data required to eventually support the contractor's costs in the final negotiations. The sooner these parameters are established, the more efficient and continuous will be the flow of modifications, negotiations, and eventual contract closeout.

Logistics

Engineers are not logisticians, and even great and dedicated engineers cannot do a day's work, let alone several months of intensive work in a hostile environment, if they cannot eat and sleep properly. The valuable database previously mentioned depended upon computers, generators, copiers, paper, cartridges, and spare parts on a daily basis, or it would not have worked and its benefits could not have been realized. In short, the success of an operation of this magnitude and duration revolves around the efficiency and effectiveness of its logistical plan. The KERO planning team, working with the Transatlantic Division, had to assume that nothing would be available for use in Kuwait, except perhaps a building shell for shelter. Since KERO was an ad hoc temporary organization that did not exist prior to this operation, there was no equipment organic to KERO, and no property book. Further, since all costs were to be borne by the Kuwaiti government, new equipment would have to be purchased on short notice with Kuwaiti funds and subsequently turned over to Kuwait upon completion of the mission. Everything needed to sustain KERO operations for at least 30 days, from vehicles to copiers to personal products to food and water, had to be purchased in Saudi Arabia in the same 10-day period during which the original contracts were awarded and then loaded on leased semitrailers for the journey to Kuwait. A complete list of all functions managed by the logistics staff is shown in Figure 12.24. The most startling and memorable example occurred when a KERO purchasing agent walked

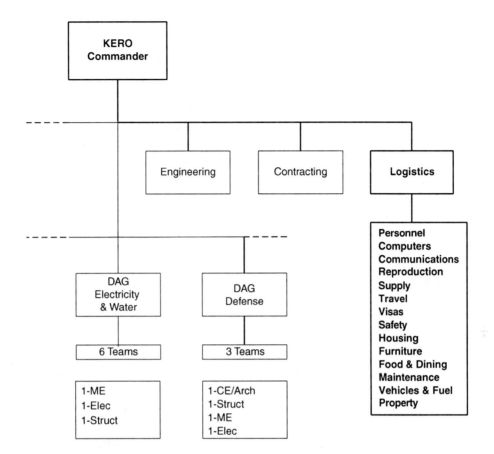

FIGURE 12.24 Logistical operations

into a Nissan showroom in Dhahran on January 18, 1991, and purchased 62 Nissan Patrol 4x4 vehicles on the spot for immediate delivery because they were the only suitable and available vehicles in Dhahran. The expression on the Nissan dealer's face was indescribable. Miraculously, despite such hasty actions, the skilled logistics staff forgot virtually nothing in over 4,000 line items of materials that were purchased.

As a result of this effort, the KERO team was able to conduct self-sustained operations almost immediately on arrival. However, since even the best-planned operation is never perfect, a rapid resupply base must be available to replenish critical items that cannot be found locally, or to satisfy new requirements that develop as the operation matures and changes. The Transatlantic Division, KERO's parent organization in Winchester, VA, provided this support using military and commercial air cargo resources through a sister office in Dhahran, Saudi Arabia, where the staging had occurred. Two aspects of this logistical tale deserve special consideration in nation assistance operations. First and foremost, the equipment utilized by the team was host nation property and had to be treated as such. Accountability, maintenance and repair, and the general condition of host government property, especially upon turnover, are all key components of the image of quality performance that the KERO team sought to leave on completion of the operation. I might add that this was a part of the vision for the KERO operation established by the commander. In his vision statement, he made it clear that the reputations of the Corps and the U.S.

Army were at stake, and therefore, delivery of a "quality" product in a "professional" manner were essential elements of success. Similarly, the host government did not want to be viewed as having engaged a ragtag outfit to reconstruct its country in full view of its own population, its Arab sister states, and the world media. From the client's viewpoint, both these factors would be seen as indicators of the weakness of the Kuwaiti government, its economy, and its ability to quickly recover from the Iraqi invasion. Translated into day-to-day KERO operations, this meant strict attention to the condition of equipment and rapid attention to repairs, housekeeping, uniforms, and the like. Another logistical consideration that deserves careful attention is the cost of doing business. Again, notwithstanding the crisis environment, the recovery operations must be sufficient to produce quality projects, but without extravagance and in keeping with U.S. and host nation government practices. In the KERO operations, the target was to hold pure overhead costs to less than 10% of all expenditures. KERO engineering salaries were charged to projects as direct costs wherever possible. Consequently, actual overhead costs came in at around 8%. This includes the salvage value of approximately $2 million worth of vehicles and equipment that were ultimately returned to the Kuwaitis. The final overhead cost, without mobilization equipment, was around 7%.

Budget Control

When KERO began work, there were no operational governmental agencies in Kuwait to accomplish the normal budgeting actions necessary to fund a large-scale national reconstruction program. Within 45 days, KERO had exhausted the original $46.5 million and was in need of additional funding authority to continue contract operations. A process had to be developed for obtaining additional funds at the appropriate time (see Figure 12.25).

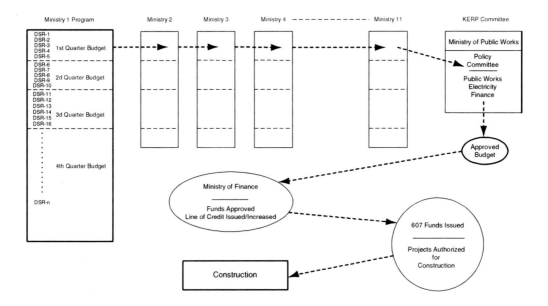

FIGURE 12.25 Budget Control Operations

Extremely cost conscious, as any host nation or sponsor can be expected to be in such a situation, the Kuwaitis requested a monthly accounting of all funds. In addition, they asked for a complete financial summary prior to approving any additional funds. An efficient method of managing and displaying project information and funding status had to be developed quickly. Similarly, KERO's more than 1,200 damage survey reports served to inventory, record, quantify, and assess the cost of the damage; however, a flexible system of managing this voluminous information was needed.

Using standard software packages and laptop computers operating on generator power, KERO engineers developed and implemented a computerized project database within days of arrival in Kuwait. A substantial achievement, this database allowed field engineers to add or modify project information as fieldwork, estimates, and construction were added or completed. This gave PMs a consistently accurate information base and the capability to manipulate the data as required to prepare budget documents and reports in minimal time. To develop the augmented budget request, the survey data contained in the computerized DSR database were arranged into sector programs by the PMs and presented to the appropriate ministry officials for decisions on priority, timing, scope, and so on. KERO PMs soon found that they had more accurate and timely information on the status of facilities than did the ministry officials, who were unable to respond quickly to such requests due to a lack of time, staffing, information, facilities, and such.

This is a normal circumstance in any disaster situation because the responsible government officials are likely to be a casualty of the disaster themselves. Working feverishly, KERO PMs therefore took it upon themselves to prioritize the DSRs and develop a line item repair program for each ministry based on their acquired understanding of the ministry's function and the known damage. The PMs then presented these programs to the respective ministry authorities for approval. Once approved, the programs were incorporated into a nation-wide KERO repair program and presented to the KERP central committee. Once approved by the KERP committee, this reconstruction budget, which totaled some $212 million, was sent to the Ministry of Finance for funding (see Figure 12.25). Using this procedure, the entire budget formulation process for 1991–1992 was accomplished in little

FIGURE 12.26 The U.S. ambassador to Kuwait and Kuwaiti ministers kept close tabs on the recovery budget

more than one week. It took about one month for the Ministry of Finance to act on the budget. Later, the KERP committee reserved the authority to adjust funding among programs but allowed KERO some flexibility in moving funds within a ministry program, provided the committee was notified. This process was summarized and reviewed at biweekly meetings with the KERP committee, which also served to monitor progress and add or delete projects, etc. In addition, written progress reports were sent to each ministry on a monthly basis. These timely and accurate reports covered all projects in the ministry and ensured that the Kuwaiti officials were never at a loss for information about the status or funding for their projects. This frequent and comprehensive communications program was a key element in gaining and maintaining the confidence of the Kuwaiti government throughout the operation, and its importance cannot be overemphasized.

To stay abreast of the extremely fast-moving project operations and funding transactions, it was necessary for KERO to hold internal in-process reviews (IPRs), or project review boards, on a weekly schedule. Again, the data that formed the basis for these reviews emanated from the project database, which was manipulated to form project management summaries, bar chart program schedules, key project fact sheets, and other management documents. This same automated system of reports was used to inform headquarters, the ambassador, Kuwaiti ministers, and other interested parties on project status and KERO operations in general. Several salient features of the budgeting operation should be noted for future operations. First of all, a computerized and flexible project management database was needed from day 1. The KERO team elected to use a simple spreadsheet program to develop this system. These spreadsheets, which could be manipulated by the PMs themselves, were extremely flexible and allowed KERO managers to quickly arrange data in user-friendly formats. This seemingly trivial feature saved a tremendous amount of time and enabled PMs to convincingly demonstrate their control of project operations and costs. Second, KERO task force engineers, who were more versed in project details than were the host government officials, were prepared to develop and defend project priorities and budget requests and subsequently to advise the ministries on how to proceed. Finally, continuous and open communication between KERO project managers and their clients was the key to good working relations and credibility.

Political Factors and Partnering with the Host Nation

There is no question that such an operation has political overtones at all levels that must be carefully managed. Working in an overseas environment, engineers must deal with vast cultural differences that, in spite of the unbridled goodwill exhibited by all participants, could easily result in disagreements and lasting misperceptions. After all, aside from the immediate humanitarian relief, the only lasting benefit for U.S. national interests is the goodwill and close working relationships generated by the work management process. These are political and humanitarian rather than engineering or construction products. The first and foremost consideration for a lasting professional image is the quality of construction provided. Early on, during the heat of the crisis phase, there was tremendous political pressure to conduct only emergency repairs to facilities so that limited funds could be spread over many project areas and substantial progress could be demonstrated. Experience has shown that, for all but the smallest repairs to facilities, this "crisis attitude," which tolerates lesser quality, will subside long before the completion date of the work. Consequently, in the post-crisis environment the user may no longer recognize, or

FIGURE 12.27 Politics and population responses are a real part of recovery

accept, such crisis or emergency scoping as quality construction and will complain bitterly about being served with shoddy work. Therefore, the standards chosen for repairs must be sufficient to stand the test of quality over time. When such questions of hindsight do arise, and they will, they need to be resolved immediately, on the ground, with the customer, and if at all possible, in his or her favor, rather than through long, drawn out administrative appeals. This is especially true for major programs that affect large groups of citizens, as these projects justifiably and expectedly receive close scrutiny and media coverage. Here, a strong, involved, and active public affairs officer can ensure balanced coverage of the work so that such problems are properly explained in the media and do not receive undue attention.

Political and humanitarian factors determine the *ultimate success of a recovery project!*

330 INFRASTRUCTURE RECOVERY INITIATIVES: A RETROSPECTIVE ASSESSMENT

A corollary of the preceding concerns project selectivity. All projects do not have equal value in light of the culture and the political and national security objectives of the host population. The same is true for the U.S. population that will undoubtedly view the assistance effort from our own national interest and priority base. Finally, progress will undoubtedly be reported by a critical and skeptical media. Consequently, every project in the Corps program must stand the test of having a clear humanitarian purpose and content. It must appear equally beneficial to an observer in rural America as it does to someone in the poorer neighborhoods of the host country. The logic and value of each project must be obvious because these observers, who shape political decisions and rate our performance based on public opinion and perception, will not understand or consider the intricacies of a complex approval process. As a general rule, U.S. personnel should always be assigned to projects that afford the highest and best use of our resources and that reflect the maximum positive image and the least risk, politically or otherwise, to our government and personnel. Private contractors working directly for the host government should not recommend projects that do not meet this criterion for construction. Close cooperation with the U.S. ambassador and his or her staff will serve to provide the political sensitivity necessary to properly screen projects.

In general, projects acceptable for U.S. government construction will meet this standard and suffice for the host nation. Projects that support major population segments, versus those that satisfy the goals of special interest groups, are usually acceptable. Planners should beware of cultural differences and the attitude that "it's their money; they can do what they want with it." Citizens of the host country may readily accept certain practices, such as the use of government labor to work on private projects, as one of the privileges or prerogatives of high office. Such practices may even be sanctioned legally, but they would never be condoned or understood by the U.S. citizens supporting your operation, and they should be avoided. For example, work on residences, especially ornate facilities, VIP facilities, and the like, should be avoided unless there is an overriding and unmistakable social value to be gained. In Kuwait, for example, KERO never succeeded in explaining the reconstruction of a former palace for use as the site of replacement offices for the Kuwaiti government because of the ornate nature of the facility and the connotation of the title "palace" to the U.S. population. By contrast, reconstruction of the national Parliament building involved very special and expensive construction and furnishings, but since the overriding value of providing a necessary and suitable facility for the return of democratic government to Kuwait was universally acceptable to both U.S. and Kuwaiti citizens, the project was instantly hailed as a great and successful work.

A second consideration for those in charge of nation assistance efforts is the political question of who is in charge of the recovery effort. Clearly, the host nation must be in charge and U.S. elements must keep this steadfastly in mind. This is not as trivial a matter as it might appear because many engineering judgments, which we make routinely and frequently without asking, are driven by our cultural imperatives and habits, both of which may not be valid in the host nation. For example, Americans value time more than money as a general rule, especially in crisis situations. Therefore, we are likely to accept a higher than normal cost in a disaster recovery situation. "Let's fix the damage quickly" is the spirit that drives many decisions. Eastern cultures are much more patient. Cost is valued as more important than time, and this difference can cause major disagreements and the misperception that Americans are not good managers.

FIGURE 12.28 Repair of 156 Kuwaiti schools became an important political priority

Leadership and Partnership

In the case of Kuwait, the host nation managers were adamant about having a partnership with the Americans. They wanted to share decisions, no matter how long that took. This stemmed from their need to feel, and be seen by their citizens, as if they were directing their own recovery. A second factor, though unspoken, was their sincere desire to learn the American way of managing such a program, so that they could do it themselves if necessary. This true partnership was relatively easy to achieve in the case of field engineering activities where Kuwaiti volunteers were assigned to KERO and worked on project teams with their American counterparts. These engineers worked side by side throughout the entire recovery operation and shared every experience, good and bad. In this way, the Kuwaitis and the Americans developed mutual understanding, respect, and trust for one another, and this was the basis for the true partnership that was formed. That is not to say that there weren't any problems. A conscious effort was made to ensure that both Kuwaiti workers and Americans followed the same rules and practices, but there were still

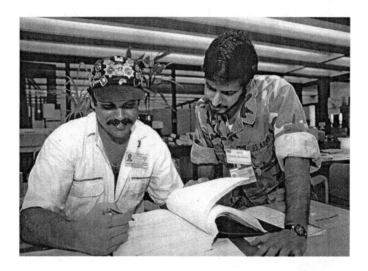

FIGURE 12.29 U.S. and Kuwaiti engineers formed a close partnership

many differences in work practices, compensation, privileges, housing arrangements, etc., all of which were potential areas for discord. However, because a basic partnership had been achieved, these differences could easily be discussed and resolved as each case arose.

At the managerial level, the partnership was equally important but much harder to achieve because of the complexity of the manager's role, the fact that managers did not work side by side, and time constraints. The individuals involved had to make a concerted effort to include their counterparts in any and all decisions pertaining to the recovery program. This usually necessitated a time delay to participate in additional discussions; however, this time was well spent as it avoided later misunderstandings and delays during construction. In fact, in almost every case where managers failed to achieve a consensus on policy prior to construction, there were misunderstandings and delays during the critical, and more costly, construction cycle.

The preceding are questions of leadership rather than managerial or technical acumen. As mentioned so often in this chapter, it falls to the project leader to establish the parameters for success of the project. Perhaps more importantly, it is up to him or her to transmit those values throughout his or her organization and infuse his or her subordinates with the will and passion to excel in all aspects of project operations, including partnership and politics. In addition to these political and partnership concerns, the task force or project leader serves as a role model for all who observe his or her conduct. There is no escaping the fact that the leader of an organization sets the moral tone for subordinates. If the leader is coldly efficient in interactions with subordinates, that pattern will be reflected in the subordinates' relationships with each other. If he or she exudes enthusiasm and a positive attitude about the project's purpose and potential benefit to society, the workforce will be similarly motivated to overcome all obstacles to achieve success. Great motivation is a tremendous "morale multiplier." It is the adrenaline of workforce behavior, enabling employees to cheerfully achieve what otherwise might have been unheard of levels of performance. The photo in Figure 12.15 was taken at 11:30 p.m. on March 5, 1991. No one had ordered those individuals to work through the night, but they understood the importance of their mission and contractor operations the following day. Such performance is the result of positive leadership and motivation.

On the other hand, if the leader is cheerless and negative in the conduct of business and in dealings with subordinates, constantly criticizing them or complaining about his or her superiors' lack of judgment or failure to provide resources, the leader will similarly create a negative attitude among the project team, sapping their physical and mental power and crippling the project.

Project management is first and foremost *a leadership activity!*

Unfortunately, engineers are most often not trained in human relations, and they may not even see this as part of their job responsibilities. However, once the engineer transitions from a purely technical role into the area of project management, he or she has entered a whole new world in which the art of human relations becomes the medium with which technical success is achieved. If the reader were to take away only one lesson from this chapter, I would ask that he or she remember the brief phrase repeated so often throughout this chapter: *Project management is first and foremost a leadership activity!*

FIGURE 12.30 Kuwaiti Central Water Pumping Station before and after Repairs

Lessons Learned

To sum up the Kuwait experience in a few words is difficult. There were so many lessons learned that this chapter can only scratch the surface. Certainly, the impact of positive leadership on team performance should be at the top of the list. In addition, several more general observations deserve to be mentioned.

First of all, as an intergovernmental operation, the Kuwait recovery project was a tremendous success. The humanitarian spirit of the participants easily bridged the cultural and professional differences and paved the way for close cooperation and good working relations between the Americans and the Kuwaitis. What resulted from this cooperation was the prospect of a long-term relationship based on trust and goodwill, which is probably more important than the operation itself. Of paramount importance to these excellent working relations was the responsive and accountable support of the new engineering

FIGURE 12.31 Key transportations systems were rapidly restored to operation

management structure and procedures developed by the KERO team. As a result, budget documents and funds accountability were precise and convincingly accurate. This is of cardinal importance to the establishment of trust with the host nation. Second, a true and honest partnership in all engineering decisions eliminated the potential misunderstandings that could easily have delayed construction and undermined the completion of key projects. Finally, free and open communications with all parties ensured that both U.S. and Kuwaiti government officials, and their constituents, understood exactly what was happening as the recovery progressed. In the end, the Kuwaitis were satisfied with the outcome of the operation and thankful to the Americans for their very timely and professional assistance.

The Kuwait recovery was rewarding and professionally exciting for all who had the opportunity to participate. As an intergovernmental experience, it holds a promise for future application not only in the aftermath of a conflict but also potentially as an instrument of foreign assistance or a method of recovery from natural disasters. The security, political stability, and goodwill that resulted from the timely restoration of the civil infrastructure in Kuwait cannot be overemphasized and warrants serious consideration as a conflict management or conflict reduction tool for future nation assistance operations.

Application to Recent Disasters

What follows is a description of the project management activities utilized in Kuwait with implications for similar recovery operations in New Orleans. There is virtually no difference between the recovery from a military or man-made disaster and that of a natural disaster. This chapter discusses the following subjects as they were applied during the Kuwait operation, with the implication that almost identical procedures would significantly improve recovery operations in New Orleans and the Gulf Coast. With this in mind, a theoretical New Orleans recovery operation has been overlaid on all diagrams and figures used to describe the Kuwait operations so that the reader can easily see how a parallel operation could be organized for the New Orleans recovery.

FIGURE 12.32 New Orleans

The following elements, considered the key elements essential to the success of any disaster recovery operation, will be discussed in comparing the Katrina recovery to the very successful Kuwait recovery operation: planning, organizing and staffing, project management, contracting, budget control, leadership, and partnership.

Planning for the Recovery Operation

A well-defined plan can save time, money, and frustration. In fact, no project can really succeed without a comprehensive plan to outline the key milestones and events that guide the day-to-day operations of the project team. The project manager's most fundamental responsibility is to define the vision for the project and clearly identify the actions the team must take to achieve success. Once the project is under way, he or she must then make timely decisions to adjust the plan for unforeseen problems that arise along the way. In the case of a natural disaster such as Hurricane Katrina, the timing of the event is unknown and therefore the time available for planning the recovery is seemingly nonexistent. For this reason, it is essential to conduct the planning cycle in the time available *before* the event occurs, basing decisions on similar events and most probable outcomes.

In the case of the Kuwait recovery, the team had only 45 days to begin full-scale recovery operations over 8,000 miles from their support base, precious little time to prepare for such a monumental undertaking. Funds were extremely tight at the outset and risk was high. No one had asked for a plan prior to funding or supporting the operation, and lastly, there was absolutely no information upon which to base a definitive plan: no scope, no site plan, and no summary of damages to be repaired. When faced with a paucity of information planners must rely on other sources for inspiration: (1) their experience, (2) their training, and (3) their imagination.

Corps personnel are continually trained in disaster recovery operations at their National Training Center in Huntsville, AL. In addition, the Corps' experience with past hurricane recovery operations in the United States was directly applicable to the scope

FIGURE 12.33 Debris Removal

and magnitude of the Kuwait operation. Finally, KERO team leaders would have to use their imagination to develop a "notional" scope of recovery operations for Kuwait. It was against this notional scope that operations were planned, with adjustments made to account for actual conditions, made on a daily basis, as they became known. The results of these early decisions are summarized in Figure 12.34.

Using Corps procedures, the KERO teams would prepare a damage survey report (DSR) and cost estimate for each damage site or potential project. The DSR would define specific elements to be repaired at each site and establish a budget for that project. The DSRs would then be aggregated into repair programs organized along functional lines, such as roads, electricity, buildings, and so on. Each program was assigned to a KERO project manager for prioritization and direction. In addition, these project managers were functionally aligned with a Kuwaiti ministry that had responsibility for that function; i.e., the KERO project manager for roads worked directly with the Ministry of Roads as the client. During the subsequent recovery phase, construction contractors would be hired to execute the repairs defined by these individual DSRs.

Because each project manager was assigned to one client, in this case a Kuwaiti minister, this gave each client a single point of contact for information and control. Project managers were instructed to become completely familiar with the culture and mission of their client's ministry so they could make informed decisions on project features, trade-offs, and schedule, just as if they were the minister. By this means, the minister was assured a constant flow of information, complete understanding of the intended outcome for each project, and therefore, complete control of the program for his ministry.

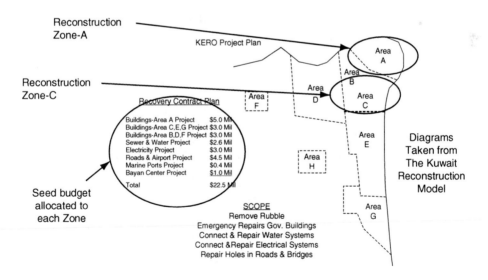

Divide New Orleans into work areas or "Reconstruction Zones"...

- By parish for housing ("geographical")
- By municipal department for infrastructure ("functional")
- Allocate a "seed" budget to each reconstruction zone

FIGURE 12.34 New Orleans Reconstruction Zones

Organization and Staffing

In the case of the KERO operation, there was not much time to organize, staff, and train the initial task force of approximately 140 Corps employees, all of whom were experienced volunteers from various districts and divisions who would rotate to Kuwait on a 3-month cycle. During the initial emergency phase of the recovery, KERO field offices were set up according to Corps emergency management practice. Labeled *damage assistance groups* (DAGs), each group had a variable number of assigned damage assistance teams (DATs), depending on the mission of the group (see Figure 12.35).

As mentioned earlier, DAGs were aligned with specific Kuwait ministry offices: buildings, roads, airport, electricity, water and sewer, ports, defense, and so on, according to pre-war Kuwaiti management conventions. Adoption of these conventions simplified the interface with the host nation, as leaders and volunteers from these Kuwaiti ministry offices were readily available to work with KERO. Additionally, this organization allowed Kuwaiti volunteers to work directly with their parent ministry. This structure was used for about 45 days while damage survey reports (DSRs) were being completed and contractors were mobilizing. Actual construction during this time was minimal; however, more than 1,200 DSRs and associated cost estimates were prepared. These reports later became the basis for all future budgets and recovery work orders.

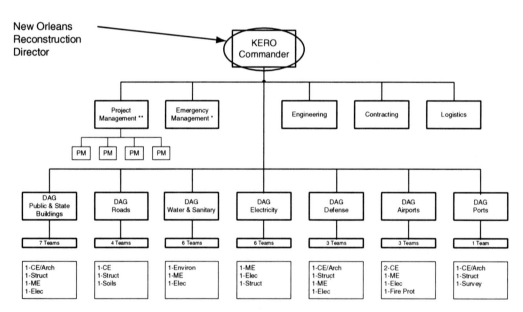

*   Provides primary direction during "emergency" phase
**   Provides primary direction during "recovery" phase

Assign a "Damage Assessment Group (DAG)" to each New Orleans Reconstruction Zone
- Appoint a "Project Manager" for each DAG
- Assign teams & skills based on magnitude & type of work
- Teams survey repair projects produce Damage Survey Reports
- Repair projects are cataloged and prioritized by the DAG

FIGURE 12.35 Damage Assessment Operations Budget Control Operations

Project Management

In 1991, the Corps of Engineers did not have a formal system of project management as part of its operating procedures. Instead, projects were traditionally managed from within the functional discipline that had the largest share of the work. Around the time of the Kuwait operation, the Corps of Engineers had been considering a new, more customer-oriented form of organization for project management. Although the Corps of Engineers had not formally adopted this new management system, KERO managers decided to implement it on a test basis, as shown in Figure 12.36. Under this arrangement, all project managers in KERO were assigned to a Department of Project Management, which reported directly to the KERO Commander.

Within the PM department, individual project managers were assigned to each Kuwaiti ministry to provide a single point of contact for all projects in that ministry. Each PM was completely responsible for ensuring that the scope, cost, quality, and schedule of all projects met the needs of his or her assigned ministry. Consequently, the client had only

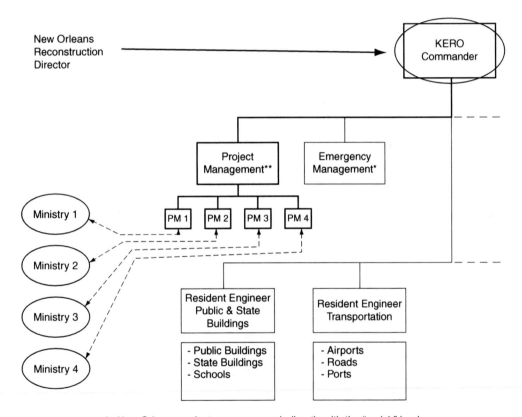

In New Orleans project managers work directly with the "parish" leaders
- Their tasks are to coordinate work with the "owners" in each zone
- To ensure responsive and frequent communications
- To ensure the "needs" of parish are being met
- PMs report daily & weekly to "Director of PM" for coordination

  * Provides primary direction during "emergency" phase
** Provides primary direction during "recovery" phase

FIGURE 12.36 Project Management Operations

one person to contact within KERO to gain complete knowledge of all his or her projects. Similarly, all functional elements in KERO were directed to look to only one PM for decisions concerning projects for a given ministry. In addition, the PMs were instructed to serve as advocates for their client's projects. As such, they had to understand the mission, operations, and culture of the ministry well enough to know how each project served the needs of that ministry. With this insightful knowledge they were able to make intelligent decisions about project features or trade-offs throughout the life cycle of project development, thereby reducing errors and saving time and resources. Within KERO, they were empowered to control all variables associated with their projects.

During the subsequent recovery phase, when construction management became the dominant operational consideration, a more conventional Corps' project management structure, with traditional resident offices to manage contractors in the field, was adopted to ensure that project delivery and contract administration were accomplished according to Corps' procedures and quality standards (see Figure 12.4). Following the same practice used during the emergency phase, project managers were assigned to coordinate work with the responsible Kuwaiti ministry officials.

Contracting

Speed and, later, control were the driving forces in all KERO contracting operations from the outset. KERO staffers planned for a 45-day competitive emergency contract award cycle to be executed from Saudi Arabia. They were eventually forced to utilize a 10-day contingency plan when the war ended more quickly than expected. Working nearly around the clock and without specific knowledge of the conditions in Kuwait, the KERO staff, with Kuwaiti counterparts, completed necessary project scoping, solicitation, prequalification, and proposal evaluation actions to award eight cost-plus letter contracts, worth approximately $25.4 million, before the war had ended. Once the recovery operations had begun, each DSR would be converted to a task order and assigned to one of these letter contracts for execution. This procedure allowed repair work to begin as quickly as possible. Each task order would later be definitized and converted to a fixed-price contract, once the actual scope, timing, and cost of the repair task were known.

Since these letter contracts were initially cost-plus instruments, a precise method of controlling and documenting the flow of work to the contractor was required once construction operations began. As described earlier, the original DAGs were converted to standard Corps resident offices for contract control and administration once mobilization was complete and construction had begun in earnest. For the purpose of passing specific work requirements to the contractors, the DSRs and associated cost estimates prepared during the emergency phase were converted into work orders. These documents, which resided in the project database, were reviewed by PMs and adjusted to define the exact scope of repair work.

The work orders were then provided to the Engineering Services office, where the original cost estimate for the work order was reviewed, and corrected if needed, using the Corps' newly automated M-CACES process. This extremely valuable tool allowed detailed estimates to be developed quickly and accurately, sometimes overnight. It also allowed the Corps to adjust the unit cost of labor and materials in accordance with changes in the market values of these items as time passed. These work orders, detailed cost estimates, and record of negotiations were incorporated into a contract modification that served as the plans and specs during construction.

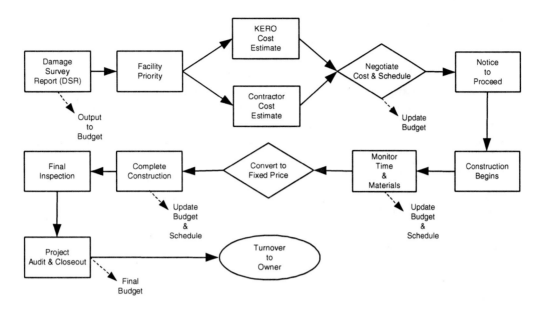

FIGURE 12.37 Contract Operations

Ultimately they, in conjunction with daily construction management logs for time and materials, became the basis for final negotiations with the contractor on the cost of each work order. As the KERO operation neared completion, these documents were audited to verify that no over–under payments were made (see Figure 12.37). As the operation matured, some larger projects were advertised individually to enhance competition and encourage participation of both U.S. and Kuwaiti small businesses.

In retrospect, some form of cost-plus contracting is inevitable, in fact, essential in such operations. The instruments must be responsive enough to meet the urgent requirements of the crisis, flexible enough to shape the tasking to the actual scope as it becomes known, and yet controllable to minimize risk and ensure cost efficiency. Staffing plans must consider these factors and provide adequate personnel for contract supervision, administration, and most certainly, one or more audits. The role of auditors very early in the contract scenario cannot be overemphasized for any cost-type contract. While engineers are supervising contract execution, the auditors can work with the contractors to shape the allowable range of contract overhead to suit the work environment. They will also specify the level of cost and pricing data required to eventually support the contractor's costs in the final negotiations. The sooner these parameters are established, the more efficient and continuous will be the flow of modifications, negotiations, and eventual contract closeout.

Budget Control

When KERO began work, there were no operational governmental agencies in Kuwait to accomplish the normal budgeting actions necessary to fund a large-scale national reconstruction program. A process had to be developed for obtaining additional funds at the appropriate time (see Figure 12.17). An efficient method of managing and displaying project information and funding status had to be developed quickly. While KERO's more than

1,200 damage survey reports served to inventory, record, quantify, and cost the damage, a flexible system of managing this tremendous volume of information was needed. Using standard spreadsheet software and laptop computers operating on generator power, KERO engineers developed and implemented a simple but extremely effective computerized project database within days of arrival in Kuwait. A substantial achievement, this database allowed field engineers to add or modify project information as fieldwork, estimates, and construction were added or completed. This gave PMs a consistently accurate information base and the capability to manipulate the data, as required, to prepare budget documents and reports in minimal time.

Working feverishly, KERO PMs used this system to prioritize the DSRs and develop a repair program for each ministry based on their acquired understanding of the ministry's function and the known damage. The PMs then presented these programs to the respective ministry authorities for approval. Once approved, the programs were incorporated into a nation-wide KERO repair program and presented to the central committee. Once approved by the central committee, this reconstruction budget, which totaled some $212 million, was sent to the Ministry of Finance for funding (see Figure 12.38). Using this procedure, the entire budget formulation for 1991–1992 was accomplished in little more than one week. To stay abreast of the extremely fast-moving project operations and funding transactions, it was necessary for KERO to hold internal in-process reviews (IPRs), or project review boards, on a weekly schedule. Again, the data that formed the basis for these reviews emanated from the project database, which was manipulated to form project management summaries, bar chart program schedules, key project fact sheets, and other management documents. This same automated system of reports was used to inform headquarters, the ambassador, Kuwaiti ministers, and other interested parties on project status and KERO operations in general.

Leadership and Partnership

In the case of Kuwait, the host nation managers were adamant about having a partnership with the Americans. They wanted to share decisions, no matter how long that took. This

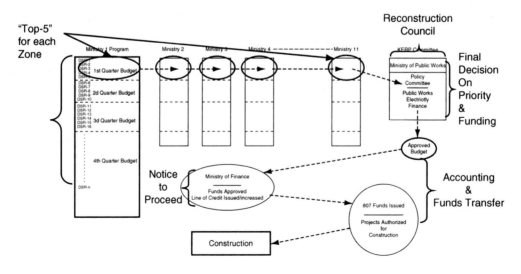

FIGURE 12.38 Budget Control

stemmed from their need to feel, and be seen by their citizens, as if they were directing their own recovery. This true partnership was relatively easy to achieve in the case of field engineering activities where Kuwaiti volunteers were assigned to KERO and worked on project teams with their American counterparts. These engineers worked side by side throughout the entire recovery operation and shared every experience, good and bad. In this way, the Kuwaitis and the Americans developed mutual understanding, respect, and trust for one another, and this was the basis for the true partnership that was formed. In the end, the Kuwait recovery was satisfying and professionally exciting for all who had the opportunity to participate. As an intergovernmental experience, it holds a promise for future application not only in the aftermath of a conflict but also potentially as an instrument of foreign assistance. The security, political stability, and goodwill that resulted from the timely restoration of the civil infrastructure in Kuwait cannot be overemphasized and warrants serious consideration as a conflict management or conflict reduction tool for future nation assistance operations.

Summary

The procedures described in this paper were not unique to the Kuwait operation. They were developed and refined over a period of 5–7 years in a variety of applications from disaster recovery operations to the routine management of public works engineering operations for a community of 21,000 inhabitants. The key elements of success are derived from the well-organized and systematic approach to program and project management, tight fiscal and budgetary control, abundant accurate information about program status, and most importantly, the direct involvement of the owner of the facilities throughout the progress of the repair operations. From what is known about the magnitude and sensitivity of the New Orleans recovery, or of any transportation infrastructure disaster that may loom before us, it is clear that these same elements, modified to suit the incident conditions, will ultimately enable the successful recovery from such events. It is unfortunate that successful operations, such as the Kuwait recovery, seem to be ignored when new disasters strike. Those responsible for the recovery seem to start with a clean slate, completely ignorant of any process or procedure that may have worked well in the past. With this in mind, these procedures, and these recommendations for a regional crisis recovery system for the United States, are respectfully submitted for consideration and application.

13

Immigration and National Security: Best Practices

Jo Ram

Objectives of This Chapter:

- Provide an overview of our ePassport, eVisa, eID, and deportation solutions
- Discuss the importance of immigration and national security issues that countries must focus on to protect their borders
- Discuss the regulations established by the International Civil Aviation Organization (ICAO), the standards board regulating ePassports and eIDs
- Discuss the evolution of the machine-readable travel document (MRTD) and its advantages
- Provide a background on data security and discuss concerns about data privacy
- Provide a background on best practices implemented in the national security arena and give examples of successful implementations

Introduction

A concern that has been brought to the forefront of the travel industry in the last decade has been national security in terms of immigration and protecting borders. There are heightened concerns regarding safety and potential terrorist threats in tourist destinations after the Bali bombing where Australian tourists were specifically targeted. As a result, governments around the world have begun to implement more secure travel documents in order to protect their borders and deter illegal immigration, fraud, and potential criminal or terrorist activities.

After the September 11, 2001, attacks in the United States, the Bali bombing, and the attacks in London and Spain, a renewed push for tighter security measures has come about. Concerns about global terrorism and the influx of illegal immigrants in the United States and Europe have resulted in a public outcry for action. A strong demand for rigorous controls over national borders exists, resulting in the worldwide adoption of stricter standards for securing travel documents such as visas, passports, work permits, and other types of identification documents.

The United States now requires that all visitors having a visa to the United States be fingerprinted before the visa is issued and at border entry. Biometric identification and authentication in the form of facial images (digitized photograph) and fingerprints are starting to play a vital role in the debate on national security and travel safety. A majority of U.S. visa-waiver countries are testing machine-readable and chip-based passports with at least one biometric identifier. These governments want their citizens to be able to travel safely and freely to the United States. The United States and participants of the U.S. visa-waiver program are also issuing machine-readable visas to those visitors who plan to visit their country. Other nations are quickly following suit, as they want their citizens to be able to enter countries such as the United States, the United Kingdom, France, Germany, and other EU countries with ease.

In early 1990, the International Civil Aviation Organization (ICAO) set standards for travel documents used for international travel, which required passports to have a machine-readable zone in the data page, a digital photo, and other security features. The new wave of ICAO regulations, which will be discussed in greater detail later in this chapter, deal with microchips, biometrics, and RFID technologies.

The term *RFID* (radio frequency identification) spans the entire spectrum of contactless identification products. There are three main categories for RFID in smart card applications; namely, for government IDs (e.g., ePassport, eID, health care ID, driver's license), personal IDs (e.g., transportation, access control) and object IDs (e.g., inventory tagging, asset tracking). RFID smart cards differ from traditional contact smart cards by not requiring physical connectivity to the card reader. The advantages of RFID include ease of use, faster transaction time, and ease of reader maintenance.

The use of RFID contactless identification technology in object IDs and personal IDs has grown prevalent in today's society and can be produced just as easily as it is obtainable in the market. However, a different story applies for its uses in identity programs in government ID projects; the sensitivity of the information stored therein brings added complexity and the need for greater assurances in the integrity of both data and system.

Since 9/11, a lot of debate has been focused on security, access control, and identification. To be able to positively identify and authenticate a person is more critical today than ever before. The identification of individuals historically was performed through some known piece of data or information, such as a Social Security number or government-issued photo ID. But in today's environment, identification is not enough. Biometrics allows governments to confirm the identity of an individual by matching human physiological characteristics such as fingerprints, eye retinas and irises, facial patterns, and so on.

In 2006, Indusa Global partnered with IRIS Berhad Corporation to provide chip-based, machine-readable travel documents that promote tighter security measures at national border crossings. Established in 1995 in Greenville, SC, Indusa Global is an enterprise solutions provider and systems integrator providing ePassport, eID, business intelligence, and data warehousing solutions and services to its clients worldwide.

IRIS is a Malaysian company known for pioneering and developing the I.R.I.S. (Image Retrieval Identification System). The Image Retrieval and Identification System provides the ability to capture textual data, live images, and human biometrics and compress this information into the limited space of a microchip. The uniqueness of the I.R.I.S technology is that even if the information is compressed, the system is able to retrieve and decompress the information without any discernible loss in resolution and quality. Despite initial skepticism, the I.R.I.S. process slowly but surely gained recognition for being a truly innovative product that eventually gave birth to the world's first electronic passport. It was this pioneering spirit and effort that managed to gain the support of the Malaysian government in implementing the first electronic passport in 1998 and the government's multipurpose smart card MyKad in 1999.

On December 22, 2006, The Bahamas Ministry of Foreign Affairs signed an agreement with Indusa Global for supplying ePassports, machine-readable visas, five types of eIDs (work permits, permanent residence cards, permit-to-reside cards, home owner's cards, and resident spouse cards), as well as a deportation and border management system. This project was a complex undertaking with teams rolling out ePassport, visa, eID, border control, and deportation systems in 38 locations throughout The Bahamas and eight overseas missions: 19 locations in The Bahamas for foreign affairs and 19 for immigration.

The new Bahamian passports are fully compliant with the standards put forth by the International Civil Aviation Organization and are completely interoperable with other countries that follow these ICAO standards. The new Bahamian passports have a machine-readable zone (MRZ) and a chip embedded in them that stores facial scans, fingerprints, and digital signatures. The passports have security in various forms, including the data page, the chip, and of course the passport paper itself.

Protecting Borders

Numerous terrorist threats within U.S. borders and attacks on U.S. embassies and American-owned interests around the world have forced the U.S. government to analyze the weaknesses in its domestic security policies. Reforms as large and complex as establishing the Department of Homeland Security (DHS), to more manageable ones like defining security measures in travel documents, have been put in practice in this past decade.

Intercepting chatter, "orange alerts" at airports, national security and immigration debates, and the need to protect borders are topics that not only are of importance to the United States government but also have increasingly become important to the average American citizen after the attacks on September 11. Being the first significant attack on U.S. soil among the 48 contiguous states, 9/11 raised concerns in the minds of many Americans about what the U.S. government was doing to protect their homeland and ensure safety and security in their daily lives. Throughout history, Americans have been fortunate to experience unprecedented democracy, freedom, opportunity, and prosperity without worrying about terrorism—either from within their national boundaries or from outside.

Each year, approximately 500,000 illegal immigrants cross U.S. borders in search of the American dream: finding a job, building a home, saving money, and creating a future for their family. This is a real problem today, as there are over 12 million illegal aliens in the United States.

The economic and social impacts of illegal immigration are the following:

- *Economic impact*: Undocumented children or U.S.-born children of undocumented parents can strain the educational and health-care systems in the United States as the taxes earned from their parents are not sufficient to support their education and medical expenses.

- The cost of law enforcement officials required to man crossings and enforce border security, the cost of implementing highly sophisticated and expensive technologies to monitor physical borders, and the cost of apprehending, processing, and deporting illegal immigrants impose a financial strain on the nation's economy.

- *Social Impact*: The most obvious social impact of illegal immigration is racial tensions caused by the belief that these immigrants are taking jobs that belong to Americans. Another controversial debate is spurred by the fact that Americans believe that illegal aliens are responsible for the low minimum wage, as they will work for less.

- Crime is a significant social impact of illegal immigration, as stated in a report from Family Security Matters: An estimated 2,158 murders are committed every year by illegal aliens in the United States, surpassing the number of American soldiers killed in the war in Iraq each year. The group says that the murders committed by illegal aliens constitute 15% of total murders committed in the United States.

The potential threat of terrorism activities from foreign groups and illegal immigration practices are the two main reasons for protecting U.S. borders. However, increasing security at border crossings, enhancing passport security, and curbing illegal immigration do not guarantee the safety of Americans. Another concern that threatens the safety of Americans comes from within U.S. borders and is called *domestic terrorism*.

The Oklahoma City bombing proved that domestic terrorism can happen as easily from within our own borders as foreign terrorist attacks orchestrated by Al Qaeda or other such groups that protest U.S. foreign policy. In fact, domestic terrorism is much harder to track as resources, funding, and technologies are focused mainly on tracking visitors and illegal persons.

In protecting the civil liberties of their own citizens, the United States and other nations are reluctant to use biometric-based technologies to track their own citizens. Groups such as the American Civil Liberties Union (ACLU) are vehemently opposed to Real ID programs, stating that these IDs violate the principles of personal privacy. The ACLU continues to lobby against such programs because they believe that the Real ID program is simply a well-disguised mask for implementing a national ID and strongly believe that having a national ID system will lead to more bureaucracy, longer lines, increased identity theft, and higher fees.

The main challenge that the U.S. government faces in protecting its borders from illegal immigration and its citizens from domestic and foreign terrorist attacks is its ability to do the following:

- Guard the borders of such a large nation, ensuring that its borders with Mexico and Canada are not porous and monitoring the activities on borders exposed to thousands of miles of ocean

- Protect civil liberties of U.S. citizens but be able to track citizens who support domestic terrorism groups or activities

- Promote foreign investment, trade, and tourism while making it impossible for foreign terrorist groups or state-led terrorists to gain entry into the United States

- Encourage other nations to invest in securing their own borders and citizens in order to promote shared data and open communication with foreign security agencies

- Educate citizens and governments about the importance of national security and its processes, loopholes, and increased costs
- Allocate budgets for research and development and for implementation of biometrics-based searching and tracking technologies

National security and immigration are even playing a key role in the debates for the 2008 presidential election. Candidates have discussed options of building physical and electronic fences to deter illegal immigration by creating a physical barrier between us and our neighbors to the south and north. Discussions on enhancing programs such as virtual fences, the US-VISIT program, and the Temporary Worker ID initiative have been heated, as each of the candidates has strong opinions on what is the most appropriate solution.

Following is a short description of the three main issues concerning the enhancement of citizen safety and protecting borders:

Virtual fences: The virtual fence developed by the Immigration and Naturalization Service (INS) originally in the 1970s was based on low-light video cameras and portable electronic intrusion–detection ground sensors deployed at border crossings. This technology evolved over time, and by 2000, INS deployed motion, infrared, seismic, and magnetic sensors. Over 13,000 ground sensors were installed, which covered only 4% of the border.

The sensor technology accompanied with cameras worked reasonably well, as the cameras effectively monitored borders within short ranges. The cameras would activate on motion sensors and capture images of illegal border crossings and/or smuggling activities. The problem with this technology was that it was costly to maintain and hard to manage in inclement weather conditions. In addition, the sensor technology led to many false alerts caused by animal triggers, as it could not differentiate between a human or an animal crossing the border.

In 2004, the then recently created Department of Homeland Security (DHS) reviewed the sensor technology and called for a more comprehensive plan for the surveillance of U.S. land and maritime borders. However, virtual fences address only part of the illegal immigration issues faced by the United States.

The U.S. shares a 7,514-mile border with Canada and Mexico, which is virtually impossible to have under surveillance 24/7. However, even if it could be monitored continuously, over 45% of the 12 million illegal immigrants residing in the United States today entered legally and were processed properly at different ports of entry. The problem is that these individuals overstayed their visa and did not return in accordance with the terms of the visa they were granted. Virtual fences are totally ineffective in resolving this type of problem.

US-VISIT program: In order to deal with the overstay issue, in 2004 the US-VISIT program was deployed as an automated biometric entry–exit system. By the end of 2005, this system was in place in all 284 U.S. air, land, and sea ports of entry.

The US-VISIT program captures biographical and biometric data, including fingerprints of travelers applying for visas in U.S. embassies around the world. Once again, when travelers enter the United States, they are fingerprinted and their names are checked against the U.S. watch list and verified to ensure that criminals, illegal workers, and terrorists are not granted entry.

By June 2006, over 60 million travelers had been processed by the US-VISIT program, and more than 1170 criminals and immigration violators had been denied access into the United States, resulting in appropriate action being taken.

Temporary Worker ID initiative: A biometrics-based electronic ID is being proposed and supported by many Republicans for the Temporary Worker ID. The goal is to provide legal status to undocumented workers with eIDs after they complete a security and health check.

This initiative is controversial and continues to be heavily scrutinized in the 2008 election debates, as promoters of the Temporary Worker ID initiative want to start the program with foreign workers but expand it quickly to include the entire 146-million-person U.S. workforce.

On the matter of protecting borders—regardless of what will be the eventual next steps taken by the administration that wins the 2008 election—foreign governments are looking for guidance from ICAO to set standards on travel safety, ePassports, eIDs, and secure visas in order to be able to protect their own borders and citizens.

ICAO Regulations

The International Civil Aviation Organization (ICAO), a United Nations agency started from the Chicago Convention in 1944, is charged with the mandate to set standards on travel safety. ICAO promotes travel safety and security through cooperative aviation regulation among its member states. ICAO council members meet every 3 years to set strategic objectives for the upcoming 5 years. According to ICAO's Web site (http://www .icao.int/), the objectives for 2005-2010 are as follows:

- Safety: Enhance global civil aviation safety
- Security: Enhance global civil aviation security
- Environmental protection: Minimize the adverse effect of global civil aviation on the environment
- Efficiency: Enhance the efficiency of aviation operations
- Continuity: Maintain the continuity of aviation operations
- Rule of law: Strengthen laws governing international civil aviation

With the high rate of illegal immigration practices and numerous terrorist threats around the world, in the United States there has been a steadily growing demand for rigorous controls over America's national borders since the 1980s, resulting in strict standards over securing travel documents, such as visas and passports. More recently, the United States began issuing ePassports and eIDs for its own citizens and requiring that all visitors having a visa to the United States be fingerprinted once when filing their U.S. visa application and again when entering the United States.

Based on a collaborative effort with countries such as the United States, United Kingdom, and other EU nations, ICAO worked closely with governments to help them enhance travel security.

In 1990, ICAO began setting standards for travel documents used for international travel. On May 22, 2003, the Air Transport committee of the council approved a four-part recommendation from the Technical Advisory Group/Machine Readable Travel Document (TAG/MRTD), which subsequently became known as the ICAO "blueprint."

Based on the New Orleans Resolution in May 2003 that states, "ICAO TAG-MRTD/ NTWG future recognizes that in addition to the use of digitally stored facial image, Member States can use standardized digitally stored fingerprint and/ or IRIS* (*subject to the resolution of intellectual property issues) images as additional globally interoperable biometrics in support of machine assisted verification and/or identification. Member States, in their initial deployment of MRTDs with biometrics identifiers, are encouraged to adopt contactless Integrated Chip (IC) media of sufficient capacity to facilitate on-board storage of additional MRTD data and biometrics identifiers," This excerpt supports the use of biometrics in travel documents and the use of microchips as a storage medium.

ICAO Document 9303 considers only three main biometric identifiers:

- Facial recognition (mandatory)
- Fingerprints (optional)
- Iris recognition (optional)

The deployment of biometric technology in passports and other travel documents, for purposes of machine-assisted identity verification and confirmation, is one aspect of ICAO's strategy to improve border clearance processes with machine-readable and associated technologies. Moreover, ICAO assembly resolution A32-18 (adopted in 1998 and updated in 2001) urges states to intensify their efforts to safeguard the security and integrity of their passports, protect their passports against passport fraud, and assist one another in these matters.

Passport integrity is a significant factor in securing global travel systems. Confidence in the integrity of a state's travel documents promotes easy and efficient facilitation of border control formalities. Biometric identification is considered an important tool in securing international travel documents, according to ICAO.

The recommendation put forth by the Technical Advisory Group entailed selection of facial recognition to be used worldwide for machine-assisted identity confirmation; use of a contactless integrated circuit (IC) chip with a minimum capacity of 32 kilobytes of data as the medium for storage of electronic data, including biometric(s), on a travel document; programming of the IC using the instructions set out in the specified logical data structure (LDS); and use of a modified public-key infrastructure (PKI) scheme for the implementation of digital signatures to secure the electronic data against unauthorized alteration.

ICAO has made recommendations on a wide range of features for the ePassport, from the symbol that identifies an ePassport to the logical data structure that ensures interoperability. In fact, ICAO even has suggested formats and guidelines for photographs, biometric properties, MRZ data, chip security, and various other elements related to machine-readable travel documents.

All ePassports that contain a 32KB or higher microchip that is encoded according to the required logical data structure should carry the symbol shown in Figure 13.1 to show their compliance with the basic ICAO standards.

The security recommendations made by ICAO on the public-key infrastructure (PKI), passive authentication, active authentication, basic access control (BAC), and extended access control (EAC) are described in greater detail in the "Data Security and Privacy" section of this chapter.

Interoperability was one of ICAO's main mandates in defining standards—enabling all nations to easily share data regardless of platform, vendor, or hardware. ICAO set

FIGURE 13.1 ePassport Symbol
Source: http://www.policylaundering.org/archives/ICAO/Biometrics_Deployment_Version_2.0.pdf

standards on logical data structures (LDS) that define the format on which the data should be stored on the contactless integrated chip.

The recommendation on logical data structures identifies the mandatory and optional data elements and their grouping within the ePassport chip. Both the format of the data elements and the groupings are important for reading the chip in the MRTD and for achieving global interoperability.

All data elements—for example, first, middle, and last name; date of birth; date of expiry; and nationality—are mapped in to larger data groups (DG1-DG16) and three potential future groupings (DG17-DG19).

Table 13.1 defines mandatory and optional data groups.

Table 13.1 Data Groups Within the Logical Data Structures

File Name	Data Groups	Mandatory (M)/Optional (O)
EF.COM	Common directory file	M
EF.SOD	Document security object	M
DG1	Machine-readable zone (MRZ) data	M
DG2	Encoded facial image	M
DG3	Encoded fingerprint	O
DG4	Encoded iris scan	O
DG5	Portrait	O
DG6	Reserved for future use	O
DG7	Signature	O
DG8	Data feature(s)	O
DG9	Structure feature(s)	O
DG10	Substance feature(s)	O
DG11	Additional personal details	O
DG12	Additional document details	O
DG13	Discretionary data elements defined by issuing State	O
DG14	Reserved for future use	O
DG15	Active authentication public-key info	O
DG16	Person(s) to notify data element(s)	O
DG17	Automated border clearance details	O
DG18	Electronic visa details	O
DG19	Travel record details	O

As specifically stated in the ICAO Document 9303 Part I, Volume II, Section III, ICAO based the design of the LDS on the following principles:

- Ensure efficient and optimum facilitation of the rightful holder
- Ensure protection of details recorded in the optional capacity expansion technology
- Allow global interchange of capacity expanded data based on the use of a single LDS common to all MRTDs
- Address the diverse optional capacity expansion needs of issuing states and organizations
- Provide expansion capacity as user needs and available technology evolve
- Support a variety of data protection options
- Support the updating of details by an issuing state or organization, if it so chooses
- Support the addition of details by a receiving state or approved receiving organization while maintaining the authenticity and integrity of data created by the issuing state or organization
- Utilize existing international standards to the maximum extent possible, in particular the emerging international standards for globally interoperable biometrics

Issuing states are granted tremendous flexibility in engaging vendors and selecting platforms and integration technologies when implementing their ePassport, eID, and machine-readable visa solutions. ICAO simply provides guidelines and recommendations.

Governments that are ICAO compliant will find it easier to share necessary data with other nations, while enhancing their own border security. Compliance with ICAO standards is a key consideration for governments embarking on deploying their own ePassport solution. It ensures that the MRTD technology being deployed is in fact interoperable and can be read easily by other nations.

Evolution of the MRTD

In many countries, *electronic passport* simply means a passport with paper-based security features that are read electronically. Many of these paper-based security features are commonly found in passports and currency notes. The emergence of the electronic passport was followed by a collective effort to promote interoperability between the 188 member states of the International Civil Aviation Organization with regards to the machine-readable travel document (MRTD).

Until the emergence of a chip-based passport, the main security features in a passport were the machine-readable zone and later the 2-D bar code. Other integrated security features were based on secure printing, available on the data page and throughout the body of the passport. The data page holds the picture of the identity holder and contains the machine-readable zone and 2-D bar code.

The illustration in Figure 13.2, from the U.S. embassy Web site in Singapore, shows an image of a passport data page with various security features.

The four main types of passport security discussed in this section are paper-based security, machine-readable zones, 2-D bar codes, and chip-based security.

Security printing was the first feature used to protect passports from fraud. Security printing is a highly specialized technique developed for use on documents of value that

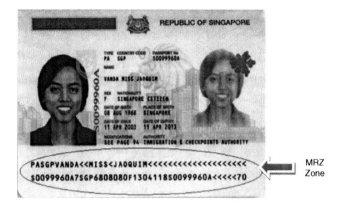

FIGURE 13.2 Data Page
Source: http://singapore.usembassy.gov/uploads/images/SUzKh9a5S1F7APOJFh77fA/ppt2.JPG.

need protection from counterfeiting and illegal reproduction. Advanced security printing features integrate various specialized printing techniques to create documents that are easily and effectively verifiable but in fact totally counterfeit-proof.

Printed security features can range from basic security that is easily identifiable by the naked eye to more sophisticated features that require specialized lights and tools to be able to view images or holographs hidden within the paper. Countries decide which printed security features to integrate into their passports based on cost and design.

A few simple and easily viewable printed security features for passports include the following:

- *Guilloche elements*: These unique, intricate, and very complex patterns are designed by mathematical formula and cannot be replicated by a digital copier due to the size of the lines and the curves. The illustration of the Guilloche design shown in Figure 13.3 can be found on the Web site of Jura JSP Gmbh, a Hungarian passport printer.

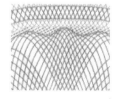

FIGURE 13.3 Guilloche
Source: http://www.jura.at/en/fe_guilloches.htm.

- *Changing security elements*: Varying background print or colors from page to page makes it far more complex to forge documents.

- *Conic laser numbering*: The passport number is perforated into every page of the passport booklet by a laser, making the holes smaller and smaller toward the end of the booklet and giving a brown edge around the holes. Neither of these features can be easily replicated by using a needle, pin, or other such tool to create the perforated passport number look.

- *Optical variable ink (OVI)*: The colors of the image and/or text will change depending on the angle at which they are being viewed.

The Austria card Gmbh's Web site shows the following optical variable ink images on a sample ID (see Figure 13.4):

FIGURE 13.4 Optical Variable Ink Print
Source: http://www.austriacard.at/main/EN/Products/IndustryAndGoverment/SecurityFeatures/index.html.

- *Laminate*: A foil is used to protect the information on the passport's data page from being easily changed. Usually hologram images are part of the security foil, which is almost fused onto the data page through the use of high heat in order to ensure that the data page cannot be easily forged. Any attempt to make changes in the data page will destroy the laminate, making it very evident that someone has tampered with it.

- *Rainbow and intaglio print*: Rainbow print brings together bands of different colors to create appealing patterns, which cannot be duplicated by color copiers or scanners as they separate these colors and do not keep the original color image. Intaglio print is a very specialized printing technique that can be achieved only with highly expensive and specialized equipment that usually is available only to security printers. Counterfeiters using high-end digital scanners and printers cannot create the look and feel of either rainbow or intaglio print. The Intaglio print sample illustrated in Figure 13.5 is from the Lithuanian passport and can be found on the Lithuanian government's Web site, www.vdtat.lt/html_english/lrpassport/4.jpg.

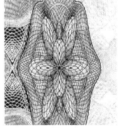

FIGURE 13.5 Intaglio Print
Source: www.vdtat.lt/
html_english/lrpassport/4.jpg.

More complex printed passport security features requiring specialized tools include the following:

- *Microtext*: The text written on the passport is so fine that it appears as a simple line to the naked eye, and even copiers and scanners are not sophisticated enough to pick up these small details. Microtext design samples shown in Figure 13.6 can be found on Hungarian passport printer Jura's Web site, http://www.jura.at/en/fe_microtext.htm.

- *UV print*: Images and text are visible only under UV light. Most of the designs can be very complex and highly counterfeit-proof as laser copiers, scanners, and printers can copy a UV print, making it forgery-proof. The UV print sample of a U.S. visa sticker

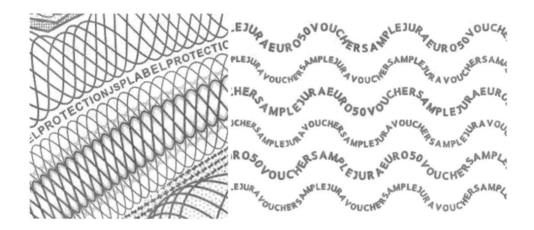

FIGURE 13.6 Microtext Design
Source: http://www.jura.at/en/fe_microtext.htm.

under a black light depicted in Figure 13.7 can be found on Crimesight's Web site, http://www.crimesight.com.au/ images/TrueScan/s7.jpg.

- *Security thread and invisible fibers*: Security thread provides a polymer thread with microtext that can be embedded in passport pages, so it is viewable only under UV light. Invisible fibers can also be embedded in the passport pages, so they become visible in various colors under UV light.

- *Temperature print*: Images and/or text appear or disappear based on the temperature applied to the document.

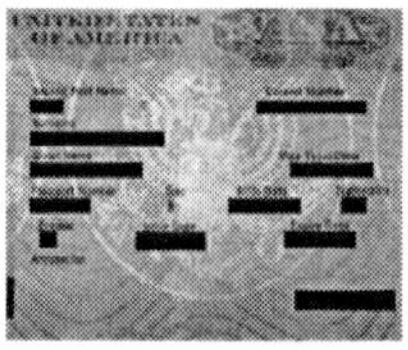

FIGURE 13.7 UV Print
Source: www.crimesight .com.au/images/TrueScan/s7 .jpg.

The aforementioned printed security features are only a few of those being used by high-end, specialized security printers. For example, there can be five or six different types of intaglio print itself with other security features mixed in to make it next to impossible to counterfeit a secure document.

Printed security is a highly specialized industry, and the top security printers in the world ensure that it is almost impossible to replicate or copy the exact security features of the documents they produce. However, hardware companies have made great advances in their scanner and copier technologies in the past few decades, resulting in counterfeiters having the ability to inexpensively forge secure documents. The forgeries are mostly inexact and can easily be detected upon close inspection or with the use of tools such as a black light. The area of concern is that only travelers who raise a security concern or have MRTDs that seem suspicious are brought in for closer inspection.

A sample of how printed security features are brought together on a passport data page is illustrated in Figures 13.8 and 13.9.

An enhancement to secure printing was machine-readable travel documents, which offer a printed code referred to as the *machine-readable zone* (MRZ). The MRZ is available on the data page and can be read by an optical reader to automatically load simple passport details (name, passport number, date of birth, etc.) into an electronic medium such as a computer file.

FIGURE 13.8 Data Page Showing Printed Security

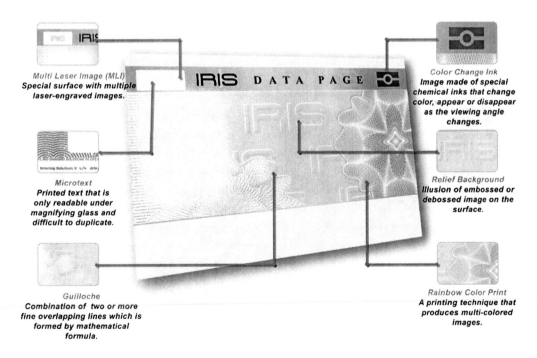

Multi Laser Image (MLI)
Special surface with multiple laser-engraved images.

Color Change Ink
Image made of special chemical inks that change color, appear or disappear as the viewing angle changes.

Microtext
Printed text that is only readable under magnifying glass and difficult to duplicate.

Relief Background
Illusion of embossed or debossed image on the surface.

Guilloche
Combination of two or more fine overlapping lines which is formed by mathematical formula.

Rainbow Color Print
A printing technique that produces multi-colored images.

FIGURE 13.9 Data Page Showing Printed Security

The codes on the MRZ provide automation in data entry but offer little in securing the travel document. Since the algorithm is public knowledge, the codes can quite easily be forged and reproduced as opposed to a chip-based ePassport containing a biometrics and meeting basic PKI requirements established by ICAO.

A further enhancement to the machine-readable zone in a passport was the two-dimensional (2-D) bar code feature. This is another type of printed security that allows a slightly larger amount of information to be stored as compared with the similar machine-readable passport (MRP) concept.

Two-dimensional bar codes are not commonly used in passports but are frequently used on IDs such as driver's licenses or company IDs. The illustration in Figure 13.10 from www.datastrip.com displays how a 2-D bar code is used on a company identification card.

Both MRZ and 2-D bar codes offer no true electronic capability in the passport itself, and as such the security benefits offered are at best similar to those of paper (printed) security. Because the bar code is also generated using a fixed algorithm, knowledge of this algorithm can result in reproduction of the bar code with a forged identity.

ICAO recognized that printed security, machine-readable zones, and 2-D bar codes, while cutting-edge technologies, were not sufficiently securing travel documents to curb illegal immigration or address the threat of terrorism.

Now that technology is available to enhance the security of travel documents to the next level, ICAO has recommended that passports be embedded with a contactless secure microprocessor chip. The United States is at the forefront of this implementation by requiring countries under its visa-waiver program to accommodate ICAO's standards.

The new chip-based electronic passports are not intended to replace secure printing, the MRZ, or even 2-D bar codes; they simply enhance and supplement security measures being put in place. The chip's increased capacity allows for storage of more information, including biometric data: facial scans, fingerprints, and iris scans. It also provides for enhanced electronic security through encrypted data and symmetric keys.

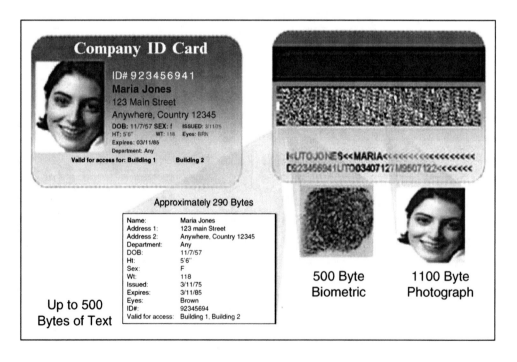

FIGURE 13.10 Two-Dimensional Bar Code
Source: www.datastrip.com/english/images/faq_2d_1.jpg.

Each security measure discussed in this chapter has its own purpose and value. By itself, each measure is just a feature of a passport, but together each measure builds on the other's strengths, securing the MRTD and protecting the identity of the passport holder.

True chip-based electronic passports encapsulating biometric features are now emerging in the marketplace. These ePassports incorporate a silicon chip (integrated circuit) embedded somewhere inside the passport. Communication between the chip inside the passport and an external electronic medium is done via special readers that have a contactless interface. The embedded chip can store a fully digitized photo, fingerprints, a digital signature, and several pages of text data. This is many times more data than is possible in a paper medium such as the MRZ or even the 2-D bar code.

The epitome of the evolutionary cycle was the birth of the first 8KB electronic passport in Malaysia in 1998, which provided comparative advantages over its predecessors such as the paper medium and 2-D bar code in preventing forgery or tampering. This was evident in the Malaysian passport because as to date there have been no instances or cases of chip forgery; in fact, the U.S. Immigration and Naturalization Service (INS), which procured ePassport readers from IRIS, was able to apprehend Chinese nationals using photo-substituted Malaysian passports to cross into the United States. Today, Indusa Global and its partner IRIS Berhad Corporation are implementing electronic passports with 64KB and 72KB chips, which not only are far more secure but also can store significant amounts of data.

The security offered by such an embedded chip is many orders of magnitude higher than paper-based security. The chip cannot be read without properly configured equipment and cannot be easily forged or duplicated. The data can be digitally encrypted and even configured to be nonerasable after encoding to prevent fraud. These vastly superior security features differentiate the true electronic passport from its lesser cousin, the passport with paper-based security features only read electronically and offering little future growth opportunities.

While security experts will debate the advantages and disadvantages of the various security features available in passports today, it is beneficial for countries to incorporate as many of the available features as possible. The chip-based electronic passport is simply a security enhancement to the earlier versions. For instance, the chip-based, biometric passports still contain paper security as well as the machine-readable zone, thus allowing countries to easily authenticate the validity of the identity holder's document on any or all three of the main types of security available.

Electronic Passport and ID Solutions

Many different companies provide electronic passport and ID solutions, and in this section one such solution is described in great detail. In fact, the selected solution was the first ePassport implementation in the world, nearly a decade before most countries even considered issuing chip-based passports.

In March 1998, Malaysia was the first country to offer an ePassport with stored electronic data and biometrics (facial scan and fingerprints) in an integrated circuit (IC) chip. The Malaysian Immigration Department, working closely with IRIS Berhad Corporation, introduced the world's first ePassport based on a contactless chip design.

Today, Indusa Global partners with IRIS to offer ePassport and eID solutions to many clients. Indusa's solutions are fully ICAO compliant with regard to the data page, the MRZ, and the data stored on the chip. These solutions are completely interoperable, providing a medium for security agencies and border control organizations to be able to securely exchange and share data.

The Malaysian ePassport solution contains both machine-readable and chip-based security features similar to the regulations established by ICAO. It is one of the only fully tested, truly proven electronic passport technologies in existence in the world today. With over 8 million ePassports in circulation at present, this ePassport solution has been implemented for almost 10 years in Malaysia and several years in other countries.

The first few generations of the Malaysian ePassport were not ICAO compliant, as the Malaysian Department of Immigration was deploying groundbreaking technology for which ICAO did not yet have regulations. However, there are many similarities between the ICAO-recommended ePassport and the Malaysian solution. In fact, ICAO studied the Malaysian implementation when developing its requirements for the MRTD. Malaysia is currently in the process of migrating its ePassport to be fully compliant with ICAO standards.

The solution offered to governments includes the following:

- Chip-based, machine-readable ePassports for citizens and certificates of identity documents that are fully compliant with ICAO's specification on machine-readable zones (MRZ), as well as electronic passport features incorporating biometric identifiers (fingerprint and facial recognition), for enhanced authentication.
- eIDs including national ID cards, medical ID cards, driver's license cards, work permit cards, residence cards, permit-to-reside cards, home owner's cards, resident spouse cards, seafarer ID cards, social program cards, and any other chip-based multiapplication ID cards. These cards incorporate chip-based and security printing technologies to secure these documents against tampering or forgeries.
- Machine-readable visas that contain paper-based security features and an MRZ.
- Enrollment, data entry, cashier, approval, production, QC, and issuance systems for ePassports, eIDs, and visas.
- Border control, deportation, and apprehension exercise systems.
- Automated fingerprint identification systems (AFIS) are optional based on customer needs and requirements.
- All systems have business intelligence–based reporting and searching capabilities to be able to track activities and find necessary information quickly and efficiently.
- Desktop, mobile, and PDA systems.
- Autogate systems for fast, easy immigration clearance for citizens, residents, and work permit holders.
- Public-key infrastructure (PKI) and key management systems to manage the security of the ePassports and eIDs.
- ICAO-compliant scanners–readers to read MRZ-enabled passports and chip-based ePassports.

The ePassport solution is based on a contactless inlay within the chip that provides a highly secure and portable medium for authentication and identification of the passport holder. It is a proven technological solution and a mature product that has been in operation in Malaysia for the last 10 years (see Figure 13.11).

US Patent ⟶

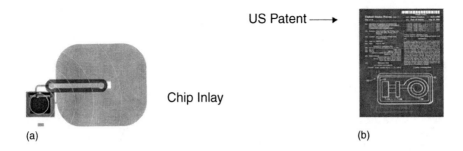

Chip Inlay

(a)   (b)

FIGURE 13.11 Electronic Passport and ID Solutions

The ePassport solution described in this chapter meets ICAO's specification on both machine-readable zones (MRZ) and electronic passport features, incorporating biometric identifiers (fingerprint, iris, or facial recognition) on a 64 KB contactless integrated circuit (IC) chip for enhanced authentication.

The basic architecture for the ePassport, eID, and secure visa solutions is designed to possess the inherent expandability and extendibility to satisfy a government's growing needs without sacrificing security. The design takes into consideration the unique and exacting requirements of secure document issuance and production systems, allowing for the seamless exchange of information between front-end, production, and issuance stations to accommodate the workload.

A typical ePassport, eID, and secure visa implementation architecture is centralized. The logical location to run any system that is meant for a national government is from a national central host (NCH). Individual site connectivity to the NCH should be provided through a secured government local area network (LAN). At a very basic level, the types of servers and subsystems that reside at the NCH include active directory servers, Web servers, database servers, key management system (KMS), public-key infrastructure (PKI) system, operations management servers, backup servers, and automated fingerprint identification system (AFIS; optional).

The illustration in Figure 13.12 depicts a high-level system architecture.

The three main functions of the ePassport, eID, and secure visa solution are as follows:

- *Front office*: Enrollment, issuance, cashier, and photography booth (if required)
- *Back office*: Interview, approvals, criminal or blacklist checks, administration, and reporting
- *Production office*: Data page or visa personalization, chip personalization, lamination, and quality control (Note: Chip personalization happens only for ePassports and eIDs.)

In the front office, all interfaces with the applicants are handled in a smooth, methodical manner. The proposed configuration seeks to reduce the time spent by the applicants throughout the application process and to provide them a prompt and efficient service. Each counter functions in a highly specialized manner to eliminate the risk of data entry errors and minimize the number of equipment handled by each staff member.

All background processes that should be kept confidential are located in a more secure part of the building, called the *back office*. The back office functions can also include external activities such as criminal, birth, or death record checks.

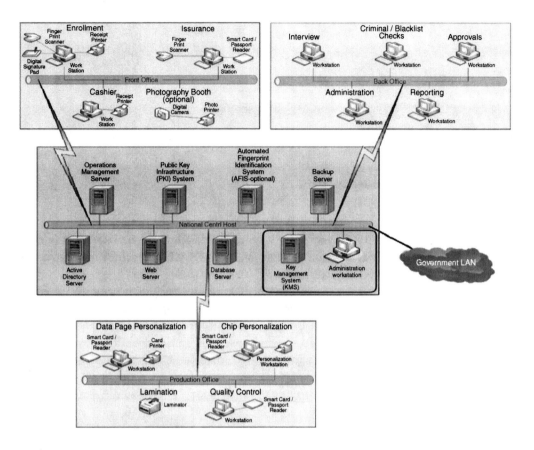

FIGURE 13.12 System Architecture

The production office is the area where the ePassport, eID, and secure visa are created, secured, and passed by quality assurance prior to issuance. This is the most critical of all processes and requires that the location be secured and even the employees follow a strict security policy. For example, no one person should have access to perform all production processes, reducing the risk of fraudulent activities.

Due to the high rates of adoption and incredible success of the ePassport system, the Malaysian government was constantly looking at ways to further improve immigration clearance for its citizens and residents. A few years after the launch of the ePassport in Malaysia, in August 2000, the Malaysian Immigration Department launched Immigration Autogate, an automated immigration clearance and border control system for travelers holding an ePassport. It is a fully automated and secure immigration clearance and border control system for travelers at immigration entry and exit points.

Immigration Autogate is a user-friendly access barrier developed for fast immigration processing of people in airports, seaports, and border crossings. The modular panel concept allows adaptation to a variety of designs and implementations. The design of Immigration Autogate and equipment installed within its modular panel structure is based on customer requirements and costs.

There are currently over 49 Immigration Autogates installed at 11 entry–exit points throughout Malaysia. Implementation of the Immigration Autogate system delivered

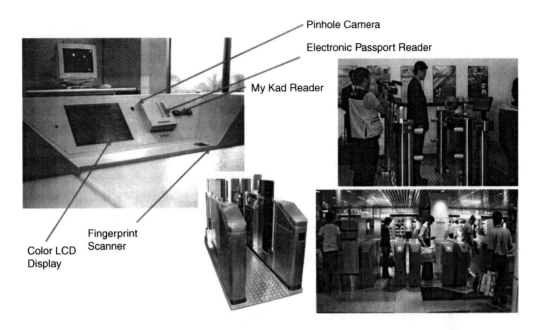

FIGURE 13.13 Immigration Autogates

significant benefits by reducing passport-processing time for Malaysian travelers to an average of 8 to 10 seconds as compared with 45 seconds for manual processing.

The Autogate contains an integrated retrieval system with automatic barriers for access control. Upon successful data retrieval and positive authentication of the electronic passport, the system provides entry to the traveler while creating a record in the central database of the traveler's entry details. In the process of verification, the Autogate system automatically performs a check with the country's stop list and deportation list and raises an alert for any positive results.

Figure 13.13 illustrates first- and second-generation Immigration Autogate systems. A typical process for immigration clearance with an Autogate is as follows:

- The visitor or resident arrives at the border checkpoint by commercial or private plane, cruise ship or private boat, automobile, or on foot.
- Visitors do not get processed by the Immigration Autogate and are sent directly to immigration officers. The immigration officer inspects the MRTD and reads the machine-readable zone and/or the chip.
- Residents and work permit holders are processed by Immigration Autogate. These individuals place their passport on the passport reader or ID in the ID-reader slot.
- The fingerprint and photograph of the traveler are verified with the data on the chip.
- The information is processed within seconds by the central host, and the identity of the traveler is checked against the country's stop list and deportation list.
- If the authenticity of the document is verified, the traveler are cleared and allowed to proceed.
- If the authenticity is not verified an alert is raised in the system, and the traveler is redirected to a service counter for further checks and interviews, where the visitor is either cleared or denied entry.

Integrated with the newer generations of the ePassport solution's being implemented in other countries is the immigration iSearch solution. Because it is vital to quickly and efficiently track and find citizen and visitor information on demand and in real time, iSearch allows governments to do just that—providing countries with the ability to quickly and effortlessly search for a visitor and pull up the image of his or her immigration card. The solution is available via a Web browser through the government intranet. Due to the private and sensitive nature of the data, extensive levels of security, redundancy, and disaster recovery are built into the product.

The iSearch solution provides immigration officials with the ability to search for citizens and visitors based on a variety of parameters, including dates of entry and departure, first and last name, country of citizenship, airline, passport number, and so on. Upon typing in the search criteria, the immigration officer can quickly and easily retrieve all relevant information regarding the visitor in question. He or she can review the visitor's personal and travel details, as well as all other information gathered on an immigration card. The actual immigration card can then be retrieved from the system as a scanned image. Detailed information about where the physical card has been stored is also included in the record details.

Other systems integrated into the ePassport, eID, and secure visa solutions are border control, deportation, and apprehension exercises. All these systems further help national governments to protect their borders as they either capture biometric data or require fingerprint authentication.

A short description of each of the systems follows.

Border Control

The border entry clearance, approval, and tracking systems are installed on each port of entry at each air (for commercial jetliners and private planes), sea (for private boats and cruise ships), and land (for cars, trucks, and by foot) arrival locations. These systems enable immigration officers to quickly and efficiently process visitors, residents, work permit holders, and citizens.

The typical process for border clearance is as follows (and shown in Figure 13.14):

- The traveler arrives at the border checkpoint.
- The immigration officer inspects the MRTD and reads the machine-readable zone and/or the chip. Fields not captured by the MRZ are manually input.
- The fingerprint and photograph of the traveler are verified with the data on the chip.
- The immigration card's bar code is scanned.
- The information is processed within seconds by the central host, and the identity of the traveler is checked against the country's stop list and deportation list.
- If the authenticity of the document is verified, the traveler is cleared and allowed to proceed.
- If the authenticity is not verified, an alert is raised in the system, and the traveler is redirected to a service counter for further checks and interviews, where the visitor is either cleared or denied entry.

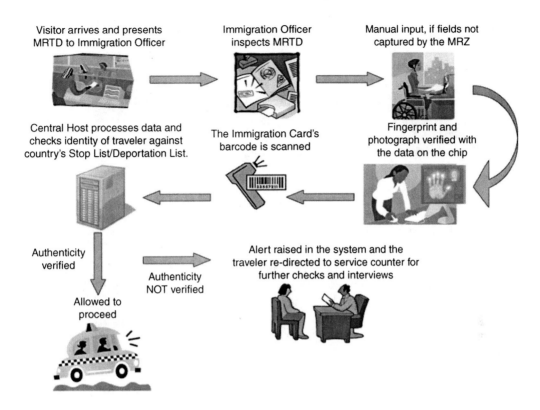

FIGURE 13.14 Border Control Process Flow

Deportation

The deportation system provides immigration departments with the ability to capture biometric information on travelers being deported. Reasons for deportation can include visitors who have overstayed their visa, visitors who have broken the law or been involved in illegal activities, illegal workers not authorized to work in the current country, or captured foreign criminals being extradited.

The overall process for deportation is as follows (and shown in Figure 13.15):

- The traveler is detained by immigration and meets the deportation criteria.
- The immigration officer inspects the MRTD and reads the machine-readable zone and/or the chip. Fields not captured by the MRZ are manually input. This can be a detailed and extensive manual process, as all details about the individual's MRTD data, visa information, contact details, employment details, incarceration records (if required), and travel details back to their home country are entered into the system.
- Documents such as birth certificates, court orders, and other relevant files are scanned into the systems.
- The fingerprint and photograph of the traveler are captured, and the name of the person is added to the deportation list. Once the traveler's name and biometrics are entered into the system, the traveler will be denied entry the next time he or she attempts to enter the country.

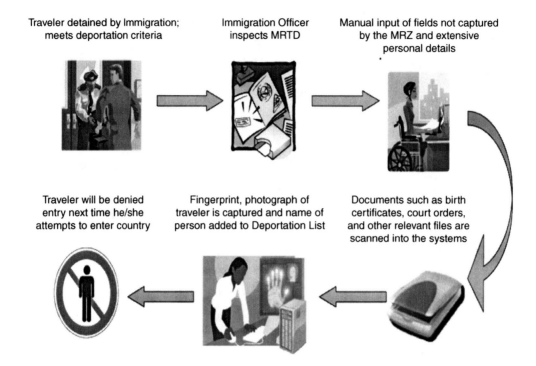

FIGURE 13.15 Deportation Process Flow

Apprehension Exercises

Apprehension exercises are conducted in the field to search for illegal workers, visitors who have overstayed their visas, or other travelers who need to be deported. The apprehension exercise software provides immigration departments with the ability to authenticate individuals being captured and detained prior to deportation. Handheld devices fitted with fingerprint readers, scanners, cameras, chip readers, and MRZ readers are used for apprehension exercises, as they take place in the field and require mobility.

The high-level process for deportation is as follows (and shown in Figure 13.16):

- The traveler or worker is captured by immigration for suspicious activity or not having proper documentation.
- Immigration officers use their handheld devices in the field to inspect the MRTD and read the machine-readable zone and/or the chip of either the passport or work permit. Fields not captured by the MRZ are manually input.
- The fingerprint and photograph of the traveler or work permit holder are verified with the data on the chip.
- The identity of the traveler or work permit holder is checked against the country's stop list and deportation list, held on the handheld device.
- If the individual is properly authenticated, he or she is released.
- If the individual is not properly authenticated, he or she is detained in a holding cell and prepared for deportation.

The chip-based, biometrics-enabled ePassport, eID, and secure visa project with Immigration Autogates, border control, deportation, and apprehension exercise systems

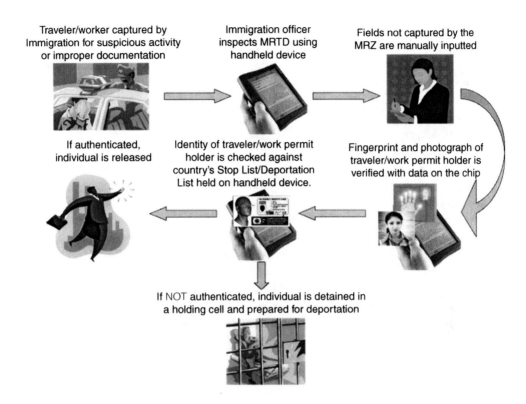

Traveler/worker captured by Immigration for suspicious activity or improper documentation

Immigration officer inspects MRTD using handheld device

Fields not captured by the MRZ are manually inputted

If authenticated, individual is released

Identity of traveler/work permit holder is checked against country's Stop List/Deportation List held on handheld device.

Fingerprint and photograph of traveler/work permit holder is verified with data on the chip

If NOT authenticated, individual is detained in a holding cell and prepared for deportation

FIGURE 13.16 Apprehension Exercise Process Flow

are based on cutting-edge technologies. Truly integrated ePassport solutions such as the one detailed in this chapter help governments improve processes, reduce costs, and incorporate efficiencies into their daily practices.

With the launch and success of the world's first ePassport solution, the Malaysian government became a driving force in developing and testing new chip-based technologies. In 2001, the Malaysian government took on yet another innovative, high-profile eID challenge to launch one of the first fully integrated multiapplication cards. This multipurpose card provides convenient transactions between the Malaysian public, the government, and the private sector through the use of a single smart card called MyKad.

The challenge faced by the Malaysian government was to develop a system that incorporated multiple government and private sector applications into a single card, along with appropriate and sufficient security features. With no known precedent for a card incorporating these requirements anywhere in the world, the implementation team had the formidable task of pioneering the development of all systems, essential procedures, and methodologies.

In an article released in April 2003 by the Center for Digital Government and the Government Technology International Group, the multiapplication smart card was devised by a highly skilled and knowledgeable eID team as a standard credit-card-sized plastic token with an embedded microchip. The initial solution had an ATMEL 32KB EEPROM (Electrically Erasable Programmable Read-Only Memory) chip. The MyKad uses the M-COS operating system (OS), which allows many different applications to be stored separately on the chip and enables different types of information to be saved securely. MyKad also incorporates a sophisticated, fully-tested public-key infrastructure (PKI), ensuring that the data on the chip are stored securely. The PKI enables MyKad holders to feel confident that when they are transacting business in public kiosks or over the Internet, they have a digital certificate ensuring that their identity can be properly authenticated without being stolen.

In order to strengthen security, biometric identification technologies including facial scans and fingerprints are used in the MyKad solution. In addition, its technology is designed with multilayered clone- and tamper-proof security features.

The smart card is equipped to accommodate both symmetric (3DES) and asymmetric (RSA-type) encryption technology. Integrated with a challenge–response type authentication mechanism, this combination of technologies ensures that the smart card is highly secure. The security is contingent only on keeping the respective cryptographic keys secret, and this process (of handling secret keys) is one already extensively deployed in the banking environment. Such a smart card–based system will be beyond the means of counterfeiters to forge and as such will defeat all currently known cloning attempts. A major benefit is the relative ease with which enforcement officers can determine the authenticity of the cards and tie the card to its rightful owner.

eIDs provided by Indusa and its partner IRIS are based on state-of-the-art memory (EEPROM) microprocessor smart cards equipped with a cryptographic coprocessor. The EEPROM is tested to be very high capacity and nonvolatile, so it can facilitate many applications to be loaded into one chip. In addition, the cryptographic coprocessor makes it possible to upload and delete applets to and from the card efficiently and dynamically.

Many types of devices are used to read the information in the MyKad chip. For instance, mobile card devices are used by enforcement agencies to read cardholder information and verify fingerprints. Similarly, Immigration Autogate facilitates checking cardholder information and fingerprint verification for automatic entry and exit at border crossings. Small key ring readers (KRR) are used by citizens and residents to access basic information stored in the chip, while enforcement officers use the larger and more powerful KRRs. Card acceptance devices used by vendors for public and private services include the desktop with credit-card readers.

The goals set for MyKad in Malaysia, as specifically described in detail in the April 2003 article published by the Center for Digital Government and the Government Technology International Group, include the following:

- *Single card for all*: The national identification card and driving license are combined in one card, which also acts as the access key to other services.
- *Efficient exit and re-entry (Autogate feature)*: Instant passport information facilitates efficient exit and re-entry of Malaysians at Malaysian immigration checkpoints.
- *Medical emergency assistance*: Health records are available for instant information during emergencies and general treatment.
- *Easy and flexible payment*: ePurse facilities enable cashless financial transactions. Its reloadable ePurse, branded locally as MEPS-Cash, is accepted at government agencies, restaurants, clinics, bookshops, and petrol stations in designated areas. It also incorporates "Touch 'n Go" functions, allowing the holder to make payment for toll, car parking, bus fares, and LRT charges. It can now also be used as an ATM card for banks providing this facility.

MyKad was conceived as a multiapplication card with five governmental applications, including identification, driver's license, immigration, medical, and ePurse. However, when it first launched in 2001, it served as an identification card, driver's license, and immigration clearance card. Over the years, many other services were made available to Malaysian citizens, who are mandated to have a MyKad, which acts also as a national ID. The idea is that as the MyKad gains in popularity and usage increases, cardholders will be given the option to select and add public- and private-sector functionalities and services to their card as they need or want.

While the Malaysian government selected the aforementioned applications to be part of its multipurpose smart card system, other governments have the ability to chose other applications more important to them and scale up accordingly. This smart card system will harmonize the efficiency of government (or public) service by providing a "one-stop shop" for delivery of government services and programs and preventing the existence of parallel smart card systems sponsored by various government ministries.

Another very successful smart card technology launched by the Malaysian government was the Touch 'n Go card. The Touch 'n Go is an electronic "purse" that can be used on all highways in Malaysia to pay tolls, in all major public transportation vehicles, and at some parking sites and theme parks. Cardholders can reload their ePurse at toll plazas, train stations, automated teller machines (ATMs), cash deposit machines, kiosks at gas stations, and at selected authorized third-party outlets.

The ePurse functionality was key to the success of the MyKad, as it is a seamless vehicle for Touch 'n Go cardholders to pay for government and commercial services. What's more, the ePurse is more than just a highly secure virtual payment and account management solution.

Following is a short list of benefits for government-issued smart cards with an ePurse capability:

- Built on tested PKI infrastructure and technologies, they are able to minimize the instances of fraud and theft and are viewed to be a highly secure payment system.
- They reduce cash-handling costs and essentially replace the use of cash very efficiently.
- They are viewed by customers as much more valuable than any other type of loyalty cards.
- They incorporate efficient and valuable processes such as enabling low-value payments to be paid directly by the card but requiring higher value transactions to be combined with a PIN.

Magnetic strip–based cards such as credit cards are not an ePurse. There are many differences between magnetic strip cards and smart cards with ePurse capabilities, the most significant being that money is loaded into an ePurse card while credit cards do not contain money. Another important difference is that, for a credit card to be brought to market, it requires a huge investment in a highly sophisticated back end, while an ePurse card does not. Basically, for an ePurse card, a handheld terminal and a smart card are the main infrastructure required. In addition, the security in an ePurse smart card is much better than magnetic strip cards, which can be quite easily copied and reproduced by criminals. In Europe and around the world, ePurse concepts are widely spread, and as far as government payment technologies are concerned, the cashless society can start with a government-issued ePurse card.

Overall, the reason why technologies deployed by the Malaysian government are analyzed in great detail throughout this section is because the Malaysian government has been a pioneer and innovator in the public and private sectors with regard to implementing chip-based biometric passports, multiapplication smart card government IDs, and consumer ePurse systems.

Data Security and Privacy

The primary advantage of implementing an ICAO-compliant electronic passport is interoperability and data security; the chip can confirm, without any doubt, the authenticity of the information stored within the passport chip. The data on the chip is "write once only" and should be read only by an authorized reader; however, counterfeiters wanting to forge secure travel documents try to skim data from travelers' passports in order to make fake documents using a legitimate person's identity. This type of activity has brought to the forefront many concerns about data security and, specifically, data privacy.

On August 3, 2006, Lukas Grunwald, a German security consultant and RFID expert working for DN-Systems said that he was able to clone ePassports. He went further to say, "the whole passport design is totally brain damaged. From my point of view all of these RFID passports are a huge waste of money. They're not increasing security at all."

Grunwald said that it took him only 2 weeks to clone the German ePassport, and much of that time was not spent on trying to duplicate the ePassport but rather on familiarizing himself with readily available ePassport standards published by ICAO. In his demonstration, Grunwald quickly created what looked like an empty fake passport and an ID card meeting ICAO standards. He put the fake ID card closer to the reader and showed that the chip closest to the reader would be read rather than the original chip inside the ePassport.

While Grunwald says that he was able to clone the tag, he was not able to change the data on the chip due to the cryptographic technology used in producing the ePassport. He also acknowledges that his plan easily would have been thwarted by immigration officers as soon as they opened and inspected the passport or read the MRZ.

During an interview in the ICAO MRTD report *Stressing Security* (2007), Barry Kefauver, formerly of the U.S. Department of State and currently a consultant chairing the ISO task force on new technologies, states, "We try to let these critics understand where the holes in their arguments are and how false the premises are that they're basing their positions on, but in the end business is business I suppose and their companies' vested interests rely on a certain level of misinformation persisting in the public domain. It's unfortunate for the technology's credibility and it does a tremendous disservice to the many IT, security and cryptographic specialists who took part in the lengthy and very diligent development stages of the ePassport."

Although Grunwald's demonstration is of serious concern to electronic passport providers, governments, and passport holders, there are many security measures being put in place to prevent passport holders who want to commit fraud from being able to clone a passport and enter a country. Such measures are noted in the following list:

- MRZ and chip information do not match.
- The picture on the passport and on the chip do not match.

- Because the chip is "write once only," it would need to be replaced, which would be very evident to the immigration official when inspecting the passport.
- If the chip were not replaced and a new inlay was inserted so that it was closer to the reader than the original inlay, it would be quite evident during the physical inspection of the passport that the passport had been tampered with.
- Changing any data on the data page to match the chip would be evident due to the printed security features and lamination available on passports.

As published in www.wired.com on August 3, 2006, in an article titled "Hackers Clone E-Passports," Frank Moss, deputy assistant secretary of state for passport services at the State Department, says that designers of the ePassport have long known that the chips can be cloned and that other security safeguards in the passport design—such as a digital photograph of the passport holder embedded in the data page—would still prevent someone from using a forged or modified passport to gain entry into the United States and other countries. "What this person has done is neither unexpected nor really all that remarkable," Moss says. "The chip is not in and of itself a silver bullet.... It's an additional means of verifying that the person who is carrying the passport is the person to whom that passport was issued by the relevant government." Moss also says that the United States has no plans to use fully automated inspection systems; therefore, a physical inspection of the passport against the data stored on the RFID chip would catch any discrepancies between the two.

In conducting a public demonstration of cloning an ePassport, one of the main concerns that Grunwald has brought to the forefront is that human interaction during border control is still very important. Many countries are considering, and some have implemented, automated processing for residents and work permit holders. However, most countries still think that the human element is crucial as it's the only true way to inspect whether the physical document was tampered with or not. Technology, although critical for securing borders, is more of an enhancement and safeguard to human inspection and paper security.

Each of the security measures put in place by ICAO is a deterrent for attackers to skim data, eavesdrop, or clone a passport. Most of the electronic travel document technologies are evolving and changing rapidly. Some have been created in the last decade, and some, like extended access control (EAC), are just being tested and debated today. While the framework for first- and second-generation ePassports is provided by ICAO, these are simply recommendations based on existing implementations, live tests, and innovative thinking. Overall, ICAO wants to make it increasingly difficult for unauthorized users to gain access to passport holders' personal data.

Formerly of the U.S. State Department, Barry Kefauver states in an interview for the ICAO MRTD report (2007), "In a real-world sense, where one deals with actual, practical security threats, these are all non-issues. We've implemented standards and recommended practices to preclude skimming and eavesdropping, and in every other credible area we've also taken any and all required measures to ensure the bearer's safety, privacy and security. The newer EAC chips coming out now in Europe would require massive amounts of long-term networked computing in order to break their cryptographic measures, and basically this is an area where security and privacy are going to be vigilantly pursued and expertly and reliably reinforced as every new threat emerges."

Based on ICAO recommendations, there are several ways that passport providers try to ensure high levels of security, from physical security for the passport booklets to electronic security on the chip. Such measures are described as follows:

- *Physical security:* Passport books with chip inlays inserted in them are delivered to the government in a locked mode to ensure that they have not been tampered with during transportation. These booklets include a manifest file to ensure that the books are numbered properly, and all books are accounted for after transportation. Upon arrival, the books are stored by the government in a temperature-controlled, secure vault. A transport code and the unique chip ID are required to unlock the chip for further processing.

- *Material security*: The primary material of the inlay is Teslin, which is selected over materials such as Polycarbonate (PC) and Durasoft because of its superior properties including durability, tamper-evident properties, and ease of manufacturing. Teslin is not easily broken or cracked, compared with many other materials that may claim to be more bendable. Teslin is much more flexible and elastic for the book-making processes compared with polycarbonate.

  The proposed inlay is designed with a proven protection capability to withstand the physical immigration stamping (cache) of passports. Slotted rollers are used as the protection feature against damage during book lamination. Special glue for bonding paper to the coated polyester (PET)-based substrates–inlays provides significant resilience and durability characteristics to the finished electronic passport book covers.

  The chip and inlay are protected against tampering and destruction in order to further ensure passport security.

- *Software Security*: The chip incorporates a privacy feature known as basic access control (BAC). BAC is a mechanism used to ensure that only authorized readers are allowed to access the information in the contactless chip.

  For second-generation passports, the European Union is pushing for extended access control (EAC). While ICAO is making a recommendation on the use of EAC for fingerprints, iris scans, and other personal data, it is leaving the implementation and technology decisions up to each country.

- *Chip security*: The security for electronic passports and IDs is governed by the public-key infrastructure (PKI). Access to the chip is protected by cryptographic keys, and data can be written to the chip only once after successful authentication with these keys. Three different 16-digit keys are generated in order to ensure that no one person or even one agency holds the entire key. A secure Key Management System (KMS) is implemented as part of each project.

  The aforementioned keys are encrypted and stored securely onto special smart cards known as Secure Access Module (SAM) cards, which are located in passport encoders and accessible only to authorized passport production personnel.

  The data on the chip are digitally signed by the government and verifiable only by the passport reader, to ensure that the data on the chip are in fact authentic.

  During the passport production phase, as data are injected onto the chip, a personalization (PERSO) SAM created by the KMS must be used to perform mutual authentication with the passport chip in order to access the Create

File, Update Binary, and other file creation and write commands. After the chip is personalized and the data are checked for correctness, it is locked against future modification. It is paramount that data can be entered into the chip only by an authorized party and remain secure against unauthorized modification.

A PKI requires four basic components to exist and operate:

- ○ A certification authority (CA) responsible for issuance of public-key certificates to the community of interest
- ○ A registration authority (RA) responsible for collecting and verifying the required identity information needed for the CA to bind to a public key in an issued certificate
- ○ A certificate repository, generally implemented as a directory service, to retain certificates and revocation lists that are made available for search and retrieval by the community of interest
- ○ A subscriber to be named as the subject in a public-key certificate and in control of the associated private key

Data privacy is a very serious issue for countries implementing electronic travel documents. All countries want to protect the private data of their citizens from unauthorized users. Therefore, ICAO, working with various countries and ePassport providers, is making security recommendations for easily readable but tamper- and forgery-proof travel documents. ICAO-recommended security measures include the following:

- *Passive authentication (mandatory)*: Data on a passport chip are digitally signed by the issuing country. These data can be read by an authorized reader to check for the authenticity of the travel document.
- *Basic access control*: As described earlier, BAC protects against eavesdropping and skimming by ensuring that the communication between the passport and the reader is encrypted. BAC forces the involvement of the passport holder, as the passport must be given to an authorized reader to be able to read the key, allowing access to the contactless chip. It uses the passport holder's data on the physical data page to derive a set of access keys in order to unlock the data on the chip as well as create a secure encrypted communication channel between the reader and the chip. The reading of the data from the chip is then done through this secure channel. This effectively protects the passport data against both skimming from an intruder with a reading device in the vicinity of the passport holder who is trying to read the person's personal passport information, and from eavesdropping by an intruder with a reading device in the vicinity who is trying to intercept data transferred between the passport and a valid reader being used by an authorized immigration officer.
- *Active authentication*: Data on a passport chip essentially cannot be copied. While a public key is stored in the readable data on the chip, the corresponding private key is locked, stored on the chip, and cannot be read.
- *Extended access control*: EAC builds on the existing technology of BAC and chip authentication (same function as active authentication) by adding terminal authentication. BAC prevents skimming and eavesdropping by obtaining and creating an encrypted symmetric key by reading the optical data on the machine-readable zone (MRZ). Chip authentication prevents the ePassport from

being cloned by verifying the chip is genuine and replacing the encryption key with a totally random key. EAC does not replace either BAC or chip authentication but rather builds on them, so that in the future, ePassport authentication at border crossings will sequentially follow all three steps.

At a high level, the terminal authentication step of EAC not only will consider the keys generated by the issuing country but also will work with keys created by the requesting and issuing countries. When a visitor arrives at a country's border crossing, immigration officials must request electronic authorization from the home country in order to be able to gain access to the visitor's fingerprints and other personal data. Visa-waiver countries, preclearance countries, and other friendly countries will have mutual agreements in place to provide information to each other. Once the key agreement is in place, both the issuing country and the requesting country will generate the same unique, private, secret key. Thus, at border crossings, when an ePassport is presented for entry, it will be authenticated using the secret key to establish a secure communication channel to prove its status as a valid or counterfeit travel document.

Security measures such as the ones previously discussed are put in place to make it time consuming, costly, impractical, and almost impossible for unauthorized users to gain access to a person's identity and use it to commit fraud or, worse, terrorist activities. No technology is foolproof, but the goal is to implement good, secure technology in order to deter identity theft. In the ICAO MRTD report published in 2007, *Stressing Security,* ICAO specifically states that, "an attacker trying to copy and clone an ePassport faces the problem of computing the microprocessor's private key given the public elements (which can always be obtained freely). Carrying out this task is commonly referred to as the Discrete Logarithm problem and requires massive computational resources even for practical key sizes." The report goes on further to describe, "A brute-force attack, where the attacker gathers as much computational power as possible and implements the fastest known discrete-log extraction algorithm (currently GNFS) would typically require 273 (respectively 2103) operations for a 1024-bit (resp. a 2048-bit) DH public key, and 2128 operations for a 256-bit ECDH public key. This represents several decades of unceasing computations over a large-scale computer network and by far exceeds the limits of practicality."

Summary: National Security Best Practices

While there is an abundance of reading materials, media coverage, and ongoing debates—pro and con—concerning MRTDs, chip-based ePassports and eIDs with biometrics are relatively new and groundbreaking technologies. Apart from Malaysia, which implemented its ePassport in 1998, most U.S. visa-waiver program countries have deployed only their first-generation eMRTD with a single biometric feature in the past 2 years. The non–visa-waiver countries are simply trying to play catch up with minimum ICAO requirements for passports: machine-readable zones by 2009 and integrated chips technology by 2010. Therefore, national security best practices for eMRTDs are now starting to be developed.

The illustration in Figure 13.17 provides highlights and an overall summary for this chapter.

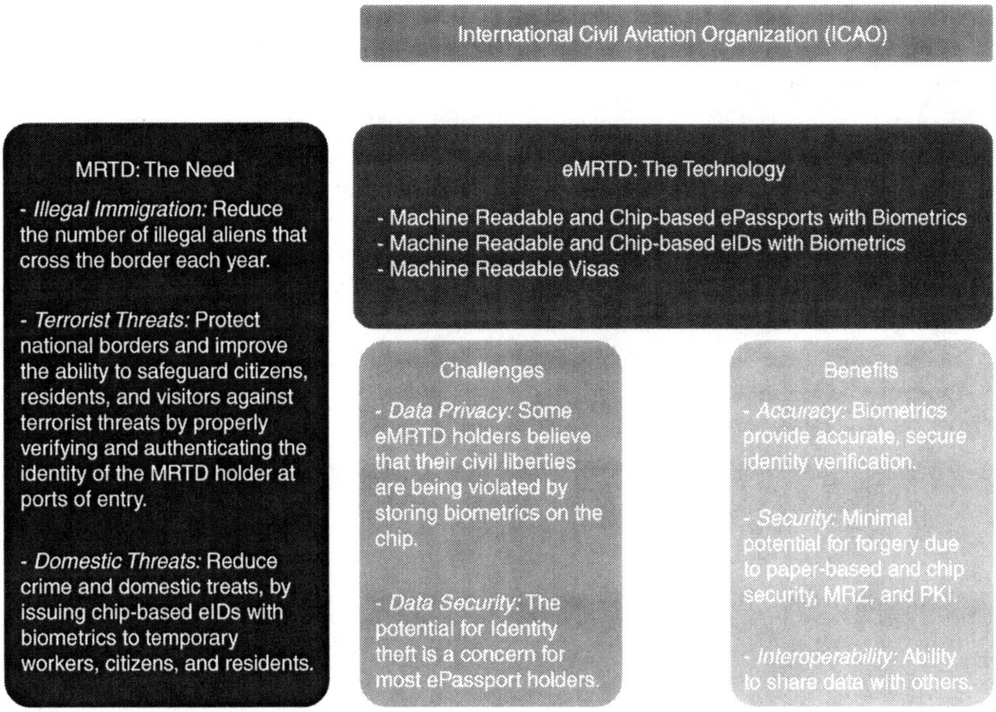

FIGURE 13.17 eMRTD Highlights

No technology is accurate, reliable, or tamper-proof 100% of the time. However, as an integrated group, each security measure works together to reduce the potential of internal and external terrorist threats and illegal immigration while promoting the need for public safety. Government leaders and lawmakers are optimistic that virtual fences on borders, secure MRTDs, properly trained TSA staff, surveillance of boats in U.S. waters, scanning of cargo, and other such security measures will collectively work together to improve overall security for our nation's citizens and residents. The overall goal is to face security challenges with a flexible but robust security plan.

References

American Civil Liberties Union (ACLU). ACLU slams draft DHS regulations on real ID, says delay fails to address privacy and civil liberties concerns. Accessed March 1, 2007, at: http://www.aclu.org/safefree/general/28755prs20070301.html.

Center for Digital Government and Government Technology International. MyKad: The Malaysian government multipurpose card. Accessed April 2003 at: http://www.centerdigitalgov.com/international/story.php?docid=49229.

GOPUSA. More Americans killed by illegal aliens than Iraq war, study says. Accessed February 22, 2007, at: http://www.gopusa.com/news/2007/february/0222_illegals_report.shtml.

International Civil Aviation Organization (ICAO). Document 9303: Machine readable travel document; Part 1: Machine readable passports; Vol. 2: Specifications for electronically enabled passports with biometric identification capability, 6th ed., 2006.

International Civil Aviation Organization (ICAO). MRTD report: *Stressing security*, Vol. 2, No. 2: Gemalto White Paper (Moving to the second generation electronic passports: Fingerprint biometrics for enhanced security and privacy), 2007.

International Civil Aviation Organization (ICAO). MRTD report: *Stressing security*, Vol. 2, No. 2: Cover story, Barry Kefauver (ePassports: The secure solution), 2007.

International Civil Aviation Organization (ICAO). Web site: http://www.icao.int/icao/en/strategic_objectives.htm.

Malaysian Government. Touch 'n go card. Accessed at: http://www.touchngo.com.my/WhatTNG.html.

Social Science Research Council (SSRC). Border battles: The US immigration debates (Immigration reforms and border security technologies), Rey Koslowski. Accessed July 31, 2006, at: http://borderbattles.ssrc.org/Koslowski/.

Wired. Hacker's clone e-Passports. Accessed August 3, 2006, at: http://www.wired.com/science/discoveries/news/2006/08/71521?currentPage=all.

14

Fast Integrated Response Systems Technology (FIRST) and Establishing a Global Center for Preparedness (GCP)

Clifford R. Bragdon, Ph.D., AICP, FASA

Objectives of This Chapter:

- Identify a security-based global model
- Show how it will enhance the world's social and economic well-being
- Describe the sector role players
- Describe the primary global concerns: natural disasters, man-made disasters, and infrastructure
- Depict a management life cycle for these concerns
- Outline primary elements in the management of threats
- Describe the creation and missions of a Global Center for Preparedness, along with the missions it will carry out
- Highlight FIRST as a means of implementation

> *"We must live by an ideal that satisfies the exigencies of civilized existence and not merely those of material subsistence."*
> —Irvin Lazzlo, author of *Essential Society*

Overview

It is generally agreed upon that natural and man-made disasters collectively are a threat to global economic and social well-being. The combined cost is equal to 5% of the world's gross domestic product, and annually it amounts to $2.5–$4.0 trillion. As noted previously, the total direct and indirect costs associated with the attack on the World

Trade Center Twin Towers have now been calculated to be $2 trillion. Hurricane Katrina is classified as the costliest natural disaster in U.S. history at $86 billion.

Transportation involving physical and electronic movement is essential for the world's economy to succeed. The movement of people, goods, and information involves a logistically based transportation network. The institutional approach to protect this network has been unacceptable for a variety of reasons and is characterized by provincialism, self-interest, stovepipe thinking, under-representation, nonintegrated perspective, technological naiveté, improper management structure, inadequate or unreliable commitments, and the lack of a consensus-based road map.

Efforts today have fallen short, whether they are a turf war in the battle against terrorism (e.g., the New York City Police Department versus the FBI), confusion as to who is in charge (e.g., military, FEMA, Corps of Engineers, Coast Guard, TSA, state, or city in the aftermath of Hurricane Katrina), inconsistencies in the allocation of resources (Congress and the executive branch), political opinion (Democrat versus Republican), or general administration and management responsiveness. The situation is now reaching significant litigation thresholds in both categories of disasters (e.g., World Trade Center 9/11–related, and the New Orleans Katrina victims). People unhappy with delays, proposed settlements, and bureaucracy are now aggressively initiating lawsuits in frustration. All presidential candidates, despite their party affiliation, are now indicating that Katrina was mismanaged by the Bush administration.

A security-based global model needs to be launched so a comprehensive, consensus-based resolution can be achieved. Figure 14.1 depicts this framework based on several interrelated principles:

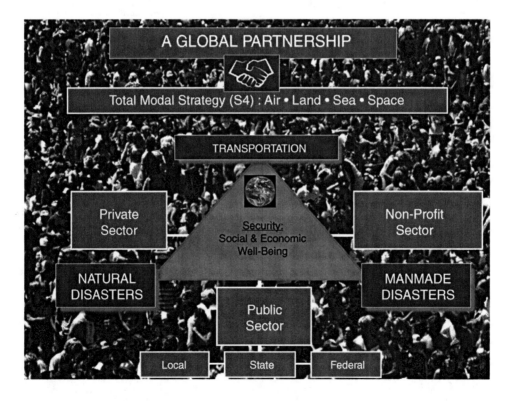

FIGURE 14.1 Global Partnership for Security

1. A global partnership is necessary since these disasters are occurring on all continents (e.g., reported terrorism has been identified as a problem in over 40 countries).

2. These disasters often defy geopolitical boundaries (e.g., the 2004 tsunami impacted three continents and 12 countries).

3. The magnitude and frequency of impact are increasing; therefore, any delay in responsiveness is unacceptable.

4. It is necessary to expand the total modal strategy to now include air, land, sea, and space due to changes in potential settlement patterns.

5. Primary global concerns should address natural disasters, man-made disasters, and infrastructure as a trio of interdependent issues, not separately or in isolation.

6. Security sector role players need to include the private sector, public sector, and nonprofit sector in a partnership network using an effective system of coordination.

Global Partnership Model: Issues

There is a concerted effort by many organizations, political scientists, government officials, and academicians to give emphasis to one particular area of interest that we should focus on, at the exclusion of others. For example, some people say that natural disasters should be the highest priority due to the number of fatalities and injuries inflicted upon the nations in the world, especially in comparison with risk. Steve Flynn (2007), Senior Fellow for National Security Studies, argues that since 90% of Americans live within a moderate- or high-risk area, natural disasters should rank as our number-one concern. He speaks of "tiers of vulnerability" with natural disasters first, followed secondly by infrastructure, and thirdly by national security terrorism threats.

However, a tiered approach breeds isolationism and separatism, which can divert efforts from seeing the bigger integrated picture. It also plays into the convenient "specialist expert model," where a generalist is considered unneeded, letting the experts deal with each of these subjects separately with their expertise. What is more beneficial is a comprehensive approach where both ideas and resources are shared, thereby ending this generalist–specialist debate. Sharing enhances education and the understanding of other positions, including the development of collaborative opinions. Often this leads to a synergistic solution where all parties benefit by cooperating.

Transportation interests, regardless of the mode, have primarily been concerned about responding to travel demand and bricks-and-mortar issues. In this instance, the planning, design, and construction have focused principally on the physical build-out (e.g., highway improvement, tunnel or bridge expansion, airport runway or terminal improvement, track extension, etc.). Issues about security or even energy efficiency are project subsets but not necessarily considered in the overall developmental scheme of things. Similarly, security issues have dealt with a mode-by-mode approach (i.e., seaport, airport, transit station) in isolation, even when these facilities physically come together at an intermodal point.

The Department of Homeland Security (DHS) frequently emphasizes hardening sites at risk, using an engineering solution, rather than an architectural concept design with an engineering component. Security sign-offs should not be the only level of interaction between the project architect, planner, engineer, and security team. Strategic security design infrastructure integration is the preferred protocol, working with an interdisciplinary planning, design, and engineering team. This holistic interdisciplinary approach is always a more beneficial one than examining just component parts.

There appear to be instances where intermodal security initiatives have been utilized during specific time periods or events supported by the Department of Homeland Security and implemented through the Transportation Security Administration (TSA). For example, DHS has initiated the Visible Intermodal Prevention and Response Team programs that are present but on a random basis at ports. Over the past 2½ years, TSA has deployed such teams on a coast-to-coast basis, utilized on approximately 200 separate occasions. These programs are specifically used for special events, or when there is a large crowd associated with a transportation system (e.g., multiple cruise ship arrivals and/or departures at a port). Port Canaveral, now ranked as the second largest cruise ship port in the United States, will be receiving greater attention by TSA, coordinated through the TSA Orlando office. This initiative will be composed of federal agencies (FBI, Coast Guard, U.S. Customs officers), the Florida Department of Law Enforcement (FDLE), and the Brevard County Sheriff's Department.

At other transportation sites, dogs are continuously used. The question of consistency and emphasis has always been one of the security surveillance issues for intermodal initiatives. Major hub airports, particularly those involving international travel (e.g., John F. Kennedy International Airport) utilize dog surveillance teams continuously, which raises the question of federal dollars expended for airports versus other transportation modes (i.e., ports, transit, utilities, and rail facilities). As previously mentioned, U.S. airports receive over 70% of all the Department of Homeland Security dollars budgeted for modal security, in contrast to all other modes.

Electronic communication and surveillance are rapidly becoming the substitute for physical movement whenever possible. For example, distance-based commuting, or telecommuting, now impacts approximately 20% of the workforce, in preference to the use of a personal conveyance for traveling from home to work on a daily basis. Today nearly 50% of the adult population desiring to pursue a degree (i.e., bachelor or master's degree) or some type of professional continuing education (i.e., certification program) obtain it online, or electronically. Information technology (IT) is a major part of tracking and command and control systems, which also include logistics and supply-chain management. The fusion of information for integrating data electronically is now the end goal for any security-based system. A white paper written on this subject by John Douglas (1996; formerly Chief of Staff, Her Majesty's Service, Northern Ireland) has had a significant impact on the data fusion process in security and military information–related matters. It has been most effectively applied by the military, as discussed in Chapter 10.

Active camera surveillance systems for transport are a standard procedure, and the level of sophistication in terms of resolution, data processing and analysis, and multitasking, as well as integrated communication linkages, is steadily improving. State-of-the-art camera surveillance technology can now sense biochemical particles, radiation, facial recognition, and detailed facial response (e.g., nervous twitching, among other physical characteristics). Camera sentry monitoring on ships is now being employed on military ships, due to their surveillance accuracy and resolution over great distances.

The size of these integrated networks can be significant. The largest surveillance system in the world operates in London and contains 1,200,000 cameras. They are used for both underground and surfaced-based transportation surveillance, including the travel demand system for entering the city of London by motorcar. Frequently, these camera systems used for security monitoring are outsourced due to their size and complexity. In Moscow one system contains a network of 37,000 cameras and is operated by a company based in the United States.

Air, land, sea, and now space venues represent total modal strategies (S^4) and must be included in safety and security inquiries on an integrated basis. This is because all methods of transport can be used for terrorism purposes (e.g., rental van and aircraft attacks on the Port Authority of New York and New Jersey Twin Towers in 1993 and 2001), as well as transportation modes themselves being recipients of terrorist attacks (e.g., Madrid, Spain, attack on both rail transit and buses).

Sector Role Players

It is absolutely essential to have all sectors represented in this partnership, including public representation linked to nonprofit organizations and the corporate sector. A consortium of all three groups interacting is very favorable, especially since the private sector business and commerce groups are typically left out of this process. Steve Flynn calls this deficiency a neglected defense against disaster management. He suggests several weaknesses by the U.S. Department of Homeland Security in dealing with business that need to be overcome. They include, among others: lack of information sharing, federal reorganization since 9/11 raising the difficulty and transaction costs for the private sector to work with federal government, adequate voluntary efforts by business going unrecognized, private sector assets not fully integrated into DHS-supported exercises, and failure to target tax incentives to promote investments in security and resiliency in the highest-risk industries. DHS now recognizes this deficiency, but time will tell if needed changes are remedied.

Businesses can be very responsive in terms of equipment, supplies, expertise, logistical assistance, and staff resources. This has been effectively demonstrated by many corporations in terms of their active participation in emergency response and recovery situations throughout the United States. A good example of this kind of involvement is the Business Roundtable, a company that represents 10 million employees and $4 trillion in assets nationally and serves disaster areas in need.

Coordination is a critical management ingredient that is often missing among the three sector role players. The anarchy of giving has to be carefully monitored and avoided also. National contributions from businesses need to be distributed through their local offices, since they are closer to the local needs and community-based requirements. Under Florida Governor Jeb Bush's administration, legislation was passed requiring a certain category of business (e.g., fuel oil, service stations, building supply stores) to install auxiliary generator sets, maintain business hours, and provide supplies and equipment needed in recovery areas. There still remains a certain level of hesitancy between businesses and government that needs addressing. In part it may be due to security issues and the lack of familiarity with protocols and procedures, besides general government red tape. All this can be overcome with strategic planning, including cooperative agreements, and multisector training in advance of any disaster.

There appear to be increasing numbers of organizations offering to assist that have integrated capabilities related to security in the business sector. This is a new breed of one-stop security company. One such organization is referred to as the PGC Consortium. Consisting of more than 40 interrelated companies, it offers agility in stand-alone security products and services from top companies, including Best of Breed, in each area of the security life cycle, as well as custom Intelligent Security Integrated Systems (ISISs).

ISIS leverages the latest, most capable and appropriate blend of technologies, environment-specific knowledge, and security expertise to all the available platforms. The PGC Consortium specializes in advanced IT and physical security sensors, products, and solutions, including Internet-based crisis management solution suites. These suites are applicable to any type of facility or environment, such as educational institutions, health-care facilities, airports, seaports, embassies, government and corporate environments, energy facilities, utilities, and borders, as well as specialty applications desired by large integrators. Their scalable solutions provide a range of "secure" levels for the most demanding and harsh environments, including advanced detection, surveillance, and tracking technologies with proprietary Intelligent Security Integrated Systems. They are setting the benchmark worldwide in facility protection capability, damage mitigation, and crisis resolution solutions, including education and training.

Companies such as Raytheon, EMC, Pearson Learning Solutions- E-Learning, among others, are merging technological capabilities with an educational engine, the Global Center for Preparedness (GCP) and Phoenix Genesis Corporation (PGC). The Global Center for Preparedness, led by Florida Institute of Technology, is now providing integrated training and educational solutions and certificates working with business and government worldwide.

In the nonprofit and trade association category, there are professional societies and organizations that relate to infrastructure, transportation, and security. The National Academy of Science's Transportation Research Board (TRB) is an independent nonprofit organization established in 1920 that focuses on transportation research involving members from government, industry, and academia. Its membership exceeds 10,000, and it hosts an annual conference in Washington, D.C., attended by representatives responsible for policy, practitioners, educators, and businesses. Other organizations that fall into this general category include the Intelligent Transportation Society of America (ITS), the Institute of Transportation Engineers (ITE), and the Intermodal Association of North America (IANA). Additionally, within each transportation mode (i.e., airports, seaports, transit, rail, etc.) there are trade associations that have an interest in transportation security as it relates to their mode, such as the American Public Transit Association.

SOLE - The International Society of Logistics—is a nonprofit international professional society composed of individuals organized to enhance the art and science of logistics technology, education, and management. Logistics is a critical element in operating a seamless system of movement in support of the economy and national preparedness. SOLE - The International Society of Logistics can play an important role. All these groups need to be collaboratively linked to this global partnership model.

The focus up to now has been on transportation and logistics–related professional societies and trade associations; however, there are others that can play an important role. By instituting such sustainable initiatives in our infrastructure, there will be a reduced dependency on imported natural resources. If we examine the infrastructure global concern, there are organizations that address sustainable building performance. Historically, our building codes, site design requirements, and zoning regulations have done little to address energy conservation or, more importantly, the larger issue of sustainability. Green buildings and green cities have been popular marketing terms, but until recently this has not been backed up with substantive energy-based standards.

The Leadership in Energy and Environmental Design (LEED), developed by the U.S. Green Building Council (USGBC), provides a suite of standards addressing environmentally sustainable construction. Begun in 1998, LEED has been involved in over

14,000 building projects throughout the United States and over 30 countries. LEED's membership includes other nonprofits, government, architects, planners, engineers, developers, builders, and product manufacturers. Studies suggest up-front energy investments averaging an additional 2% will yield 10 times the investment (20%+), over a building's life-cycle operating costs. In the area of LEED certification for new construction and major renovations, a numerical point system has been established at four levels. This certification ranges from basic Certified, to Silver, Gold, and Platinum levels. New York City's first green-certified office tower award was presented to 7 World Trade Center on May 26, 2006 (i.e., gold status certification category, 39–51 points).

Another nonprofit that has addressed this energy conservation subject in detail has been the Association of Energy Engineers (AEE). Established in 1977, this nonprofit professional society today has 8,500 members located in 77 countries. Its mission is to promote scientific and educational interests in those engaged in the energy industry and to foster action for sustainable development. The focus of AEE has been on increasing energy efficiency, utilization of innovative energy service options, enhancing environmental management programs, upgrading facility operations, and improving equipment performance.

There are many other sector role players that are important affiliates in assisting with natural, man-made, and infrastructure subject areas. What is important to emphasize is that there are evolving interest groups that could become key players due to their technology, expertise, or innovation. A continuous surveillance of these potential partners is a step toward bringing them into the process. Their eventual entry is through the public, private, or nonprofit portals.

The last organizational group that can contribute to this security-based global model is an informed public that gets involved in some cause, stimulated by a constructive movement or societal concern. This has recently been demonstrated by a social advocate group called Earth Hour. Supported in part by the World Wildlife Foundation (WWF), this organization initiated a international campaign, for the second year, to turn back candlepower (i.e., nonessential lights and electronic goods) for 60 minutes. It occurred on March 29, 2008. The group's purpose in turning off the lights was to make a statement, to help find new ways to reduce human impact on the environment, and to start a movement that ends with a solution to the common challenge we all face: climate change.

Earth Hour 2008 was a notable global event that mobilized over 100 million people from five continents. These countries ranged from Albania to Zimbabwe, Bosnia to Uzbekistan, Canada to Uruguay. In the United States, over 40 metropolitan areas participated along with states. Reports indicated that social boundaries were broken because people from so many backgrounds, cultures, and geographies came together to press for change. Earth Hour's Web site (2008) remarked, "Never before have governments, NGOs, businesses and average people called upon each other and the world to find a new direction." An inspiring movement, if it can be focused, could bring about a reinforcing purpose to all the role players in this area of global safety and security for social and economic well-being. A 2009 campaign focus on March 29 could bring about an important new coalition incorporating the interests of safety, and security and sustainability internationally.

Disaster Management Life Cycle

Regardless of the security issue related to a natural or man-made disaster and the infrastructure consequences, there is a life cycle involved (see Figure 14.2). It consists of eight interrelated steps that many agencies and organizations often do not consider in totality.

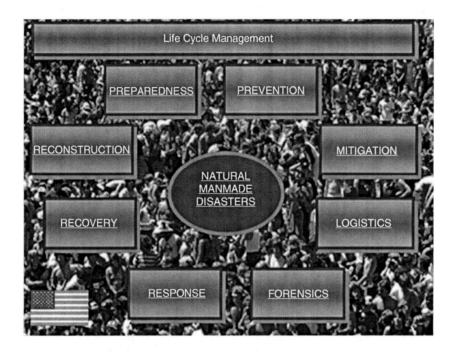

FIGURE 14.2 Life Cycle Management

Step one is preparedness. It is the philosophy that, whatever the impending incident, you are prepared to address it. Step two is the assumption of prevention, or what can be done to prevent an act from occurring. This can range from an accidental or intentional act (e.g., lightning triggering a fire, or an arsonist who purposely sets a fire that might be terrorist motivated). Eco-terrorism is beginning to appear as an act of protest in wealthier countries and may grow in both frequency and magnitude of impact. Step three focuses on mitigation: which among the abatement strategies, or what combination thereof, can be effective? Here we are recognizing that the event will potentially occur but are focused on what can be done to minimize the level of impact. Step four is having a logistics plan in place for the mobility of people, goods, and information that can be instituted in a supply-chain delivery system. Step five is often the most overlooked area, forensics. When there is a problem or dysfunction, what is the cause and how can it be overcome next time? It is learning from your experiences, both successes and mistakes. Step six is the method of response to a disaster and your comprehensive and strategic approach. An integrated response has been one of the greatest weaknesses in the security management process, as evidenced by the lingering problems with the Katrina hurricane nearly 3 years later. Step seven deals with recovery, which administratively is a major transition in handling a crisis. Initial responders have now left the scene and you are dealing with a different group of individuals, who may not possess the history and understanding of the client base they are trying to serve. The final step (Step eight) is reconstruction, which can be very difficult. Many times there is a polarity between those who want to rebuild on the same footprint and others who are advocating a start-over process, with new construction. Each of these options has merit and must be decided in a conflict resolution professional manner based on facts, not opinions.

Today there are still 38,000 families remaining in FEMA-provided travel trailer and mobile homes along the Mississippi Gulf Coast, due to the ravages of hurricanes Katrina and Rita that occurred in 2005. These dwelling units were recently found to be toxic because of potentially hazardous levels of formaldehyde gas, according to government tests conducted by the Centers for Disease Control and Prevention (CDC) in Atlanta, GA. Hazardous levels ranged from 5 to 40 times higher than those levels found in a typical home.

Formaldehyde is used in a variety of products, including the composite wood and plywood panels installed in these trailer and mobile homes purchased by FEMA. It now appears that the FEMA governmental purchases of travel trailers and mobile homes has involved formaldehyde health issues beyond those units deployed to assist those people in Mississippi and Louisiana. The geographical problem area appears to be even larger than those trailers FEMA used after Hurricane Katrina. More recent information now suggests those units in Arkansas, among other states, could be contaminated as well. In addition to causing respiratory ailments, formaldehyde is considered a human carcinogen.

This chemical health urgency announced in February 2008 has sped up the concerns and the process. People still remain in these FEMA-supported temporary housing units containing formaldehyde. However, strong differences of opinion still remain in certain neighborhood areas of New Orleans. The ultimate neighborhood community decision should be based on the level of potential risk of a hurricane reoccurring, the preventive design to mitigate any future damage (based on a category-5 hurricane), and the costs that may be involved if a hurricane of this magnitude ever repeats itself.

What many people forget is that this financial burden is being shared by all taxpayers. Consequently there are limited dollars to be used on behalf of all taxpayers in the United States. They should be used only once, unlike many previous instances where the federal government has come back again and again (e.g., government guaranteed flood insurance programs in high risk residential areas are continuously used as a financial bail out, despite their flood prone location). Our financial resources are finite and should be conserved.

Disaster Prevention Planning and Management

In terms of threat analysis, disaster prevention planning and management's three basic principles (see Figure 14.3) should be addressed:

1. Space is three-dimensional.
2. Time is a 7/24 phenomenon.
3. Assets include both population and infrastructure.

Urban space is finite, but it consists of three dimensions: aerial, surface, and subsurface (see Figure 14.4). However, we use only two-dimensional surface or "land use" planning principles in the design of our urban habitat. Land use is a spatially flat, passé term, since it treats property as a two-dimensional (2-D) surface. This myopic approach is referred to as "Titanic planning" (see Figure 14.5), since nothing below or above the surface is generally considered. The concept is antiquated and out of date, because we have air rights as well as surface, subsurface, water, and land development opportunities. *Space* (a 3-D concept) is a preferred term to *land* (a 2-D concept), consequently we should adopt a process of space use, not land use, planning, and design.

Subterranean zoning, subsurface site development plans, "earth scrapers," utilidoors, and underground cities, as well as rooftop utilization and water rights development, are

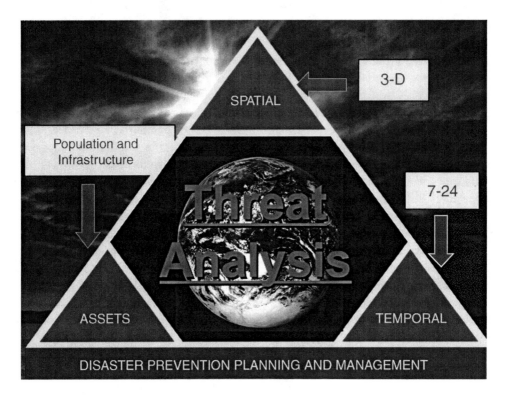

FIGURE 14.3 Disaster Prevention Planning and Management

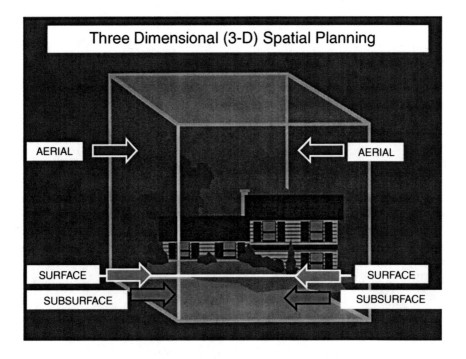

FIGURE 14.4 Three-Dimensional (3-D) Spatial Planning

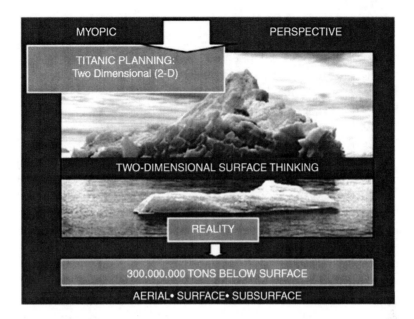

FIGURE 14.5 Two-Dimensional (2-D) Titanic Planning

principles that have disaster prevention opportunities. These are examples of the 3-D aerial and subsurface space use principles applied to planning that have security implications. The North American Air Defense Command (NORAD) was built under the Cheyenne Mountain in Colorado to sustain any possible attack (e.g., nuclear). There are many other command and control centers and bunkers constructed below the surface for safety and security purposes. Walt Disney, in his original plans for Disney World, recognized this subterranean potential. A major portion of these facilities is totally underground to handle certain administrative, staff, and accessibility issues within the Disney World, Orlando, Florida complex. Administrative offices, employee facilities, banking, and security centers are all below ground, with vertical entry points into amusement areas for only the Disney employees.

Developers in Dubai, UAE, are building an entire underwater hotel resort, and they continue to develop Palm Jumeirah, a three-palm-shaped island complex covering 12 square miles in the Persian Gulf. The ultimate water-based build-out will house 60,000 residents. The Japanese government planned a heliport 500 system to facilitate passenger movement by air, including 150 water based heli-ports, with aquatic-based farming below their landing platforms. They have also planned an "earth scrapper" in Japan, totally underground for document storage and studios for recording music. This spatial theme appears to be catching on with the public, as evidenced by the popularity of the History Channel TV series *Cities of the Underworld* (History.com-Cities of the Underworld).

The design and building of smart, sustainable, secured structures are now also being advanced. Buildings that reflect the state-of-the-art technological principles in the path of any natural or man-made disaster (i.e., self-contained, impenetrable, intelligent buildings that are threat-proof) are now on the drawing boards. These design concepts are gaining increasing global attention. More advanced building materials (e.g., security-enhanced blast-resistant exterior curtain walls and window systems) that can absorb higher explosive charges without deformation are now being installed in sensitive building locations on a worldwide basis. Such structures are attractive to certain businesses and governments, financial institutions, and data processing and information storage organizations, as well as corporate and government record-keeping agencies. Even residential developers are examining the benefits of smart buildings that go well beyond those classified as green. Sustainability now extends into the areas of safety and security, as opposed to simply being environmentally compatible.

To date the focus has been on specific buildings, complexes, or centers incorporating safety and security amenities, as a primary design feature. However the scale of things is beginning to change, to include even larger developments. The Kingdom of Saudi Arabia (KSA) is now advancing this approach by proposing redeveloping safe and secure infrastructure and services for its country, based on economic development principles. This is being referred to as King Abdullah Economic City (KAEC) which may become a change agent for the rest of the nations of the world to model themselves after.

Besides spatial planning and management, a second principle is temporal planning. Anything designed should be based on the principles of time use, where the 24-hour day is optimized. Both natural and man-made disasters represent potential 24-hour threats, 7 days a week, so our population and infrastructure must be on constant alert. Furthermore the optimization of mobility is based on using the entire 24-hour day.

Certain logistics-based industries use off-peak times to support the daily business cycle (e.g., FedEx, UPS, and DHL employ the ground–air network for time-based package deliveries less than 70 pounds). A network of feeder airports is used by Federal Express, connecting its fleet of aircraft through the Memphis International Airport and acting as its "superhub" distribution center throughout the night. The FedEx cargo fleet of aircraft is the world's largest airline, operating a fleet of 672 planes. Starting operations in 1973, this company each day delivers packages and freight to 220 countries. Its newest hub, Asian-Pacific, opens in 2008.

New York's Mayor Giuliani was the first mayor of a major U.S. city to begin operating the Office of the Mayor on a 24-hour basis, using the ombudsman program. Other cities have followed. Boston has instituted the 24-Hour Mayor Constituent Service program. By calling a number 24/7, callers receive a response from the city to a problem in one of six areas: building, health, housing, code enforcement, environmental services, and weights and measures. The logistics system of delivering grocery products at off-peak times (late evening through early morning hours) brought about the initiation of extended hours at grocery stores. In some cities such as Las Vegas, second- and third-shift workers come close to matching the number of employees working normal daytime hours. Restaurants, along with banks through automatic teller machines (ATMs), and increasingly liberal drinking laws are opening cities to expanded hours for drinking, entertainment, and dining opportunities. Such cycles are similar for police, fire, and emergency response services.

Even certain modes of transport are now responding with 24-hour service. Kansai International Airport (KIX), located near Osaka, Japan, operates 24 hours a day, handling both passenger and air cargo flights. Constructed on an artificial island in the Osaka Bay, KIX was designed to operate continuously, without noise restriction, due to its water-based, rather than land-based location. This airport island is connected to Kobe and Osaka by a rail, bus, ferry, and vehicle intermodal transportation network.

A Multisensory Environment

Although we are born with five senses (i.e., visual, acoustical, aroma, tactile, and taste), our urban environment is primarily examined and designed in terms of the visual, or what the human eye can see. This is very limiting, since we perceive our environment through all our senses, not just sight; consequently, noise should be replaced by sound, aroma supplant smell, and tranquility replace shock and vibration. The result will be a more sensory-balanced (kansai) urban environment in which to live and enjoy. "Sensory-impaired"(i.e., blind, deaf) architects, planners, and engineers would be beneficial additions to the profes-sional design community and should be actively recruited. Their sensory intuitiveness would expand the sensory experience of the built environment, making it more enriching.

We should design our human habitat in terms of our combined senses (see Figure 14.6) in order to create a harmonic balance, which is referred to as *kansai*. Although we are born with a unique sensory system that combines an integrated network of five senses, we do not utilize this social–physical attribute. Society has created a sensory imbalance toward visual dominance. Our language even reinforces this imbalance with certain words that give prominence to the visual; for example, words such as *insight, foresight, visionary, envision-ing,* and *insightful* suggest that this is the preferred way of thinking. Consequently, "out of sight, out of mind," and *shortsightedness* refer to a person who doesn't have the big picture; "the blind leading the blind" suggests paralysis of thought or a dead end. If you are from Missouri, the "Show-me state," surely you can grasp the subject, but can others?

The term *virtual reality* currently emphasizes "virtual *visual* reality." This means that we are really operating on just one of five cylinders (20% capacity) and thereby minimizing our sensory potential, including underutilizing the operational efficiency and enjoyment of our cities (see Figure 14.7). One-dimensional urban places relying on one sense, vision (e.g., beautification), alone are not sufficient. City planners, architects, urban designers, landscape architects, and engineers have the proclivity to use the visual

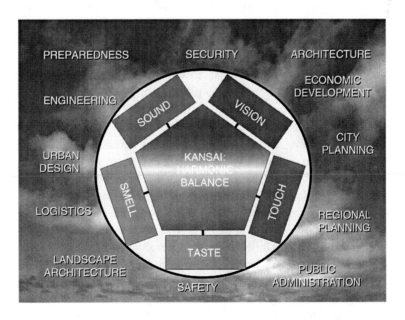

FIGURE 14.6 Kansai

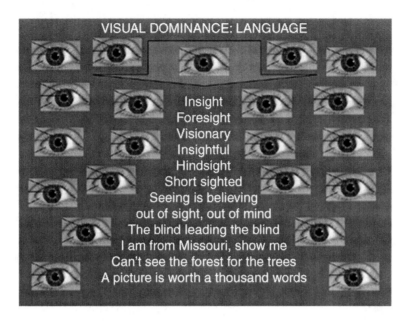

FIGURE 14.7 Visual Dominance

as their primary mode of thinking and create solutions to the problems they "observe." What is needed is multisensory-based teams of professionals composed of sighted, as well as visually, hearing, and olfactory impaired, who bring various perceptual sensitivities to the built environment.

The concept of the visual world has become dominant at the exclusion of other senses. Television, computer screens and navigation systems, VCRs, and video phones are just some of the technologies that dominate the cognitive process. Educators bemoan the fact that students learn by what they see more than what they hear. This means that visual-based thinking is held in high esteem. What is occurring in a parallel manner is that our language is building a visual bias into our vocabulary. These words and phrases are reinforcing the use and selection of visual-based words. *Insight, foresight, visionary, insightful,* and *hindsight* all suggest that there is a positive way to problem-solve, and vision is the proper means by which to pursue what one wants. After all, "seeing is believing"; "I am from Missouri, show me"; and "a picture is worth a thousands words."

Sensory integration is on the uptake. There are now major advances where a sensory-merging process is beginning to involve such areas as aromacology, geo-sonics, tactile imagery, psycho- or bio-acoustics, and so on. It is interesting to note that most senses are considered from the negative side (e.g., noise and its control, as opposed to sound to enhance the acoustic environment) rather than as part of a continuum. Again it appears that terrorism management technology is also examining smarter ways to merge the senses in the efforts to make the world a safer place to live. In some ways this technology is now setting a new pattern of sensory integration that is highly applicable.

Virtual border crossings are being merged with physical barriers, ranging in height from 15–30 feet. A 28 mile virtual border fence designed by Boeing and erected in Arizona has not been successful as initially reported by Secretary of Homeland Security, Michael Chertoff. Both the U.S. General Accounting Office (GAO) and Arthur Rotstein of the Associated Press (April 23, 2008) have reported that this $20 million virtual fence has failed to alert border patrol agents in a time sensitive manner. Consequently DHS is now going to spend an additional $45 million in more effective communications software and hardware interfacing with the border patrol.

Multisensory master planning (S^5) using the sensory palate of all human senses is needed to provide the full environmental experience. Effective acoustical detection through the medium of air, earth, or water is important in determining human movements, including those of illegals or terrorists. Underwater acoustics is being deployed in the vicinity of ports (e.g., hydrophones, buoys, and robotic devices) for detection and surveillance purposes.

Acoustical mapping of noise contours for transportation sources has been developed in response to the environmental impact process. Based on the regulations issued as part of the Federal Interagency Committee on Noise (FICON) and the Federal Noise Abatement and Control Act, noise emission standards are in place to protect the general health and welfare of the population, with an adequate margin of safety. Transportation noise has been identified by the EPA as the largest contributor to community noise. Large-scale noise abatement and control programs have been developed for highways by the Federal Highway Administration (FHA), and by the Federal Aviation Administration (FAA) for civilian airports. Noise standards for military installations were issued by the U.S. Department of Defense (DoD).

The use of noise as an offensive weapon is gaining attention by the DoD, and these devices are now being integrated into the military system for deployment purposes, as well as other types for civilian purposes (for command and control or other security operations). The Long-Range Acoustic Device (LRAD) uses a technology of focused acoustic energy as a nonlethal intervention option to enforce infrastructure security perimeters on land and sea and enhance control compliance, among many other uses (Figure 14.8). The LRAD produces an exceptionally clear, tight beam of sound, audible and intelligible up to 1,000 meters. It was originally developed for the U.S. Navy to fulfill a capability gap following the *USS Cole* attack. Providing audible alerts to locations at significant distances can be helpful under emergency-type conditions (i.e., mass notification of disasters, hazardous emergencies, and fires). These systems are also programmable for creating

Airports and Government Facilities

Clear, intelligible voice heard at 700+ meters

LRADtm Long Range

Acoustic Device

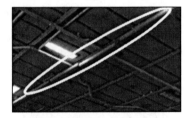

Subway or Transit Station

Maritime or Port Facility

FIGURE 14.8 Long-Range Acoustic Device

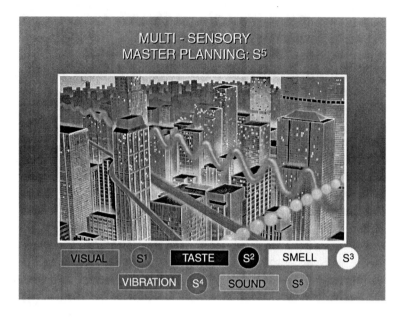

FIGURE 14.9 Multisensory Master Planning: S^5

messages in several different languages to communicate effectively in a rapid deployment manner. The American Technology Corporation is the developer of LRAD.

Rather than suppress the negatives, Figure 14.9 provides a sensory map of how the senses can be positively integrated into the urban infrastructure. Not only is there interest in creating a balanced sensory world in which to live, but also there is interest in understanding the dynamics of sensory generation and human response. A Stanford University professor, Paul Rhodes (2007), is working on a process to reverse-engineer the brain, with the hope of turning a machine into a dynamic sensing device that can sort a host of odors. His research could assist the process leading toward a goal of creating an expert olfactory system. From a security perspective, olfactory threats that are detected can result in a response mechanism to protect a population potentially at risk. The other extreme is to use sensory weapon systems that can modify human behavior. If this olfactory mechanism can be perfected, the opportunities in surveillance can be significant. This could include advanced security screening at airports and rapid detection of bacteria or freshness to avoid food poisoning, among many others.

The collective recognition of these three problems (i.e., natural disasters, man-made disasters, and unplanned growth) and designing solutions for the built environment are not as yet recognized at the policy level. Such a concern was raised and addressed, possibly for the first time, in the intermodal transportation study initiated for the town of Southampton, Long Island (C. R. Bragdon, Intermodal Transportation Study for Southampton, Long Island, New York, 2003). This unique study incorporated real-time interactive virtual simulation within the Southampton town hall. It examined solutions addressing spatial and temporal conservation, optimization of multiple modes of transportation, and the question of emergency response–evacuation due to physical (e.g., hurricanes, storm surge, and flooding) and man-made incidents.

Figure 14.10 brings together the general subject areas discussed in this section of Chapter 14 that integrate sensory (five senses), spatial (three dimensions, using the x, y,

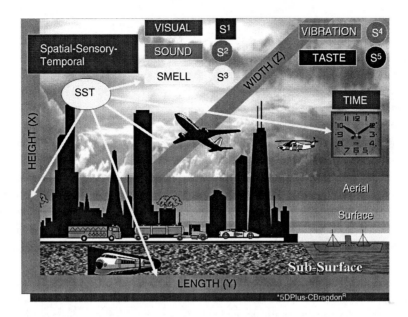

FIGURE 14.10 Spatial, Sensory, and Temporal (SST) Integration

z axis), and temporal (24-hour time) features. In the effective development of a global partnership and security plan, these factors of space, time, and the human senses all have an important role to play.

The Global Center for Preparedness (GCP)
Need

As our planet of nearly 200 countries approaches a population of 10 billion, and with an ever-increasing urbanized world now exceeding 50% of Earth's surface, it is time for the restoration of world order, to sustain our civilization before the 22nd century arrives. By the year 2100 demographic experts are predicting a U.S. population approaching 1 billion persons. Survival is the desire of all humanity, but the tenets of civilization must be built on the principles of a safe, secure, and sustainable way of life for our families and all future generations. The economic impact of disasters is rising, with the cost of damages from natural disasters appearing to double every 10 years, especially along populated coastlines. It is estimated that there may be a $500 billion hurricane during the decade of 2020 focused on the northeastern U.S. (reported at the 30th Annual National Hurricane Conference, April, 2008).

The challenges are formidable, including global warming, issues of food supply and distribution, health and environmental conditions, energy consumption, and spatial excess. The onset of disasters and their level of destruction and loss of human life must be addressed on a worldwide basis. Our scale of thinking must now be focused on world concerns, because rapidly we are becoming a global culture. As author Thomas L. Freidman has observed, the world is flat, and together our civilization faces a challenging time, but the rewards of conflict resolution are highly beneficial: global peace and respect for all humankind (Friedman, 2007). Economic and social well-being must prevail in a compatibility-based world.

We must organize ourselves, and being prepared for all eventualities is an important goal, beginning with the individual and extending to the family and then the community, region, nation, and ultimately our human biosphere. Preparedness is the key, but it also the gateway for success. The foundation of this idea and book began with the planning for the First Annual Conference on National Preparedness (2007). This author met with a dedicated group of professional soldiers at the Pentagon, affiliated with the Joint Chiefs of Staff and the military services.

Major General Mike Sumrall, assistant to the chairman of the Joint Chiefs of Staff for National Guard matters, was a pivotal person in the development of this concept of preparedness for this author. I have found General Sumrall to be a highly motivated and dedicated professional soldier who cares about this country as a citizen. He wants to ensure that we are properly prepared to deal with any natural and man-made event that could imperil this country and our citizenry. The Guard's primary mission has been to provide interface between the military and our civilian communities in times of both natural and man-made disasters, when a federal response is called for.

It is with this vision and motivation that the First Annual Conference on National Preparedness was planned and held on December 14, 2007. That conference subsequently became the primary stimulus for establishing the Global Center for Preparedness (GCP). It was at this Conference in 2007 that it was announced, with support of all those in attendance, that a Center with this theme would be established. A 2nd Annual Conference is now being planned for December 8–9, 2008, to further advance the global focus on the problem and needed seamless integrated solutions worldwide.

Philosophy

The philosophical approach was to deal with the world as one entity and disregard a more localized focus (e.g., national center for preparedness), which is relevant but does not have the global and international context. Homeland security is also an important focus, and there are now over 60 academic degree programs in our colleges and universities with that as their title and orientation. Again, that concept in most instances deals with how the "homeland," typically the United States, should address this important subject area of emergency preparedness and response, including both natural and man-made disasters. A third concern was to consider more than what the federal government is doing through the network of some 40 federal entities, led by the U.S. Department of Homeland Security. The private sector companies, along with the nonprofit participants such as foundations, are major players, as well as the international community.

Lastly, the concentration on natural and human disasters, focusing on response and recovery alone, does not tell the whole story. Infrastructure, smart sustainable growth management, and urban planning are essential as well. The story does not end with the excellent assistance and leadership from the National Guard in the military–civil initiatives, so important in a first response effort. Nor does it end with the coordinated efforts of FEMA, the Coast Guard, state and local governments, and nonprofits such as the Red Cross, Salvation Army, etc. These elements do not address the whole system of issues.

Consequently, the Global Center for Preparedness (GCP) gives the opportunity to explore in totality our global human habitat. This should include lessons learned throughout the world, with the important interplay of international organizations such as the United Nations with the World Bank, the International Monetary Fund (IMF), the Inter-American Development Bank (IDB), the International Red Cross, and foundations (e.g., Bill and Melinda Gates Foundation, etc.) with business and government. Economic development and public–private partnerships linked with a focus on the global scale have many benefits. Within the United Nations itself are at least 10 organizations of interest, including the following:

1. Inter-Agency Committee on Sustainable Development
2. International Civil Aviation Organization (ICAO)
3. International Maritime Organization (IMO)
4. International Monetary Fund (IMF)
5. International Strategy for Disaster Reduction
6. UN Development Program (UNDP)
7. United Nations Educational, Scientific, and Cultural Organization (UNESCO)
8. UN Human Settlements Program (UN-HABITAT)
9. World Bank
10. World Health Organization (WHO)

Aiming for an international focus, the Global Center for Preparedness does not rely solely on the experience or influence of the United States. It is really influenced by how our society as a whole through a network of countries, regions of the world, organizational stakeholders, and social and political leadership addresses the problem and enacts solutions.

Framework

Stephen Flynn, considered one of the leading authorities on national security in the United States, believes we should use a tiered approach when examining vulnerability. He considers the same three areas that the Global Center for Preparedness addresses (natural disasters, human disasters, and infrastructure). However his approach differs because it is sequential and layered. His first priority is natural disasters, followed by infrastructure, and lastly terrorism. Figure 14.11 depicts the approach of the GCP, which utilizes a

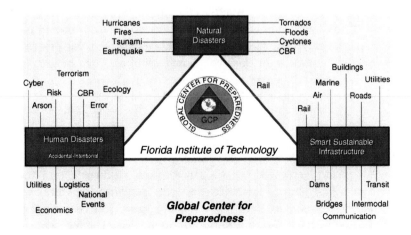

FIGURE 14.11 GCP Framework

parallel approach to all three areas, assigning no priority because of their interrelatedness and the urgency to be holistic and not compartmentalized. Vulnerability and risk do not justify singular action.

Compartmentalization is exactly what our governmental bureaucracy has perpetuated, and this stovepipe approach is still evident today in our homeland security initiatives. A flexible and comprehensive approach, not based on bureaucratic rigidity but adopting public–private partnerships with an opportunity for participation, is essential. Yes, isolationism is reduced as we recognize the necessary integrative process that is needed. However, isolationism persists. The Global Center for Preparedness represents an emergence of integrated thinking and parallel processing that is not dissected by the tiered focus suggested by Stephen Flynn and many security experts.

Certain regions of the world are picking up on this integrative concept, due to economic development realities and their present noncompetitiveness. The Inter-American Development Bank (IDB), led by Louis Alberto Moreno, has established a disaster risk management policy coupled with a $12 billion critical infrastructure fund for South America and Latin America. It understands that intermodal transportation is the backbone of the economy and recognizes the fact that logistics costs in South America and Latin America are twice that of any than other region in the world. Inadequate infrastructure limits the region's competitiveness as well as its responsiveness to any disaster that may occur. One of the major economic indices now being observed by the investment community is the question of downtime and recovery from any human or man-made disaster, as well as the logistical supply-chain management efficiency. New investment dollars are going to locations where downtime and recovery are a minimal disruption. Economic development and post-disaster economic strategies tie back to global competitiveness. A country or region of the world can ill afford to have a strategic policy and plan that do not concurrently address natural disasters, human disasters, and infrastructure.

IDB-supported programs have been launched in Brazil (e.g., Embraport, Meto-QuartoSA), Panama (e.g., Panama Canal expansion), and Chile (e.g., urban transportation reform). Additional areas of focus, supported by the IDB, address environmental issues, rural development, and disaster risk management. Partnerships are necessary to carry out these ambitious plans and future ones in Latin and South America. The Bill and Melinda Gates Foundation has now agreed to work in this region with the Inter-American Development Bank and the countries addressing new transportation (i.e., physical and electronic), health care, and water systems. This partnership for private and public collaboration and philanthropy in this region will reinforce this integrative need that merges natural disaster and human disaster recovery with infrastructure. This is the approach and philosophy of the Global Center for Preparedness.

Natural Disasters

Wind (tsunamis, tornados, cyclones, and the like), fire (spontaneous, lightning, and so on), rain (flooding, etc.), geologic changes (earthquakes, volcanoes, etc.) and chemical, biological, and radiological accidents that can occur constitute this broad category of natural disasters. These events take place in terms of acts of God, or natural occurrences, without direct human intervention. It is important to note that the frequency and severity of impact of these disasters is growing. This is due partially to global population growth, increasing human density patterns, expanding infrastructure, and value of human assets.

Health factors, using the United Nations' definition of health, will be included in assessing the impact of natural disasters on the population at risk. In virtually all natural disasters involving a population exposed there are health-related issues that must be addressed.

Human Disasters

Human disasters fall into two distinct categories: accidental, such as human performance–related factors including errors; and intentional actions, such as terrorism and arson. In both these categories, the human being has entered into the equation. Certainly one of the most globally significant is the warming of our planet Earth. However, there are others of major concern, including cybersecurity and the threat of chemical, biological, and radiological incidents that, on the largest geographical scale, could involve international warfare. Nonsustainable activities, such as the combustion of fossil fuels and pollution of the human habitat, relate to ecological effects. The potential risk applies to the area of exposure and the scale of the event or episode. Health factors, using the United Nations' definition of health, are included.

Smart Sustainable Infrastructure

Infrastructure considerations include all the man-made artifacts (i.e., buildings, roads, dams, bridges, utilities, transportation systems) supporting transportation activity. They also encompass the vehicles, aircraft, ships, boats, mass transit, and rail, operating in the modal or intermodal system of transport. IT and communications systems are also part of this inventory (i.e., towers, substations, transformers, switching systems, cabling, plants, pipelines, etc.), since electronic components and mobility are part of the overall picture of the transportation network (defined earlier in Chapter 1 as transcommunication, Figure 1.1). Transcommunication represents the integration of both physical and electronic movement. What is needed is an intelligent and sustainable infrastructure that is imbedded with sensory responsiveness, materials science, and adaptation to the onset of threats. Such an infrastructure needs to support the backbone of our economy: transportation which involves the movement of people, goods, and information by air, land, and sea.

Missions

It is important to establish missions for the Global Center for Preparedness that will be initiated to be operational on a worldwide basis. The GCP began operating in the spring of 2008. There are 10 primary missions being pursued globally through its consortium members and their activities (see Figure 14.12).

The growing membership represents business, government, academia, and nonprofits as affiliates of the Center, which is headquartered at the Florida Institute of Technology in Melbourne, Florida. The organizational missions include the following:

1. *Repository for best professional practice*: There are important things that we can learn from one another in the subject area of global preparedness and response, i.e., city to city, state to state country to country. Frequently, these practices are never inventoried and saved in a consistent manner due to the level of activity. Furthermore, there is no clearinghouse or repository for these experiences. . . . We quickly go from one subject to another, with little regard for

GCP Missions
1. Repository for best professional practice
2. Assessment including forensic logistical analysis
3. Information technology and database management
4. Training certification and academic instruction
5. Sustainable planning and management
6. Modeling and simulation
7. Basic and applied research
8. Consultation services
9. Educational and informational dissemination
10. Performance assessment and quality assurance

FIGURE 14.12 GCP Missions

what we can learn from the experiences of others. Case studies and lessons learned, based upon organizational experiences and what can be shared, will be inventoried, codified, and accessible to the members of the Global Center for Preparedness.

2. *Assessment, including forensic logistical analysis*: The movement of people, goods, and information by air, land, sea, and space is critical for our economic and social well-being. Forensics is important, since it can be an effective assessment tool in understanding the dynamics of the logistics process in order to validate standards, operations, and procedures to make improvements. It can also be utilized in establishing quality control measures (referred to as TTQM: total transportation quality management), since transportation security is a fundamental element in preserving mobility.

3. *Information technology and database management*: The role of IT in global preparedness is essential for data fusion. Command and control and the information systems in place must be protected and effectively distributed to all stakeholders and clients in a time-efficient and secure manner. Interoperability is an absolute necessity. Protection of the database management system from a cybersecurity perspective is also becoming an increasingly important factor and is a mandatory requirement in a cyber-secure environment.

4. *Training, certification, and academic instruction*: Continuous learning in this technological age of safety and security is vital to all stakeholders. Access to education and training opportunities including resident instruction, distance learning, or blended programs will be available on a worldwide basis, regardless of physical location. Levels of certification are also needed and will be a major aspect of the GCP educational program. New areas of academic development (e.g., program in humanitarian and disaster relief logistics, HDRL) and training will be pursued on a collaborative basis, among many other activities.

5. *Sustainable planning and management*: Energy independence must be aggressively pursued internationally, as most natural-based renewable resources are in the decline since global demand and fossil fuel consumption are growing unabated. The demand for petroleum products is growing on all continents. Despite improved energy resource management practices, there are no large scale effective energy efficient substitutions. In response this is impacting the cost of food, distribution, and finished goods. Sustainable initiatives are going to be a major part of the GCP mission.

6. *Modeling and simulation*: Computer-based simulation, which has evolved from the military and the entertainment industry, now is making important inroads in real-time visualization and situational awareness in the public and private sectors. Interactive real-time 3-D is gaining popularity in all aspects of tabletop training, exercises, gaming, and role playing, including immersive avatar role playing. The merging of all senses (multisensory) into this virtual world to assist in providing solutions is on the cusp of the next important plateau.

7. *Basic and applied research*: Both basic and applied research is an important factor for identifying, understanding, and developing solutions as an input to a global preparedness plan. The Global Center for Preparedness, through its consortium members, will be in a position to respond to any request for proposal (RFP) or request for quotation (RFQ) that might have potential application to its mission. The disciplines of engineering, computer science, architecture, urban planning, information technology, CIS and MIS, logistics, health, science, ecology, marine biology, business, accounting, finance, marketing, public administration, and the behavioral sciences constitute the areas of expertise within GCP. The GCP will become the education and training engine for all of the public, private, and non-profit stakeholders.

8. *Consultative services*: An inventory of resources by discipline and subject area will be developed as a potential pool of intellectual capital within the GCP that can initiate consultative scopes of work and projects that are needed to solve practical problems. These can also involve nonprofit ventures through Florida Tech with private sector partners or other affiliated organizations and professional societies. The ulitmate goal of these consultancies is to advance the safety and security solutions to preserve and enhance the economic well-being of our society.

9. *Educational development and information dissemination*: Education and information dissemination will be a primary means by which the GCP communicates to its members, affiliate organizations, and the general public. An interactive Web site, with streaming digital video among other technologies, will be in place using www.globalcenterforpreparedness.com. A digital-based library science export technology will be used for maximum high-speed connectivity and response, as well development tools to enhance the educational learning process. The GCP will be initiating a Journal for Global Preparedness with international sponsorship.

10. *Performance assessment and quality assurance*: The products developed and delivered must be reliable and scientifically accurate. GCP will initiate standards of excellence criteria, protocols with necessary assurances, and quality control measures. The name and logo of GCP will be used in its transactions, as an intellectual property.

It is the intention of the Global Center for Preparedness to be a resource to those agencies and organizations mandated by law to deal with homeland security, situational awareness, and emergency response and recovery. This should involve civilian, military, and private sector relationships. There is no desire for the GCP to become a formal part of the federal government; rather, it will act as an independent resource or advisor to provide assistance to federal, state, and local governments, as well as industry when requested. This role of the GCP is more aligned to being a "practical think tank" that desires to assist in solving real-world problems of security and national preparedness by combining private sector, academic, and nonprofit capabilities.

Administrative Structure

Administratively, the Global Center for Preparedness operates through an executive board. The board membership is primarily composed of larger-scale businesses allied to the subject areas of homeland security preparedness, natural and human disaster prevention, and response and recovery. There is a chairman of the GCP executive board, Dr. Clifford Bragdon, who is responsible for providing oversight and representing the board members in terms of policy and overall direction, with the board's input. The executive board members will play a central role in the direction and support of GCP, as major players in the field of global security.

Supporting this executive board is an advisory board. This board will consist of technical experts with an academic and research background related to this subject area. In addition, there will be representation of companies and small businesses that will participate in the mission of the Global Center for Preparedness through a committee structure network. The third group consists of auxiliary, or volunteer, members who have expressed an interest, due to their life experiences, in becoming involved with specific assignments or tasks. Their backgrounds and contributions are substantial, and they want to volunteer their time on a pro-bono basis. Collectively these professionals are referred to as Friends of the GCF.

The day-to-day operations will be led by an executive director, who has an administrative staff to carry out the duties and responsibilities of the daily activities tied to the mission statement of the GCP. At the outset, the three areas of activity will relate to the development and implementation of the Global Center for Development's strategic, administrative, and financial plans. In principal, the Global Center for Preparedness will act as the educational engine working with all its partners as a conduit to achieve the 10 basic missions as previously outlined.

National Security and Transcommunication-Based Initiatives

The interstate highway system was authorized by the passage of the Federal-Aid Highway Act of 1956. It represented the largest public works project in the nation's history. President Eisenhower was one of its strongest supporters, in part based on his experiences in observing the use of the autobahn during Word War II. In his 1954 State of the Union address, President Eisenhower declared that a national highway system was essential to "protect the vital interest of every citizen in a safe and adequate highway system." He signed this into law (Public Law 84-627) on June 29, 1956, and from this act the interstate highway system was referred to as the "Dwight D. Eisenhower National System of Interstate and Defense Highways."

The interstate highway system was designed as a dual-purpose system, primarily to serve the public. Today it exceeds 46,000 miles of dedicated roadway. Concurrent with public use, this interstate system was intended to provide rapid access using key ground transport routes for military supplies and troop deployments in an emergency. Subsequent legislation, including the Transportation Equity Act (TEA-21, enacted as Public Law 107-178 on September 30, 2003) and the Safe, Accountable, Flexible, Efficient Transportation Act (SAFETEA-LU, Public Law 109-59, enacted on August 10, 2005), extended the application of funding to include highway safety and transit. This gave an expanded direction to the application of highway trust funds for nonhighway projects (e.g., intermodal purposes) and the development of intelligent transportation systems (ITS).

Nearly 52 years have passed since passage of the act for the national system of interstate and defense highways. There has been a broader interpretation of what constitutes a highway-related project with the passage of SAFETEA-LU. Highway trust fund dollars now are allocated for nontraditional highway programs, including transit and recreational rights-of-way (joint corridors) and other intermodal themes. However, this evolution has been incremental, and the amount of funding in these newer areas has been a small percentage of the overall Federal Highway Administration (FHWA) budget.

There is growing interest in developing a more innovative approach to transportation at the federal level. Kevin Johns (2004), director of Economic Development for Palm Beach County, suggests that it is time for creation of "Eisenhower II," describing this as an innovative approach to transportation having an impact similar to that of passage of the act for the interstate highway system in 1956. Mr. Johns, an urban planner, believes that this country should create an intermodal transportation spine incorporating economic development opportunities for all modes (i.e., airports, seaports, rail and transit stations, ferry terminals, and pipeline terminals). Joint-use principles with public–private partners incorporating seamless intermodal networks are essential for reducing transportation gridlock, along with establishing tax credits for electric and hybrid vehicles. He predicts this would reduce dependency on foreign oil by 18–20%. In one of the most congested corridors in the United States, Kevin Johns has proposed the South Florida Multi-Modal Logistics Center. This is an important next-generation approach to integrated transportation, but it needs to be turned up another notch to include more emphasis on security and preparedness.

Re-Examination of Modal Elements

We should be inventorying all the present modes of transport so as to see how they come together currently and into the future. Although it may be difficult to perceive, as our thirsty fossil fuel–based vehicles vie for a globally competitive supply of gasoline, a "car-henge–based" society is on the verge of a monumental conversion to energy independence. This will bring about a significant change in international relationships and responses. Physically this means gasoline stations, fewer in number, will be replaced by "energy centers." But what impact does that have on urban infrastructure and the delivery of needed services? There are many other subject areas that need to be explored, especially with the introduction of advancing transportation-based technology, including new modes of movement and infrastructure design.

FIGURE 14.13 Transcommunication Modes: Present and Future

Figure 14.13 delineates many of the transportation technologies that are variations of present modes or new modes of transport. They all explore the three-dimensional spatial opportunities of area, surface, and subsurface. For example, a subterranean high-speed pneumatic freight tunnel (e.g., tube express) is now being designed for highly congested corridors involving truck transport. At least one pilot project is being analyzed as a potential border crossing between Texas and Mexico by Texas A & M University through a research grant. Current international border crossing, passage of commercial tractor trailers can take 2–8 hours. Installation of the freight cargo tube could positively change the dynamics of border security and surveillance, besides the benefits of an improved supply chain and just-in-time delivery.

Water taxi systems are rapidly increasing in urban areas, as the road network becomes more congested. In many cities such as Fort Lauderdale, they are coordinated with other modes (i.e., buses and their arrival and departure schedules) to create a more coordinated movement system for pedestrians. Within a decade water-borne travel could be complimented by wing-ship high-speed transit and cargo corridors. Wing in ground effect technology allows air-water–borne travel using physics' principles that significantly conserve energy. Pacific Seaflight, if certified by the Coast Guard and the FAA, will begin using a 12-passenger Blue Dolphin wingship by 2009. Wing-ship technology for emergency response, if it would be coordinated with maritime–landside connections, could rapidly improve the dwell time. Such aircraft are now designed and being tested in various countries (e.g., the United States, Germany, and Sweden).

In 2001 Dean Kamen unveiled a Segway PT, which is a two-wheeled, self-balancing electric personal transportation system. This electronic personal assistive mobility device has been adopted into the transportation network in both civilian and military applications. It is used at military bases, warehouses, corporate campuses, industrial sites, and educational campuses. Chicago exemplifies the broad-scale government use, where they are using 30 Segway GTs for police, fire, and airport operations as well as emergency management. Segway GT has an important use in individual safety and security operations as a network component, and this company is looking to expand its technology into different configurations.

The military version of the Bell-Boeing V-22 tilt rotor aircraft has recently received federal government approval to restart its military research and testing program, now based at Patuxent Naval Air Station in Maryland. There is also a civilian tilt rotor version developed by Bell Helicopter Textron, Inc., and referred to as the Bell Agusta (BA609). The BA609 holds considerable commercial promise if it can be flight certified by the FAA. Due to its aircraft engine and mechanical design, the plane initially lifts off as a helicopter and at an appropriate altitude converts its propulsion system to that of an airplane. At that point it could cruise at speeds in excess of 150 miles per hour. The potential use in interurban flights, especially in high-density city environments, could be beneficial for both passengers and small commercial cargo. This could become the next-generation Greyhound bus station system installed on a network of commercial rooftops. There are several tilt rotor ports, or vertiports that have been built in anticipation of this new technology in both Dallas and New Orleans, (Note: One was planned by this author for use during the summer 1996 Atlanta Olympics, but it was ahead of its time.) In the interim, these rooftop facilities are used for helicopter operations and can play an especially important role in emergency medical response.

Tesla Motors, Inc. introduced the world's most fuel efficient production vehicle with the 2008 Telsa Roadster. A San Carlos, California–based company it began delivery of its first 900 production two-passenger roadsters in March 2008. This electronic-powered sports car has impressive performance characteristics:

a. 0 to 60 mph in 3.9 seconds
b. Top speed of 125 miles per hour (limited for safety)
c. EPA rated at 135 miles per gallon
d. Electric regenerative power
e. Cruising range, between overnight charges, 220 miles
f. 100,000 miles on the original Energy Storage System (ESS)
g. Operating cost per mile, 2 cents
h. Purchase price: $109,000 (12–18 month waiting list for the 2009 Telsa Roadster)

In a recent conversation with the author, Elon Musk, former CEO of Tesla Motors and its chief investor, stated that there will be a five passenger sedan in production by 2010, anticipated to sell for $59,000. Certainly this is stimulating the automotive industry, since 63% of U.S. oil demand is consumed by transportation and we are relying on foreign oil for nearly 60% of our supply. A $60,000 refundable membership fee is required to secure your position when ordering a 2009 Roadster.

Every civilian-based transportation mode, depicted in Figure 14.13, has implications for both national preparedness and response. Important modal characteristics include the following:

- Response time
- Capacity
- Energy efficiency
- Availability
- Accessibility and spatial efficiency
- Convertibility to emergency response needs

- Initial and life-cycle costs
- Reliability and safety
- Supporting infrastructure
- Government regulations and requirements
- Public and political acceptance

None of these transportation modes should be examined individually or in isolation. They need to be part of an intermodal or total modal network. The city of Dallas is a good example, because its downtown convention site is transportation integrated. Besides microwave relay systems, the rooftop is also used for helicopter service and is designed to handle future tilt rotor operations. In addition, the Dallas Area Rapid Transit (DART) serves the convention center with both light rail and bus service. Taxi service is also accessible. Public parking is available for those choosing to drive. Lastly, there is a commercial rail service adjacent to the convention center. Due to the travel distances between the Dallas–Fort Worth International Airport and the Metroplex, time and energy costs are significant. Soon they will open the DART light rail system directly to the airport, which will facilitate air and ground transportation interface. It will also be another important factor in situational analysis and emergency response planning and logistics.

Spatially based intermodal transportation needs to be an integral part of three-dimensional planning for every city in the United States, if not the world. Within this planning process, situational awareness, national preparedness, and response need to done together. This was part of the objectives in preparing the Southampton Intermodal Transportation Study (SITS), conducted on Long Island. For example, concerns about emergency evacuation, efficiency of transportation networks (i.e., roadways, waterways, rail–transit, and airports), and sites for modal connections were major elements of the SITS. Those familiar with the Hamptons know that transportation access from Manhattan is severely limited. The east–west road network to the South Fork is very restricted, and the travel demand, especially on weekends, results in gridlock conditions. Furthermore, commercial truck service dominates, with 99% of all goods moved by road; the Long Island Railroad right-of-way is in place but basically serves the commuting population between the city and several destination stops in the Hamptons.

Recognizing these transportation problems, among others, the Southampton Intermodal Transportation Advisory Committee assessed these issues. Assisting the committee and town of Southampton in the study was their consulting team of Dr. Cliff Bragdon and Yossi Kaner. A virtual simulation studio was designed and installed in the town hall to initiate various transportation scenarios that could be visualized and manipulated with the use of a mouse displayed on a flat plasma screen. This was the first time that this type of real-time virtual technology was installed in a local government office for conducting planning and transportation studies in support of elected governmental officials. This studio office was called by some the "Steven Spielberg studio for transportation planning decision making." Figure 14.14 represents the portion of a simulated analysis to improve east–west access by creating a joint-use, intermodal corridor. It was spatially designed to fit within the existing Long Island Railroad right-of-way, thereby minimizing any disruption to existing land uses. In general, the Hamptons consist of upscale single family residences, with limited specialty commercial uses and an extensive network of oceanfront beaches, all in a historical setting. (The town of Southampton was established in 1640.)

FIGURE 14.14 A Joint-Use Intermodal Corridor for the Hamptons

There were several intermodal transportation planning principles incorporated into this concept, including the subjects of safety and security. These principles included the following:

- Develop a joint-use corridor that would minimize the use of land, purposely restricting this corridor to the present Long Island Railroad right-of-way.
- Because present freight movement east–west was 99% truck traffic on the limited road network, a dual-track rail bed was proposed, allowing both commercial freight and commuting rail to occur simultaneously (presently there is a single-track rail bed passing through the South Fork of the Hamptons).
- Construct a two-lane submerged roadway that would be restricted by use and direction, depending of time of day, using a contraflow model. By virtue of its submerged design, there would be no increase in the ambient sound level, attributable to this submerged roadway in the adjacent community.
- Design this joint-use corridor to enhance freight movement and create a better balance between rail and truck freight services, reducing the high percentage of trucks using the road system and reducing the energy costs to both operators and consumers.
- For emergency response, recovery, and possible evacuation, manage this devoted joint-use corridor to speed up the logistical process, since at peak hour the level of service on portions of the Montauk highway is at service level D.
- Minimize the visual introduction of any additional transportation modes, since the roadway would be submerged (approximately 10 feet below the rail track beds). No road traffic could therefore be observed.

- Ultimately connect this joint-use corridor with north–south intermodal nodes (located in the villages of Southampton and West Hampton), which assist connectivity to the North Fork and west to New York City.
- Create a fiber-optic spine (e.g., broadband and T1 connections) to enhance communications, as well as establish a utilidoor (a utility-based corridor) as an improved economic development feeder to residences and businesses. This design would improve esthetics by limiting design, which would limit vertical protrusions associated with unsightly cellular and utility towers.

Ultimately this proposal was not adopted by the Southampton board, in part because of citizen opposition that would bring about change. However, this simulation technology allowed all the stakeholders to observe what was being considered in depth. Before development of this studio, everything was done on paper, using two-dimensional flat surfaces with recommended solutions fastened to walls or mounted on easels, for the decision-makers to see.

Fast Integrated Response Systems Technology (FIRST)

The problem of effective and coordinated response to both natural and human disasters remains a concern throughout the world. Brigadier General (Ret.) Ralph Locurcio (2006) indicates this is not a new problem: "It is an inescapable problem that occurs whenever established organization units are integrated for a combined operation, and it is made even worse in a political climate such as our American democracy…the problem is greatly exacerbated when emergency actions must be accomplished in a compressed time frame." Analytically, there appear to be three classic organizational problems that require management attention, and they deal with communication, structural, and cultural problems.

Like the rest of the world, the United States is in an organizational crisis in terms of dealing with large-scale, intergovernmental (federal–state–local) problems that are related to either natural or man-made disasters. The Fast Integrated Response Systems Technology (FIRST) team's goal would be to assist in integrating collective resources of a region, including using 5-D simulation to produce a dynamic visualization working with an integrative professional team, including planning, architecture, engineering transportation, logistics, business, finance, medical, and social science disciplines. The 5-D team would build constructive scenarios to develop practical design options and real-time training and implementation tools, dealing especially with infrastructure and habitat redevelopment. This would involve conflict-resolution–based techniques that these members have initiated on a broad scale for the military as well as for governments.

The impact of most disasters extends well beyond local and state capabilities. Financially, the magnitude of impact grows especially in populated coastal regions in the United States that now represent 90 million people. This requires regional approaches, often including multistate initiatives. A regional response network is often necessary. The Fast Integrated Response Systems Technology (FIRST) team approach is intended to utilize accurate and reliable databases and geospatial information to create a platform for effective modeling and simulation (see Figure 14.15). Members of this team have been responsible for rebuilding Kuwait ($800 million); five-dimensional simulation for deepwater ports; commercial and military airports; public transit; maritime security training for DHS–TRA; intermodal safety, security analysis, and solutions for the U.S. Department of Transportation; hurricane and disaster recovery and reconstruction; urban and regional

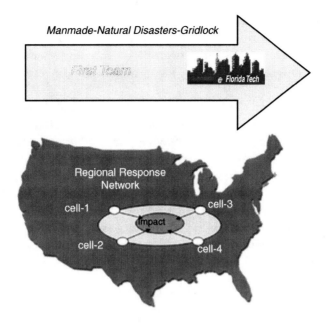

Fast Integrated Response Systems Technology

FIGURE 14.15 Fast Integrated Response Systems Technology—FIRST

planning; development of training manuals and protocols; patents; and distance learning and delivery mechanisms to maximize effective communication.

The FIRST team considers six interrelated tasks as needed for reinforcing regional assistance, including post-disaster economic development strategies and supporting the urban habitat and infrastructure. These tasks include the following:

1. Inventory recent natural–man-made disaster responses, including forensic logistics and redevelopment initiatives.
2. Develop a regional response integration model that includes all public and private sector and nonprofit stakeholders.
3. Integrate planning, engineering, housing, medical and communications technology, cybersecurity, transportation, logistics, and human factors in the model formation.
4. Validate the model using data from selected scenarios and then test this model with regional disaster planning agencies.
5. Develop an implementation methodology for specific regions, followed by training, ranging from response to recovery and including post-disaster economic development strategies.
6. Advise regional response cells as requested.

Conclusion

National preparedness and security are a problem facing the entire world. It should not be considered a specific concern for a single country, region, or continent. International cooperation is the only means to address this problem. Risk assessment and vulnerability analysis have placed improper emphasis on addressing natural disasters as the number-one priority at the expense of also considering human disasters and infrastructure. Our global challenge today is to address these three concerns on a collective basis, in part because any solutions will involve the sharing of organizational resources and there is a common source of financial support: government. Joint problem-solving results in more efficient use of resources.

There is a need to blend public and private sector stakeholders with nonprofits, foundations, and most importantly, an aroused and involved public at large. Participation by all interest groups is a positive course for developing responsibilities and collaborative solutions. Disaster prevention, planning, and management require activation of skill sets including spatial and sensory approaches to problem solving. Intermodalism or transcommunication must be central to any resolution. The Global Center for Preparedness can be of assistance to those organizations mandated by law to lead this initiative. Ultimately the safe, secure, and sustainable movement of people, goods, and information by air, land, sea, and space must be preserved to ensure that our civilization endures and achieves its goal: social and economic well-being.

References

Ackerman, Diane. *A Natural History of the Senses*. New York: Random House, 1990.

Ascher, Kate. *The Works: Anatomy of a City*. New York: Penguin Press, 2005.

Associated Press. Amtrak to step up security measures. March 14, 2008.

Association of Energy Engineers. Established 1977; accessed at: www.aeecenter.org.

Berkowitz, Carl, and Clifford Bragdon. Advanced simulation technology applied to port safety and security. *Conference proceedings, the American Society of Civil Engineers*, June 2006.

Berkowitz, Carl, and Clifford Bragdon. Advanced analysis: How can newly developed technology be applied to improving passenger terminal safety and security? *Passenger Terminal World*, November 2006.

Bragdon, Clifford R. Aromacology and the built environment. Scent Society: Yesterday, Today and Tomorrow, Olfactory Research Fund, New York, Carnegie Hall, 1995.

Bragdon, Clifford R. Emergency systems used in transportation systems. *Proceedings, Segunda Feria Internacional de Transporte Masivo* (2nd International Mass Transit Fair), Bogota, Columbia, November 8–9, 2006.

Bragdon, Clifford R. Homeland security: Developing a national center for preparedness. Brevard Community College, Melbourne, FL, January 22, 2008.

Bragdon, Clifford R. Integrated mobility-based transportation system for logistical optimization. *Logistics Spectrum*, 40(3), July–September 2006.

Bragdon, Clifford R. Kansai: The harmonic balance of the human senses: Developing a sensory master planning system (S^5). *Proceedings, Summit 2000: The Globalization of the Senses*, Plaza Hotel. New York, October 2, 1996.

Bragdon, Clifford R. Intermodal and multimodal transportation issues for the 1996 Summer Olympics. *Proceedings, Urban Regional Information and Assessment Conference*, Atlanta, GA, 1993.

Bragdon, Clifford R., and Stephen Lee Morgan. New visualization technologies for port security. *Cargo Security International*, 2(6), December 2004.

Bragdon, Clifford R. *Southampton intermodal transportation study*. Town of Southampton, June 2003, 184 pages.

Bragdon, Clifford R. *Urban cultural evolution: An historical perspective.* Master's thesis, Michigan State University, 1966.

Bragdon, Clifford R., King, David J., and Mathew Hyner. Virtual multi-sensory planning and technology applied to intermodal transportation safety and security. In *The New Challenge of International Transportation Security*, Institute of Traffic and Transportation, National Chiao Tung University, 2003.

Bureau of Transportation Statistics. *Pocket Guide to Transportation 2007*, U.S. Department of Transportation, Washington, D.C.: Government Printing Office, January 2007.

Cochran, John E., Jr., and Clifford R, Bragdon. Intermodal transportation safety and security: A Center for Advanced Simulation and Technology short course. Auburn University and Dowling College, Gulf Shores AL, August 15–16, 2002.

Cohen, Charles, and Eric Werker. The political economy of natural disasters. Harvard Business School, January 16, 2008.

Dantzig, George B., and Thomas L. Saaty. *Compact City: A Plan for a Livable Urban Environment.* San Francisco: W. H. Freeman and Company, 1973.

Douglas, John. The fusion of information. White paper, Chief of Staff, Her Majesty's Service, Northern Ireland, 1996.

Driving devices make scents. *New York Daily News*, March 19, 1998.

Earth Hour 2008. Accessed at: www11.earthourus.org/; 60 Earth Hour.

Fairlie, John. *Municipal Administration.* New York: Macmillan Company, 1908.

Farazmand, Ali. Learning from the Katrina crisis: A global and international perspective with implications for future crisis management. *Public Administration Review*, Special Issue, pp.148–158, 2007.

First Annual Conference on National Preparedness. *Proceedings*, Global Center for Preparedness, Florida Tech, December 14, 2007.

Flynn, Stephen. *The Edge of Disaster: Building a Resilient Nation.* New York: Random House, 2007.

Flynn, Stephen, and Daniel B. Prieto. Neglected defense: Mobilizing the private sector to support homeland security. Conference on Foreign Relations, Washington, D.C., 2008.

Freedman, Mitchell. Scouting for options less traveled Southampton. *Newsday*, July 15, 2001.

Friedman, Thomas L. *The World Is Flat* (2nd Revised Edition). New York: Farrar, Strasuss and Giroux, 2007.

Garten, Rothkopf. A blueprint for green energy in Americas. IDB, Washington, D.C., 2007.

Government Accounting Office (GAO). *Transportation security research and development.* Washington, D.C.: GAO-04-890, 2004.

Habitat II Conference on Human Settlements. Second United Nations Conference, Istanbul, Turkey, June 1996.

Howitt, Arnold, and Jonathon Makler. *On the Ground: Protecting America's Roads and Transit Against Terrorism.* Washington, D.C.: Brookings Institute, April 2005.

Hickey, K. The National Aviation Transportation Center (NATC) introduces intermodal transportation simulation system that is clearly not a game. *Traffic World*, 259(7), August 16, 1999.

Hubler, Eric. The fittest and fattest cities in America. *Men's Fitness*, pp. 85–91, March 2008.

Johns, Kevin. The Eisenhower II national transportation plan: A national economic, transportation and energy policy white paper. Director of Economic Development, Palm Beach County, FL, November 28, 2004.

Kay, Jane Holtz. *Asphalt Nation.* New York: Crown Publishers, Inc., 1997.

Kim, Ryan. Can a computer have a sense of smell? *San Francisco Chronicle*, March 10, 2008.

Lan, Lawrence W. (Ed.). *The New Challenge of International Transportation Security.* Institute of Traffic and Transportation, National Chiao Tung University, 2003.

Leadership in Energy and Environmental Design (LEED). Established 1994.

Lindstron, Martin. *Brand Sense.* New York: Free Press, 2005.

Linzer, Dafna. Turf war in the battle against terrorism. *Washington Post*, March 21, 2008.

Locurcio, Ralph V. Fast Integrated Regional Systems Technology (FIRST): A regional approach to disaster response and recovery. *Logistics Spectrum, 40*(3), July–September 2006.

Mass transit system threat analysis (unclassified). Transportation Security Administration, Office of Intelligence. Washington, D.C., February 29, 2008.

Moreno, Luis Alberto. Highlights of the IDB sustainability review 2007. Inter-American Development Bank, Washington, D.C., 2008. Inter-American Development Bank Web site related to this subject: www.iadb.org/sustainability.

National Commission on Terrorist Attacks Upon the United States (9/11 Commission). *The 9-11 Commission Report: Final Report of the National Commission on Terrorist Attacks Upon the United States,* official government edition. Washington, D.C.: U.S. Government Printing Office, 2004.

National Hurricane Conference. Thirtieth Annual, Orlando, FL, March 31–April 4, 2008.

Pacific Seaflight: Wing in ground craft website: wwwpacificseaflight.com (Blue Dolphin Wingship).

Perelman, Lewis J. Infrastructure risk and renewal: The class of blue and green. Public Entity Risk Institute (PERI) Internet Symposium, Washington, D.C., January 2008.

Powell, Andrew Phillip, and Juan Franciso Martinez. Environment, rural development and disaster risk management. IDB Business Seminar Series, February 7, 2008.

Raynor, Jessica. Port receives more security: Federal, local agencies work to coordinate new effort. *Florida Today*, March 31, 2008.

Reuters. Hurricane price tags soar on crowded coasts: $500 billion storm by 2020s predicted by one study. Thirtieth Annual National Hurricane Conference, Orlando, FL, April 6, 2008.

Rotstein, Arthur. Virtual border fences in Arizona a Failure. *Boston Globe*, April 23, 2008.

Rhodes, Paul. Neuronal components are connected with stochastic and dynamic synapses: Implications for biological and synthetic neural computation. Lecture presented May 18, 2007.

Safdie, Moshe. *The City After the Automobile.* Toronto: Stoddart Publishing Company, Ltd., 1997.

Slater, Rodney E. The National Highway System: A commitment to America's future. *Public Roads, 59*(4), Spring 1996. U.S. Department of Transportation, Federal Highway Administration,Washington, D.C.

Snell, Bradford C. American ground transport. *Third Rail Press*, 1974.

Strugatch, Warren. A dreamer's vision vs. reality on road and rail. *New York Times*, October 29, 2000.

Texas A&M University. Texas Transportation Institute (TTI) mission is to solve transportation problems through research, to transfer technology and to develop diverse human resources to meet the transportation challenges of tomorrow.

Tesla Motors, San Carlos, California; website: www.teslamotors.com

Transportation for tomorrow. Report of National Surface Transportation Policy and Revenue Study Commission, Washington, D.C., January 2008.

Transportation systems security management: Coming to your senses with regard to safety and security (Dr. Clifford R. Bragdon, course administrator). *Proceedings*, Florida Institute of Technology, University College, October 23–24, 2006.

Tri-State Transportation Campaign. Mobilizing the region: Transportation reform in the New York/New Jersey/Connecticut Metropolitan Region.

Williams, Stephen W. 21 Inventions for the Next Century. *Newsday*, June 13, 1999.

Younes, Bassem, and Carl Berkowitz. Guidelines for intermodal connectivity and the movement of goods for Dubai. *Logistics Spectrum, 40*(3), July–September 2006.

Zllotnik, Hania. 2007 revision of world urbanization prospects. United Nations, UN Population Division, February 2008.

Index

Note: Page numbers followed by 'f' indicate figures.

CPSIA information can be obtained at www.ICGtesting.com
Printed in the USA
BVOW01*1751040914

365234BV00004B/18/P